REVIEWING BIOLOGY

The Living Environment

With Sample Examinations

THIRD EDITION

Carl M. Raab
Former Director of Academic Initiatives for the
Office of School Programs and Support Services
New York City Board of Education, New York
Former Assistant Principal, Supervision Science
Fort Hamilton High School, Brooklyn, New York

Revised by
Michael F. Renna
Former Principal, Preparatory Academy
for Writers MS/HS, Queens, NY
Former Assistant Principal, Supervision Science
Hillcrest High School, Queens, New York

Amsco School Publications, Inc.
315 Hudson Street, New York, N.Y. 10013

The publisher wishes to acknowledge the helpful contributions of the following reviewers:

Marilou Bebak
Biology Teacher
Nardin Academy High School
Buffalo, New York

Annick O'Reilly
Biology Teacher
Bainbridge-Guilford High School
Bainbridge, New York

Bart Bookman, Retired
Assistant Principal, Science
Adlai E. Stevenson High School
Bronx, New York

Barbara Poseluzny, Retired
Assistant Principal, Science
A. Philip Randolph High School
Manhattan, New York

Michael Comet
Biology Teacher
South Lewis High School
Turin, New York

Lane Schwartz, Retired
Principal, Glen Cove High School, New York
Former Assistant Principal, Science
John Jay High School, Manhattan, New York

Vincent Kreyling
Biology Teacher
Roslyn High School
Roslyn Heights, New York

Martin Solomon
Science Teacher
Daniel Carter Beard JHS 189Q
Queens, New York

Text Design: Nesbitt Graphics, Inc.
Cover Design: Meghan J. Shupe
Composition: Nesbitt Graphics, Inc.
Artwork: Hadel Studio
Cover Photo: *Heliconia rendulata* plant, courtesy of fotosearch.com.

Science, Technology, and Society features: Christine Caputo

Regents Exams included are 6/10, 8/10, 1/11 & 6/11

Please visit our Web site at: *www.amscopub.com*

When ordering this book, please specify:
Either **R 7410 P** *or* **REVIEWING BIOLOGY:
The Living Environment, Third Edition**

ISBN: 978-1-56765-907-8

Note to the Teacher

This newly revised third edition of *Reviewing Biology: The Living Environment* offers an innovative format that comprehensively reviews the National Science Standards-based Core Curriculum for NYS and follows the new scope and sequence outlined by the NYC Department of Education. The book is readily correlated with the standard textbooks for high-school-level biology and includes material beyond the core for those teachers who want more in-depth coverage. The books of this series are specifically geared to meet the needs of students who want to review the material in preparation for final exams.

The material in *Reviewing Biology: The Living Environment, Third Edition* is divided into ten chapters, each of which is subdivided into major topic sections. The book is abundantly illustrated with clearly labeled drawings and diagrams that illuminate and reinforce the subject matter. Important core science terms are **boldfaced** and defined in the text. Other important science terms that may be unfamiliar to students are *italicized* for emphasis. In addition, the large work-text format and open design make this book easy for students to read.

Within each chapter are several sets of Part A, Part B-1, Part B-2, and Part C questions that test the students' knowledge and reasoning while provoking thought. These Regents and Regents-type questions include multiple-choice, short-answer (constructed response), diagram-analysis, and essay (extended constructed-response) formats. Topical reading-comprehension passages with question sets appear at the end of Chapters 1-9, giving students further opportunity to refine their science reading and writing skills. Tables, graphs, and diagrams that aid in interpreting, reviewing, and testing the material often accompany the questions. The nearly 900 questions found in the text can be used for topic review throughout the year, as well as for exams and homework assignments. The four sample examinations at the back of the book can be used as final exams or as practice for the final exam.

Chapter 10 (Laboratory Skills and Part D Labs) reviews the scientific skills and methods that all students should master in the course of completing one year of biology instruction at this level. Also included in this chapter is a description of each of the new mandated labs that students are required to know for the Living Environment exam. *Reviewing Biology: The Living Environment, Third Edition* also contains a full Glossary in which students can find concise definitions of all boldfaced scientific terms. Students can use the extensive Index to locate the text discussions of these and other important biological terms and concepts.

An Appendix, comprised of Part D Sample Lab Questions, has been added to this edition to give students additional practice in answering actual Regents questions on the four mandated Part D lab activities. Also included in the book are nine completely new Science, Technology, and Society features that explore current controversial issues in biology. Several constructed-response questions are presented at the end of each feature to test reading comprehension and to encourage students to evaluate the issues and make their own decisions about the impact of biological science and technology on society, the environment, and their lives.

Contents

CHAPTER 1

Ecology

Ecology is the study of the relationships between organisms and between organisms and their physical environment. No organism exists in nature as an entity apart from its environment.

ECOLOGICAL ORGANIZATION

In ecology, the relationships between organisms and the environment may be considered at various levels. The smallest, least inclusive level in terms of ecological organization is the population; the largest and most inclusive level is the biosphere.

Levels of Organization

All members of a species living in a given location make up a **population**. For example, all the water lilies in a pond make up a population, and all the goldfish in a pond make up a population. Together, all the interacting populations in a given area make up a **community**. For example, all the plants, animals, and microorganisms in a pond make up the pond community.

An **ecosystem** includes all the members of a community along with the physical environment in which they live. The living and nonliving parts of an ecosystem function together as an interdependent and relatively stable **system**. The *biosphere* is the portion of Earth in which all living things exist. The biosphere, which is composed of numerous, complex ecosystems, includes the water, soil, and air.

QUESTIONS

PART A

1. All the different species within an ecosystem are collectively referred to as the (1) niche (2) community (3) consumers (4) population

2. Which term includes the three terms that follow it? (1) population: community, ecosystem, organism (2) community: ecosystem, organism, population (3) ecosystem: organism, population, community (4) organism: ecosystem, community, population

3. Which sequence shows increasing complexity of levels of ecological organization? (1) biosphere, ecosystem, community (2) biosphere, community, ecosystem (3) community, ecosystem, biosphere (4) ecosystem, biosphere, community

4. The members of the mouse species *Microtus pennsylvanicus* living in a certain location make up a (1) community (2) succession (3) population (4) phylum

5. Which term includes all the regions (on land and in water) in which life exists? (1) marine biome (2) climax community (3) biosphere (4) tundra

PART B-2

6. Define each of the following ecological levels: *ecosystem, population, community, biosphere.*

7. List the four ecological levels of organization in their order of increasing complexity.

CHARACTERISTICS OF ECOSYSTEMS

Ecosystems are the structural and functional units studied in ecology.

Requirements of Ecosystems

An ecosystem involves interactions between its living factors and nonliving factors. It is a self-sustaining unit when the following two conditions are met.

First, there must be a constant flow of energy into the ecosystem, and there must be organisms within the ecosystem that can use this energy for the synthesis of organic compounds. The primary source of energy for most ecosystems on Earth is sunlight; the organisms that can use this energy for the synthesis of organic compounds are green plants, algae, and other photosynthetic autotrophs. Second, there must be a recycling of materials between the living organisms and the physical, nonliving parts of the ecosystem.

Until recently, it was thought that all life forms depend, either directly or indirectly, on solar energy to carry out their life activities. However, this is not true. There are some organisms that survive in the deepest parts of the ocean, where no light penetrates. They live around hot-water vents on the seafloor and rely on bacteria that use chemicals—hydrogen sulfide and carbon dioxide—in the seawater to produce energy-rich sugars. These bacteria, which function as the food producers in this ecosystem, carry out *chemosynthesis* rather than photosynthesis. They use the energy of chemical reactions, rather than the energy of the sun, to produce carbohydrates. Survival of the other organisms in that ecosystem, such as worms, clams, shrimp, crabs, and octopuses, depends on the food energy that the bacteria produce.

Abiotic Factors of Ecosystems

The components of an ecosystem include nonliving, or **abiotic**, factors, and living, or **biotic**, factors. The abiotic factors of the environment are physical factors that sustain the lives and reproductive cycles of organisms. These factors are: intensity of light; temperature range; amount of water; type of soil; availability of minerals and other inorganic substances; supply of gases, including oxygen, carbon dioxide, and nitrogen; and the pH (acidity or alkalinity) of the soil or water.

Abiotic factors vary from one environmental area to another. The abiotic conditions in any particular environment determine the types of plants and animals that can exist there. Thus, abiotic factors are *limiting factors*. For example, the small amount of available water in a desert limits the kinds of plants and animals that can live in that environment.

Biotic Factors of Ecosystems

The biotic factors of an ecosystem are all the living things that directly or indirectly affect the environment. The organisms of an ecosystem interact in many ways. These interactions include nutritional and symbiotic relationships.

Nutritional Relationships. Nutritional relationships involve the transfer of nutrients from one organism to another within the ecosystem.

Autotrophs are organisms that can use energy from the environment to synthesize their own food from inorganic compounds. Most autotrophs are photosynthetic, using energy from sunlight along with carbon dioxide and water from the environment to synthesize organic compounds.

Heterotrophs cannot synthesize their own food and must obtain nutrients from other organisms. Depending on their source of food, heterotrophs are classified as saprophytes, herbivores, carnivores, or omnivores.

Saprophytes are organisms that obtain nutrients from the remains of dead organisms. Types of saprophytes include bacteria and fungi. Animals that feed exclusively on plants are called **herbivores**. Animals that consume other animals are called **carnivores**. The carnivores include **predators**, which kill and eat their **prey**, and **scavengers**, which feed on the remains of animals that they have not killed. **Omnivores** are animals that consume both plant and animal matter.

Symbiotic Relationships. Different kinds of organisms sometimes live together in a close association. Such a close relationship, or **symbiosis**, may or may not be beneficial to the organisms involved.

A type of symbiotic relationship in which one organism benefits while the other is neither helped nor harmed is called *commensalism*. Barnacles living on whales, remora living on sharks, and orchids living on large, tropical trees all obtain favorable places to live without doing any noticeable harm to the other organism (Figure 1-1).

Figure 1-1. The shark and remora have a type of symbiosis known as commensalism.

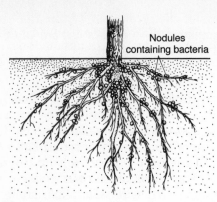

Figure 1-2. Nitrogen-fixing bacteria live in the roots of legumes in a type of symbiosis known as mutualism.

A symbiotic relationship in which both organisms benefit is called *mutualism*. For example, certain protozoans (unicellular organisms) live within the digestive tracts of termites. Wood eaten by the termites is digested by the protozoans, and both organisms benefit from the nutrients that are released. Another example of mutualism is found in lichens, which are made up of both algal and fungal cells. The algal cells carry on photosynthesis, which provides food for the lichen, while the fungal cells provide moisture and minerals, and anchor the lichen to a surface.

Nitrogen-fixing bacteria live in the roots of legumes (such as the peanut plant). The relationship between these organisms is mutualistic because the bacteria provide nitrogen compounds for the plant, while the plant provides the bacteria with nutrients and a good place to live (Figure 1-2).

A symbiotic relationship in which one organism, the **parasite**, benefits while the other, the **host**, is harmed is called *parasitism*. Examples include the athlete's foot fungus that can live on humans, and tapeworms and heartworms that can live in dogs.

QUESTIONS

PART A

8. Different species of animals in a community would most likely be similar in their (1) physical structure (2) size (3) abiotic requirements (4) number of offspring produced

9. All the living things that affect each other and their environment are considered (1) biotic factors (2) inorganic substances (3) physical conditions (4) chemical factors

10. A study was made over a period of years in a certain part of the country. It showed that the area had a low amount of rainfall, a wide seasonal variation in temperature, and short periods of daylight. These environmental factors are (1) abiotic factors of little importance to living things (2) abiotic factors that limit the type of organisms that live in the area (3) biotic factors important to living things in the area (4) biotic factors that are affected by the abiotic factors

11. The presence of nitrogen-fixing bacteria in nodules on the roots of legumes (such as the peanut plant) illustrates an association known as (1) commensalism (2) mutualism (3) parasitism (4) environmentalism

12. At times, hyenas will feed on the remains of animals that they have not killed themselves. At other times, they will kill other animals for food. Based on their feeding habits, hyenas are best described as both (1) herbivores and parasites (2) herbivores and predators (3) scavengers and parasites (4) scavengers and predators

13. Which is an abiotic factor in the environment? (1) water (2) earthworm (3) fungus (4) human

14. The organisms that prevent Earth from becoming covered with the remains of dead organisms are known as (1) herbivores (2) parasites (3) autotrophs (4) saprophytes

15. A particular species of fish has a very narrow range of tolerance for changes in water temperature and dissolved oxygen content. For this fish, the temperature and oxygen content represent (1) autotrophic conditions (2) a community (3) limiting factors (4) symbiosis

16. An example of a parasitic relationship would be (1) tapeworms living in the intestines of a dog (2) algal and fungal cells living together as a lichen (3) barnacles living on a whale (4) wood-digesting protozoa living in the gut of a termite

17. Parasitism is a type of nutritional relationship in which (1) both organisms benefit (2) both organisms are harmed (3) neither organism benefits (4) one organism benefits and the other is harmed

18. For an ecosystem to be self-sustaining, it must (1) contain more animals than plants (2) receive a constant flow of energy (3) have a daily supply of rainwater (4) contain only heterotrophs

19. Heterotrophs include (1) autotrophs, saprophytes, and herbivores (2) omnivores, carnivores, and autotrophs (3) saprophytes,

herbivores, and carnivores (4) herbivores, autotrophs, and omnivores

20. The primary source of energy for most ecosystems is (1) radioactivity (2) sunlight (3) animal proteins (4) carbon dioxide

21. An ecosystem that does *not* depend on sunlight for its energy source is found (1) at the tops of mountains (2) on tropical islands (3) in hot deserts (4) near deep-ocean vents

22. A particular species of unicellular organism inhabits the intestines of termites, where they can live protected from predators. The unicellular organisms digest wood that has been ingested by the termites, thus providing nutrients to the termites. The relationship between these two species can be described as (1) harmful to both species (2) harmful to the host (3) beneficial to both species (4) beneficial to the parasite only

23. A partial food web is represented in the diagram below. Letter *X* most likely represents (1) autotrophs (2) carnivores (3) decomposers (4) parasites

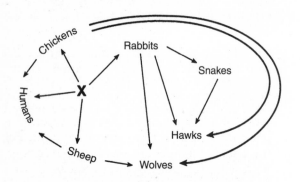

24. Identify the two main conditions that must exist for an ecosystem to be self-sustaining.

25. Explain why abiotic factors are considered to be limiting factors; give an example of one.

26. In his biology class, a student claims that all living things depend on the sun for survival, either directly or indirectly. Provide one example that shows why this student's claim is incorrect.

Energy Flow Relationships

For an ecosystem to be self-sustaining, there must be a flow of energy between organisms. The pathways of chemical energy from food through the organisms of an ecosystem are represented by food chains and food webs.

Food Chains

The transfer of energy from green plants through a series of organisms with repeated stages of eating and being eaten is described as a **food chain** (Figure 1-3). Green plants obtain energy for their life processes from the radiant energy of sunlight (that is, **solar energy**), which they convert to usable chemical energy (glucose) by photosynthesis. For all other organisms in the food chain, energy is obtained from the breakdown of food. The organisms in a food chain are described in terms of the following categories.

Green plants and other autotrophs are the **producers** in the food chain. All the food energy for a community is derived from the organic compounds synthesized by the producers (for example, grass in a savannah, algae in the sea, or chemosynthetic bacteria near a deep-ocean vent).

All the heterotrophic organisms in a community are **consumers**. They must obtain energy from the food that they eat. Animals that feed on green plants and algae are called *primary consumers*, or herbivores. Animals that feed on primary consumers are called *secondary consumers*, or carnivores. Omnivores may be either primary or secondary consumers; that is, omnivores may feed on plants and/or animals. Humans are a good example of an omnivore, since we eat both plant and animal foods.

Saprophytes are **decomposers**, the organisms that break down the remains of dead organisms and organic wastes. Decomposers return substances in the remains and wastes of plants and animals to the environment, where other living organisms can use them again. Most decomposers are either bacteria or fungi. This recycling of materials is critical to the survival of an

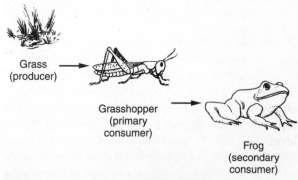

Figure 1-3. A food chain (not drawn to scale).

ecosystem; it ensures that the limited supply of materials can be used over and over again. In addition, the remains, or **residue**, of dead organisms do not accumulate in the ecosystem, since they are broken down to simpler compounds.

Food Webs. In a natural community, most organisms eat more than one species and may be eaten, in turn, by more than one species. Thus, the various food chains in a community are interconnected, forming a **food web** (Figure 1-4). Food webs have the same levels of organisms (producers, consumers, and decomposers) as food chains, but the flow of energy and materials is much more complex.

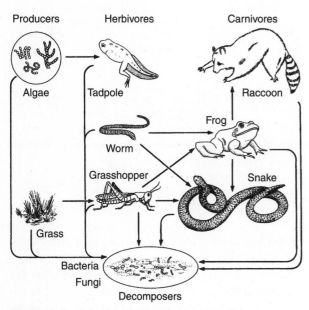

Figure 1-4. A food web (not drawn to scale).

Pyramid of Energy. The greatest amount of energy in a community is present in the organisms that make up the producer level. Only a small portion of this energy is passed on to primary consumers; and only a small portion of that energy in the primary consumers is passed on to secondary consumers; and so on. An **energy pyramid** can be used to illustrate the loss of usable energy at each feeding, or *trophic*, level (Figure 1-5).

At each consumer level in an energy pyramid, only about 10 percent of the ingested nutrients are used to synthesize new tissues, which represent the food available for the next feeding level. The remaining energy is used by the consumers for their life functions and is eventually converted to heat, which is lost from the ecosystem. Thus, an ecosystem cannot sustain itself

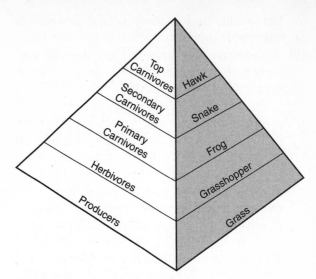

Figure 1-5. A pyramid of energy.

without the constant input of energy from an external source. In most ecosystems, this energy source is the sun.

Pyramid of Biomass. In general, the decrease in available energy at each higher feeding level means that less organic matter, or *biomass*, can be supported at each higher level. Thus, the total mass of producers in an ecosystem is greater than the total mass of primary consumers; and the total mass of primary consumers is greater than the total mass of secondary consumers; and so on. A *biomass pyramid* can be used to illustrate this decrease in biomass at each higher feeding level (Figure 1-6).

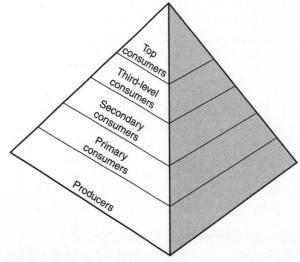

Figure 1-6. A pyramid of biomass.

If, for some reason, biomass becomes greater at a higher trophic level than at a lower level, the ecosystem will become unbalanced and unsustainable. The consumers at the higher trophic level will have exceeded the capacity of the ecosystem to support them. Eventually, they will begin to die off as their food supply runs out, and the ecosystem will return to a state of balance.

QUESTIONS

27. Which occurs within self-sustaining eco-systems? (1) The producers have a limited source of energy. (2) Consumers eventually outnumber producers. (3) Carnivores usually outnumber herbivores. (4) Organisms recycle materials with each other and the environment.

28. Which food chain relationship illustrates the nutritional pattern of a primary consumer? (1) seeds eaten by a mouse (2) an earthworm eaten by a mole (3) a mosquito eaten by a bat (4) a fungus growing on a dead tree

29. Which term describes both the bird and the cat in the following food chain?

    sun → grass → grasshopper → bird → cat

    (1) herbivores (2) saprophytes (3) predators (4) omnivores

30. Organisms from a particular ecosystem are shown below. Which statement concerning an organism in this ecosystem is correct? (1) Organism 2 is heterotrophic. (2) Organism 3 helps recycle materials. (3) Organism 4 obtains all of its nutrients from an abiotic source. (4) Organism 5 must obtain its energy from organism 1.

31. The elements stored in living cells of organisms in a community will eventually be returned to the soil for use by other living organisms. The organisms that carry out this process are the (1) producers (2) herbivores (3) carnivores (4) decomposers

32. In the food chain below, what is the nutritional role of the rabbit?

    lettuce plant → rabbit → coyote

    (1) parasite (2) saprophyte (3) primary consumer (4) primary producer

33. Which level of the food pyramid shown below represents the largest biomass? (1) bass (2) minnows (3) copepods (4) algae

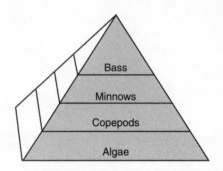

34. Fly larvae consume the body of a dead rabbit. In this process, they function as (1) producers (2) scavengers (3) herbivores (4) parasites

35. Which diagram best represents the usual relationships of biomass in a stable community? (1) 1 (2) 2 (3) 3 (4) 4

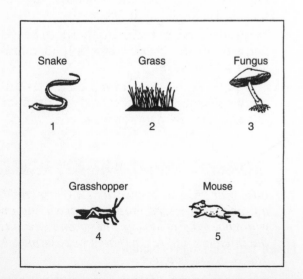

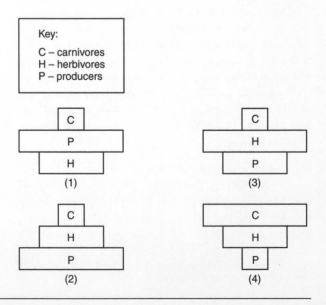

36. Which level in an energy pyramid has the highest amount of available energy? (1) highest level consumers (2) secondary consumers (3) primary consumers (4) producers

Base your answers to questions 37 through 40 on the food web below and on your knowledge of biology. The diagram represents different species of organisms that may interact with each other in and around a pond environment.

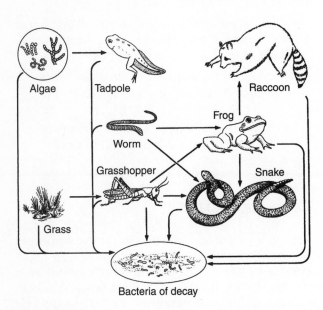

37. The adult frog represents a type of consumer known as a (1) producer (2) carnivore (3) saprophyte (4) parasite

38. Which organisms are classified as herbivores? (1) algae, tadpole, raccoon (2) worm, snake, bacteria (3) tadpole, worm, grasshopper (4) grasshopper, bacteria, frog

39. Which statement about the algae and grass is true? (1) They are classified as omnivores. (2) They are parasites in the animals that eat them. (3) They contain the greatest amount of stored energy. (4) They decompose nutrients from dead organisms.

40. The interactions among organisms shown in this diagram illustrate (1) a food web (2) geographic isolation (3) abiotic factors (4) organic evolution

Base your answers to questions 41 through 43 on the following food chain and on your knowledge of biology.

rosebush → aphid → ladybird beetle
→ spider → toad → snake

41. Which organism in the food chain can transform light energy into chemical energy? (1) spider (2) ladybird beetle (3) rosebush (4) snake

42. At which stage in the food chain will the population with the smallest number of animals probably be found? (1) spider (2) aphid (3) ladybird beetle (4) snake

43. Which organism in this food chain is a primary consumer? (1) rosebush (2) aphid (3) ladybird beetle (4) toad

Base your answers to questions 44 through 47 on the diagram below, which represents four possible pathways for the transfer of energy stored by green plants.

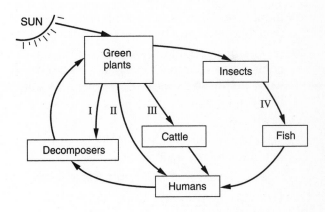

44. The pathway labeled IV represents (1) a food chain (2) a population (3) an ecosystem (4) an abiotic factor

45. Through which pathway would the sun's energy be most directly available to humans? (1) I (2) II (3) III (4) IV

46. In this diagram, humans are shown to be (1) herbivores only (2) carnivores only (3) omnivores (4) parasites

47. The cattle in the diagram represent (1) primary consumers (2) secondary consumers (3) producers (4) autotrophs

Base your answers to questions 48 through 50 on the food web and graph on page 8 and on your knowledge of biology. The graph represents the interaction of two different populations, A and B, in the food web.

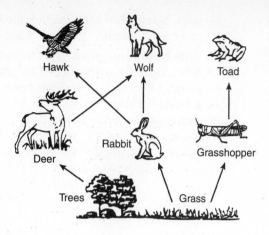

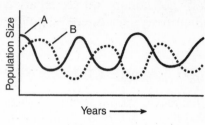

48. Population *A* is made up of living animals. The members of population *B* feed on these living animals. The members of population *B* are most likely (1) scavengers (2) autotrophs (3) predators (4) parasites

49. Identify one specific heterotroph from the food web that could be a member of population *A*.

50. An energy pyramid is shown below. Which organism shown in the food web would mostly likely be found at level *X*? (1) wolf (2) grass (3) deer (4) toad

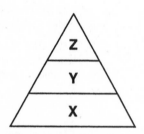

PART C

51. Draw, using specific organisms as examples, a pyramid of energy that has three trophic levels.

52. Distinguish between a pyramid of energy and a pyramid of biomass. Explain the relationship between them.

53. Explain why there is a "loss" of energy as one goes step-by-step up an energy pyramid. Discuss where the seemingly "lost" energy actually goes.

54. Explain why an ecosystem could not sustain itself without the constant input of energy from an outside source.

55. Consider the following food pyramid: corn → mice → snakes → hawks. If the total amount of energy captured by the corn in is 1,000,000

calories per day, and only about 10 percent of this energy is passed on at each higher trophic level, calculate:

• how much energy (in calories) would be available per day at each higher level (for mice, snakes, and hawks);

• how many hawks this ecosystem could support, if the hawk population needs 500 calories per bird per day.

Carrying Capacity

In every ecosystem on Earth, there are limited amounts of available resources. These resources include food, water, energy, minerals, and space (territory). Even though some of these resources may be recycled through the actions of bacteria and fungi, the pace of recycling may not keep up with the demand for these materials. The amount of resources available limits the number of organisms that an ecosystem can support. The maximum number of organisms of a particular type that can be supported in an area is known as the **carrying capacity**. In a stable ecosystem, a population of organisms will fluctuate slightly (due to seasonal and other factors), as shown in Figure 1-7. If the population increases significantly above its carrying capacity, many individuals will die off because there are insufficient resources available to support them.

Competition

Different species living in the same environment, or **habitat**, may require some of the same resources for their survival. Since resources (such as food, water, space, light, and minerals) are usually limited, **competition** occurs among the various species. Competition is the struggle between different organisms for the same limited resources.

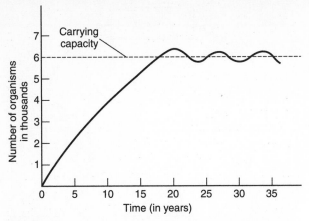

Figure 1-7. The carrying capacity for a population of organisms will fluctuate slightly in a stable ecosystem.

The more similar the needs of the species, the more intense the competition. For example, lions, leopards, and hyenas may compete to consume the same type of antelope. In addition, because their requirements are most similar, the strongest competition for resources often occurs among members of the same species. For example, competition for antelope prey (or water, mates, territory, and so on) may be more intense between neighboring prides of lions than between lions and other nearby large predators.

Each species occupies a particular ecological niche in a community. A **niche** is the role that the species fills in its habitat. A species' niche includes the type of food, or nutrients, it requires; where and how it lives; where and how it reproduces; and its relationships with other species in the area. When two species compete for the same niche, the one that is more successful at utilizing the available resources will out-compete the other, thereby maintaining just one species per niche in the community.

QUESTIONS

PART A

56. Carrying capacity is best thought of as the amount of (1) abiotic factors present in an ecosystem (2) light available for photosynthesis (3) organisms the ecosystem can support (4) producers compared to consumers in the ecosystem

57. A stable ecosystem is characterized by (1) a greater number of consumers than producers (2) population sizes at or near the carrying capacity (3) a greater need for energy than is available (4) a lack of decomposers to recycle materials

58. In a freshwater pond community, a carp (a type of fish) eats decaying matter from around the bases of underwater plants, while a snail scrapes algae from the leaves and stems of the same plant. They can survive at the same time in the same pond because they occupy (1) the same niche but different habitats (2) the same habitat but different niches (3) the same habitat and the same niche (4) different habitats and different niches

59. The role a species plays in a community is called its (1) habitat (2) biotic factor (3) territory (4) niche

60. When two different species live in the same environment and use the same limited resources, which interaction will usually occur? (1) competition (2) cooperation (3) commensalism (4) mutualism

61. One biotic factor that limits the carrying capacity of any habitat is the (1) availability of water (2) level of atmospheric oxygen (3) activity of decomposers (4) amount of soil erosion

Base your answer to the following question on the information and diagram below.

A population of chipmunks migrated to an environment where they had little competition. Their population quickly increased but eventually stabilized, as shown in the graph below.

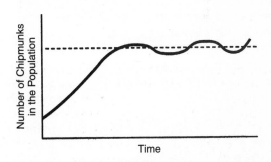

62. Which statement best explains why the population stabilized? (1) Interbreeding between members of the population increased the mutation rate. (2) The population size became limited due to factors such as availability of food. (3) An increase in the chipmunk population caused an increase in the producer population. (4) A predator species came to the area and occupied the same niche as the chipmunks.

63. The size of a mouse population in a natural ecosystem tends to remain relatively constant due to (1) the carrying capacity of the environment (2) the lack of natural predators (3) cycling of energy (4) increased numbers of decomposers

64. Purple loosestrife plants are replacing cattail plants in freshwater swamps in New York State. The two species have very similar environmental requirements. This observation best illustrates (1) variations within a species (2) dynamic equilibrium (3) random recombination (4) competition between species

65. Some bacteria can reproduce once every 20 minutes. As a result, their populations can double several times an hour. Even at this phenomenal rate of reproduction, bacteria do not overrun the planet. Give a brief, valid explanation for this fact.

66. Explain why competition between individuals of the same species is often more intense than competition between members of different species.

Cycles of Materials

In a self-sustaining ecosystem, various materials are recycled between the organisms and the abiotic environment. The recycling process allows materials to be used over and over again by living things.

Carbon-Hydrogen-Oxygen Cycle. The elements carbon, hydrogen, and oxygen are recycled through the environment by the processes of respiration and photosynthesis (Figure 1-8). During aerobic cellular respiration, plants and animals use oxygen (O_2) from the air and release carbon dioxide (CO_2) and water (H_2O) via the breakdown of glucose. During photosynthesis, plants use carbon dioxide (CO_2) from the air and water (H_2O) from the environment in the synthesis of glucose ($C_6H_{12}O_6$) and oxygen (O_2) is given off as a by-product.

Water Cycle. In the water cycle, water moves between Earth's surface and the atmosphere (Figure 1-9). The main processes involved in this cycle are *evaporation* and *condensation*. Liquid water on Earth's surface changes to a gas by the process of evaporation and enters the atmosphere in the form of water vapor. As a result of condensation, water vapor is returned to

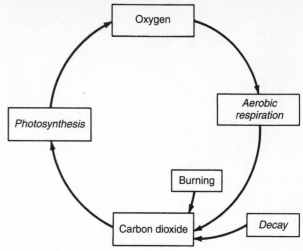

Figure 1-8. The carbon-hydrogen-oxygen cycle.

the liquid state (precipitation) and falls to Earth. Some water vapor is added to the atmosphere by aerobic respiration in plants and animals and by transpiration in plants. Water is also an essential nutrient for all living things, allowing them to carry out essential life processes and chemical reactions.

Nitrogen Cycle. The element nitrogen is needed by all living things because it is part of the structure of amino acids and proteins. Plants absorb nitrogen-containing compounds from the soil; animals obtain nitrogen in the form of proteins in the foods they eat. These proteins are broken down by digestion to amino acids, which are then used in the synthesis of animal proteins.

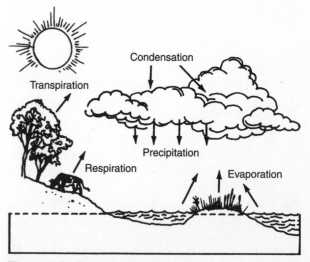

Figure 1-9. The water cycle.

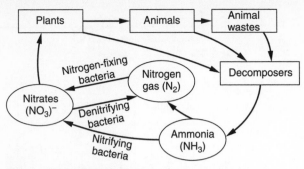

Figure 1-10. The nitrogen cycle.

The nitrogen cycle involves decomposers and other soil bacteria. Figure 1-10 shows the various components of the nitrogen cycle, which are described below.

Nitrogen-fixing bacteria, which live in nodules inside the roots of some plants (refer to Figure 1-2), convert free nitrogen (N_2) from the air into nitrogen-containing compounds called nitrates (NO_3). *Nitrates* are absorbed from the soil by plants and used in protein synthesis. Animals that eat plants convert the nitrogen-containing plant proteins into animal proteins. The nitrogenous wastes of living animals, and the nitrogen compounds in the remains of dead plants and animals, are broken down by decomposers and converted to ammonia (NH_3). *Nitrifying bacteria* in the soil convert ammonia into nitrates, which can be used again by plants. *Denitrifying bacteria* break down some nitrates into free nitrogen (N_2), which is released into the atmosphere as a gas.

QUESTIONS

PART A

67. Carbon dioxide is added to the atmosphere by (1) photosynthesis in plants (2) evaporation of water (3) respiration in animals only (4) respiration in plants and animals

68. Oxygen (O_2) is added to the atmosphere by (1) evaporation and photosynthesis (2) respiration in plants (3) photosynthesis only (4) denitrifying bacteria

69. Which of the following processes is *not* involved in the water cycle? (1) condensation (2) nitrification (3) evaporation (4) transpiration

70. The processes involved in the recycling of carbon, hydrogen, and oxygen are (1) evaporation and condensation (2) photosynthesis and respiration (3) nitrification and denitrification (4) respiration and transpiration

71. Nitrogen is both removed from the atmosphere and returned to the atmosphere by the activities of (1) plants only (2) animals only (3) plants and animals (4) bacteria

72. Animals obtain their nitrogen from (1) proteins in their food (2) nitrates in the soil (3) gas in the atmosphere (4) bacteria in their intestines

PART B-1

Base your answers to questions 73 through 75 on the diagram below, which represents a cycle in nature, and on your knowledge of biology.

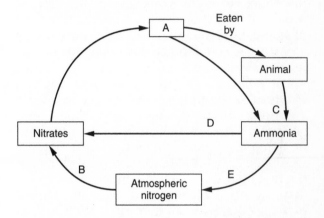

73. The cycle represented by the diagram is the (1) nitrogen cycle (2) carbon cycle (3) water cycle (4) oxygen cycle

74. Nitrifying bacteria in the soil are represented by the letter (1) *A* (2) *E* (3) *C* (4) *D*

75. The letter *B* most likely represents (1) bacteria of decay (2) denitrifying bacteria (3) a legume (peanut plant) (4) nitrogen-fixing bacteria

PART B-2

76. Describe how carbon dioxide and oxygen are recycled by the processes of respiration and photosynthesis.

77. Explain why both nitrogen-fixing bacteria and nitrifying bacteria are important for the survival of plants.

ECOSYSTEM FORMATION

Ecosystems tend to change over a long period of time until a stable one is formed. Both the living (biotic) and nonliving (abiotic) parts of the ecosystem change.

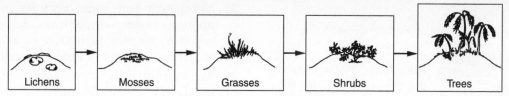

Figure 1-11. Ecological succession occurs, over time, on a new island.

Succession

The replacement of one kind of community by another in an ecosystem is called ecological, or biological, **succession**. *Ecological succession* is usually a long-term process, happening over the course of many years (and many generations of different plants and animals). The kind of stable ecosystem that eventually develops in a particular geographical area depends on the region's climate.

Pioneer Organisms.   Depending on climate and other abiotic environmental factors, succession on land can begin in an area that has no living things and end with a forest. Succession begins with *pioneer organisms*, which are the first plants, or plantlike organisms, to populate an area. Lichens and algae are typical pioneer organisms on bare rock, such as that found on a newly emerged volcanic island (Figure 1-11).

Starting with pioneer plants, each community modifies the environment, often making it less favorable for itself and more favorable for other kinds of communities. One sequence of plant succession in New York State might be lichens → grasses → shrubs → conifers (pine trees) → deciduous (beech and maple) woodlands.

Since plants are the basic source of food for a community, the types of plants present in a community determine the types of animals that can live in the community. As the plant populations change, the animal populations also change.

Climax Communities.   Succession ends with the development of a *climax community* in which populations of plants and animals exist in balance with each other and with the environment. In New York State, for example, the oak-hickory and hemlock-beech-maple associations represent two climax communities. In the Midwest, where there is less rain, grasslands are the typical climax community.

The climax community remains stable until a catastrophic change, such as a volcanic erup-

tion or forest fire, alters or destroys it. Thereafter, succession begins again, leading to the development of a new climax community. This new community may be of the same type as the previous one or, if the catastrophe has changed the environment in some important way, it may be of another kind.

Biodiversity

In addition to the factors mentioned above, a stable community or stable ecosystem requires **biodiversity**. This term refers to the presence of a wide range of different species of organisms living and interacting with each other and with their nonliving environment. These organisms play a variety of roles that contribute to the overall stability of an ecosystem. For example, green plants and algae act as producers; fungi and bacteria act as decomposers, recycling vital materials; and animals act as consumers. Some roles are readily apparent while others may not be so obvious. Nevertheless, the removal of any one species from its natural environment may have profound negative effects on the overall health of the ecosystem.

Biodiversity also increases the probability that at least some organisms would be able to survive a catastrophic environmental event, such as climate change or a volcanic eruption. In time, the surviving organisms could reestablish a healthy community.

In addition, stable ecosystems that are rich in species, such as tropical rain forests, contain a wealth of genetic material that may have beneficial uses in medicine, agriculture, or other areas. Today, tropical forests, wetlands, coral reefs, and other ecosystems that are rich in biodiversity are being destroyed at an alarming rate, mainly due to human activities. Once species are lost to extinction, they can never be recovered. Careful protection of diverse habitats and their living resources is critical to preserving the biodiversity of Earth, not only for the needs of humans or individual ecosystems but also for the health and stability of the entire planet.

QUESTIONS

78. The natural replacement of one community by another until a climax stage is reached is known as (1) ecological balance (2) organic evolution (3) dynamic equilibrium (4) ecological succession

79. In an ecological succession in New York State, lichens growing on bare rock are considered to be (1) climax species (2) pioneer organisms (3) primary consumers (4) decomposers

85. Stable ecosystems are characterized by (1) only two major species interacting with each other (2) an infinite amount of available resources (3) a variety of different species interacting with one another (4) very little recycling of materials between the biotic and abiotic components

86. Stage *D* in the diagram below is located on land that was once a bare field. The sequence of stages leading from bare field to stage *D* best illustrates the process known as (1) replication (2) recycling (3) feedback (4) succession

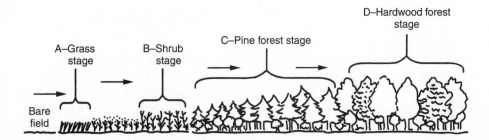

80. One of the first organisms to become established in an ecological succession leading to a pond community would be (1) grasses (2) algae (3) minnows (4) deciduous trees

81. Ecological succession ends with the development of a (1) climax community (2) pioneer community (3) ecological niche (4) abiotic community

82. Which two groups of organisms are most likely to be pioneer organisms? (1) songbirds and squirrels (2) lichens and algae (3) deer and black bears (4) oak and hickory trees

83. After a major forest fire occurs, an area that was once wooded is converted to barren soil. Which of the following sequences describes the most likely series of changes in vegetation after the fire?

(1) shrubs → maples → pines → grasses
(2) maples → pines → grasses → shrubs
(3) pines → shrubs → maples → grasses
(4) grasses → shrubs → pines → maples

84. Biodiversity in an ecosystem is important because it (1) allows one species to dominate the others in its habitat (2) slows down the pace at which species evolve (3) provides stability to the ecosystem (4) limits the amount of variation among organisms

87. A greater stability of the biosphere would most likely result from (1) decreased finite resources (2) increased deforestation (3) increased biodiversity (4) decreased consumer populations

88. List the stages that precede a beech-maple forest in New York State. Identify the pioneer organism and the climax community in this succession.

89. Compare a natural meadow with a cornfield in terms of biodiversity. In your answer, be sure to address the:

- number of species that live in each habitat;
- number of interactions among species that occur in each habitat;
- relative ability of each habitat to survive a natural disaster that might occur.

Base your answers to questions 90 through 92 on the diagram on page 14, which represents the changes in an ecosystem over a period of 100 years, and on your knowledge of biology.

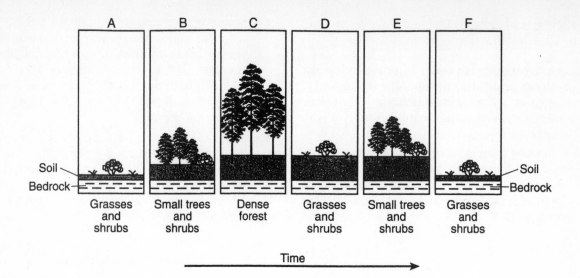

A — Grasses and shrubs
B — Small trees and shrubs
C — Dense forest
D — Grasses and shrubs
E — Small trees and shrubs
F — Grasses and shrubs

Soil
Bedrock

Time

90. State one biological explanation for the changes in types of vegetation observed from *A* through *C*.

91. Identify one human activity that could be responsible for the change from *C* to *D*.

92. Predict what would happen to the soil *and* vegetation of this ecosystem after stage *F*, assuming no natural disaster or human interference.

BIOMES

Earth can be divided into broad geographic regions by climate. The kind of climax ecosystem that develops in these large climatic areas is called a **biome**. Biomes may be terrestrial (land biomes) or aquatic (water biomes). The stretch of tropical rain forests around the equator is a land biome. The ocean is an aquatic biome.

Terrestrial Biomes

The major plant and animal associations (biomes) on land are determined by the large climate zones of Earth. These climate zones are, in turn, determined by geographic factors, including *latitude* (distance north or south of the equator) and *altitude* (distance above or below sea level). Other major geographic features, including large bodies of water, mountains, and deserts, modify the climate of nearby regions.

Climate includes the temperature range and the amounts of precipitation and solar radiation received by a region. The presence or absence of water is a major limiting factor for terrestrial biomes and determines the kinds of plant and animal communities that can be established.

Kinds of Terrestrial Biomes. Land biomes are described in terms of, and sometimes named for, the dominant kind of climax vegetation found there. Table 1-1 lists the major land biomes, their

Table 1-1

The Major Terrestrial Biomes on Earth

Biome	Characteristics	Plants	Animals
Tundra	Permanently frozen subsoil	Lichens, mosses, grasses	Snowy owl, caribou
Taiga	Long, severe winters; summers with thawing subsoil	Conifers	Moose, black bear
Temperate forest	Moderate precipitation; cold winters; warm summers	Deciduous trees (maple, oak, beech)	Fox, deer, gray squirrel
Tropical forest	Heavy rainfall; constant warmth	Many broad-leaved plant species	Snake, monkey, leopard
Grassland	Variability in rainfall and temperature; strong winds	Grasses	Antelope, bison, prairie dog
Desert	Sparse rainfall; extreme daily temperature fluctuations	Drought-resistant plants and succulents	Lizard, tortoise, kangaroo rat

characteristics, dominant plant life, and some representative animals.

Effects of Latitude and Altitude. At the equator, the temperature and amount of rainfall remain relatively constant throughout the year. With increasing distance from the equator, temperature and rainfall show more variation during the year.

Increasing altitude may have the same effect on climate as increasing latitude. Thus, the temperature and kind of climax vegetation found at the top of a high mountain near the equator may be very much like that of a sea-level region far north of the equator. This relationship is shown in Figure 1-12.

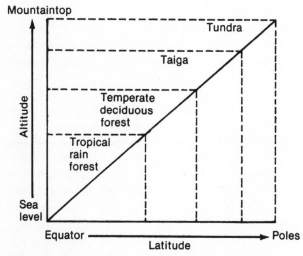

Figure 1-12. Relationship between latitude and altitude and terrestrial biomes.

Aquatic Biomes

Aquatic biomes make up the largest ecosystem on Earth. More than 70 percent of Earth's surface is covered by water; the majority of living things on Earth are water-dwellers.

Aquatic biomes are more stable than terrestrial biomes; they show less variation in temperature because water has a greater capacity to absorb and hold heat. The kinds and numbers of organisms present in an aquatic biome are affected by various abiotic factors, such as the water temperature, amounts of dissolved oxygen and carbon dioxide, intensity of light, and the kinds and amounts of dissolved minerals and suspended particles in the water (Figure 1-13).

Aquatic organisms are well adapted for the removal of dissolved oxygen from water. They also have adaptations for maintaining a proper water balance in their cells. (Water balance is affected by the concentration of salts in the water.)

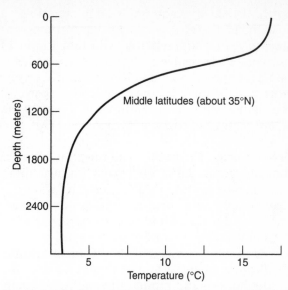

Figure 1-13. The relationship between ocean depth and water temperature: These factors, among others, have an effect on the types of organisms found in an aquatic biome.

In aquatic biomes, most photosynthesis takes place near the surface of the water, since light intensity is strongest there. At greater depths, where sunlight does not penetrate, there is no photosynthesis. However, as discussed earlier in this chapter, another type of food-making reaction takes place on parts of the ocean floor; chemosynthesis supports entire communities of organisms very different from those found elsewhere in the ocean or on land.

Marine Biome. The marine, or saltwater, biome includes all the oceans of Earth, which actually make up one continuous body of water. Most of the water on Earth is contained within the saltwater biome (Figure 1-14). The most important characteristics of the marine biome are that it: (a) is the most stable environment on

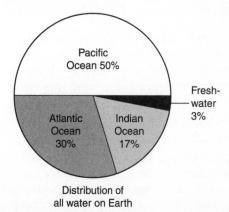

Figure 1-14. Most of Earth's water is contained within the marine biome.

Earth; (b) absorbs and holds large quantities of solar heat, thereby stabilizing Earth's temperature; (c) contains a relatively constant supply of nutrients and dissolved salts; (d) serves as a habitat for a large number and wide variety of organisms; and (e) includes the area in which most of the photosynthesis on Earth occurs (in coastal waters, along the edges of landmasses).

Freshwater Biomes. The freshwater biome includes ponds, lakes, and rivers. Because these are separate bodies of water, they vary widely in size, temperature, oxygen and carbon dioxide concentrations, amounts of suspended particles, current velocity, and rate of succession.

Ponds and lakes tend to fill in over time. Dead plant material and sediment accumulate on the bottom and around the banks, gradually making the body of water shallower and smaller (Figure 1-15). Thus, in all but the largest lakes, there is a gradual succession from a freshwater to a terrestrial climax community.

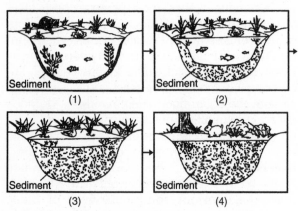

Figure 1-15. Over time, ponds tend to fill in as natural materials accumulate on the sides and bottom.

QUESTIONS
PART A

93. In which of the following biomes does most of the photosynthesis on Earth occur? (1) forests (2) oceans (3) deserts (4) grasslands

94. Drastic changes in air temperature would be *least* likely to affect which biome? (1) tundra (2) temperate forest (3) marine (4) tropical forest

95. Land biomes are characterized and named according to the (1) secondary consumers in the food webs (2) primary consumers in the food webs (3) climax vegetation in the region (4) pioneer vegetation in the region

96. The largest and most stable ecosystems are the (1) aquatic biomes (2) terrestrial biomes (3) high-altitude biomes (4) high-latitude biomes

97. Which is the most common sequence of major land biomes encountered when going from the equator to the polar region? (1) tundra, taiga, temperate forest, tropical forest (2) tropical forest, temperate forest, taiga, tundra (3) temperate forest, tropical forest, taiga, tundra (4) tropical forest, temperate forest, tundra, taiga

98. Which biome is characterized by its ability to absorb and hold large quantities of solar heat, which helps to regulate Earth's temperature? (1) desert (2) marine (3) grassland (4) taiga

99. Generally, an increase in altitude has the same effect on the habitat of organisms as an increase in (1) latitude (2) moisture (3) available light (4) longitude

PART B-1

For each description given in questions 100 through 103, select the biome from the list below that most closely matches that description.

(A) Desert

(B) Grassland

(C) Taiga

(D) Temperate deciduous forest

(E) Tundra

100. This area has a short growing season and low precipitation, mostly in the form of snow. The soil is permanently frozen and the vegetation includes lichens and mosses. (1) A (2) B (3) C (4) D (5) E

101. This area has 25 to 50 centimeters of rainfall annually. The growing season does not produce trees, but the soil is rich and well suited for growing crops such as wheat and corn. Grazing animals are found here. (1) A (2) B (3) C (4) D (5) E

102. There are many lakes in this area and the vegetation is coniferous forest composed mainly of spruce and fir. There are many large animals, such as bear and deer. (1) A (2) B (3) C (4) D (5) E

103. This area has broad-leaved trees, which shed their leaves in the fall. Winters are fairly cold,

and the summers are warm with well-distributed rainfall. (1) A (2) B (3) C (4) D (5) E

PART B-2

104. How are latitude and altitude similar in terms of how they affect the types of organisms that can live in a biome?

105. Describe the two main types of aquatic biomes; list four important abiotic factors that affect the kinds of organisms that live in them.

106. Explain why a coastal city may experience less fluctuation in temperatures during the winter and summer than a city farther inland, even though both cities may be at the same latitude.

READING COMPREHENSION

Base your answers to questions 107 through 110 on the information below and on your knowledge of biology. Source: Science News *(May 28, 2005): vol. 167, no. 22, p. 350.*

Pesticide Makes Bees Bumble

A naturally derived pesticide previously considered safe for insect pollinators may hamper the foraging of wild bees, researchers report.

Bumblebee larvae raised on pollen spiked with spinosad, an insecticide mixture of chemicals made by bacteria, grow up to be slow, clumsy foragers, say Lora Morandin of Simon Fraser University in Burnaby, British Columbia, and her colleagues in the July *Pest Management Sciences*. As adults, the bees suffer from muscle tremors and take longer to penetrate complex flower structures than do bees nourished as larvae with untainted pollen, the researchers found.

Previous studies on bees hadn't focused on sublethal effects of pesticides or on larvae, says Morandin. Moreover, researchers had looked mostly at domesticated honeybee colonies, which farmers can move before spraying a field. Wild bee colonies don't relocate during spraying, and so these bees probably suffer higher exposures to pesticides.

At least one-third of the food produced in developed countries relies on bees, birds, and other pollinators. Wild bees probably make a significant contribution, though their role hasn't been quantified. In a forthcoming *Ecological Applications*, Morandin shows that canola plants in fields with large wild-bee populations produce more seeds than do plants in fields with fewer wild bees.

Mark Winston, also of Simon Fraser University and a coauthor of the pesticide study, notes that bees were affected by spinosad concentrations that insects might reasonably encounter in a crop field. Even so, he adds, the pesticide, which farmers use on many crops against a wide range of insects, shouldn't necessarily be scrapped. Rather, he says, "timing, dose, and formulation may need to be managed in order to use this pesticide properly."

107. What effect does spinosad appear to have on wild bees' foraging (food-getting) ability?

108. Why are wild bee populations more susceptible to this insecticide than domestic bee colonies are?

109. Why is the fact that "at least one-third of the food produced in developed countries relies on bees, birds, and other pollinators" important to this article?

110. According to one of the researchers, what are some ways in which this pesticide problem can be handled?

CHAPTER 2

Cellular Processes of Living Things

CONCEPT OF LIFE

Scientists have not yet agreed on a single definition of life. Thus, life is often defined in terms of certain activities, or life functions, that are performed by all living things.

Life Functions

All living things, or **organisms**, carry on several basic life functions. First of all, *regulation* involves the control and **coordination** of the life functions. The process of **nutrition** provides all the substances that are used by an organism for the growth and repair of its **tissues**. Nutrition among heterotrophs includes the activities involved in *ingestion* (obtaining food from the environment), **digestion** (processing food for use by the organism), and *egestion* (removal of solid wastes). *Transport* includes the absorption of materials through cell membranes and the **circulation**, or distribution, of materials to all the cells of the organism. After the materials are delivered to the cells, the process of **respiration** can occur. Respiration includes the chemical activities that release energy from organic molecules for use by the cells. During respiration, the chemical bonds of **glucose** are broken down, and the energy released is stored in the compound **ATP (adenosine triphosphate)**. An organism uses the energy in ATP to perform its life functions. ATP functions much like a rechargeable battery—when it gets "run down," it is recharged by the breakdown of glucose.

Other chemical reactions are involved in building, rather than breaking down, **molecules**. During **synthesis** reactions, small molecules combine to form larger ones. *Growth* is an increase in size brought about by increases in cell size and cell number. The products of synthesis are the raw materials that are used for growth. The process of **excretion** includes all those activities that are involved in the removal of cellular waste products from the organism. These wastes include **carbon dioxide**, water, salts, and nitrogen-containing compounds. The life function of **reproduction** results in the production of new individuals. However, since each organism has a limited life span, reproduction is necessary more for the survival of each **species** (a group of like organisms) than for the individual organism itself.

Metabolism. All the chemical activities that an organism must carry on to sustain life are its *metabolism*, or **metabolic** activities. The breaking apart of glucose molecules to release their energy and the growth and repair of tissues to maintain a functioning body are both examples of metabolic activities.

Homeostasis. The maintenance of a stable internal environment in spite of changes, or **deviations**, in the external environment is known as **homeostasis**. An example of homeostasis is the maintenance of a constant body temperature in spite of temperature fluctuations in the external environment. The ability to maintain homeostasis is critical to survival. If maintenance of homeostasis fails, the organism becomes ill and, in some cases, may die.

QUESTIONS

PART A

1. The tendency of an organism to maintain a stable internal environment is called (1) homeostasis (2) nutrition (3) reproduction (4) synthesis

2. The energy available for use by the cell is obtained from the life function of (1) reproduction (2) respiration (3) transport (4) synthesis

3. The chemical process by which complex molecules of protein are made from simple molecules is called (1) regulation (2) respiration (3) synthesis (4) excretion

4. Which life function includes the absorption and circulation of essential substances throughout an organism? (1) transport (2) excretion (3) ingestion (4) nutrition

5. Which term includes all the chemical activities carried on by an organism? (1) regulation (2) metabolism (3) digestion (4) respiration

6. Which life activity is *not* required for the survival of an individual organism? (1) nutrition (2) respiration (3) reproduction (4) synthesis

7. In an ameba, materials are taken from its environment and then moved throughout its cytoplasm. These processes are known as (1) absorption and circulation (2) food processing and energy release (3) energy release and synthesis (4) coordination and regulation

8. In an organism, the coordination of the activities that maintain homeostasis in a constantly changing environment is a process known as (1) digestion (2) regulation (3) synthesis (4) respiration

PART B-2

9. Identify a life function that provides the substances an organism uses for its growth and for repair of its tissues.

10. Why are such different things as single-celled amebas and multicelled humans both considered to be organisms?

PART C

11. You are working as a biologist in a laboratory. An unknown specimen is brought in for analysis. Describe the steps you would take to determine if the specimen is a living organism or simply a collection of nonliving molecules.

CELLULAR STRUCTURE OF LIVING THINGS

All living things are composed of **cells**. Some organisms consist of only one cell, while others may consist of billions of cells. The processes that are essential for the survival of an organism are performed by its cells. Most of the cells in a multicellular organism are capable of performing all the life functions independently, as well as with the other cells of the body.

The Cell Theory

The *cell theory*, which is one of the major theories of biology, can be stated as follows: (a) Every organism is made up of one or more cells; (b) the cell is the basic unit of structure and function in all living things (for example, cells make, or synthesize, proteins and release energy); and (c) all cells come only from preexisting cells (that is, new cells are formed when previously existing cells divide).

Development of the Cell Theory. During the last four centuries, improvements in the microscope and the development of other techniques have made it possible for biologists to observe and study cells. The cell theory was developed from the work of a number of scientists. First, *Anton van Leeuwenhoek* (1632–1723) made powerful simple microscopes (magnifying glasses) that he used to study living cells; he was the first person to observe sperm cells, bacteria, and protozoa. Then, *Robert Hooke* (1635–1703) made compound microscopes (microscopes with two or more lenses) that he used to observe thin slices of cork; he used the term "cells" to describe the small compartments that make up cork tissue. In 1831, *Robert Brown* concluded from his studies that all plant cells contain a nucleus. Later, in 1838, *Matthias Schleiden* concluded that all plants are made up of cells and, in 1839, *Theodor Schwann* concluded that all animals are made up of cells. Finally, in 1855, *Rudolph Virchow* concluded that all cells arise only from preexisting cells. These last three ideas formed the basis of the cell theory.

Exceptions to the Cell Theory. Recent discoveries have led scientists to identify several exceptions to the cell theory. For example, mitochondria and chloroplasts, which are cell organelles, contain genetic material (DNA) and can duplicate themselves within living cells. Another exception is the **virus**, which is not a living cell. It consists of an outer coat of protein surrounding a core of DNA or RNA. A virus can reproduce while it is inside a living host cell; but outside the host organism, it shows no sign of life. As such, viruses are not included in any of the six **kingdoms** of living things. Recently, scientists have discovered tiny particles known as *viroids* and *prions*, which also are not living cells. However, like viruses, these infectious particles have the ability to reproduce themselves and cause diseases when they enter a living host organism.

There are exceptions to the cell theory among multicellular organisms as well. For example, some tissues in multicellular plants and animals do not appear to be made up of clearly identifiable cells. In humans, skeletal muscle tissue does not show distinct boundaries between the cells. In plants, some tissues found in seeds are also not clearly cellular. A group of protists known as *slime molds* have tissues that do not appear to be made up of individual cells, either. Finally, the very first living cells on Earth must have developed from non-cellular matter (that is, not from preexisting cells).

Cell Structure.   Cells contain a variety of small structures, called **organelles**, which perform specific functions (Figure 2-1).

The **cell membrane**, or *plasma membrane*, surrounds and protects the cell and separates the cell contents from the environment. The membrane consists of a double lipid layer in which large protein molecules float. The cell membrane is *selectively permeable*; this means that some substances can pass through it, while others cannot. In this way, the **membrane** regulates the passage of materials into and out of the cell and controls the cell's chemical makeup.

The **cytoplasm** is the fluidlike material that fills the space between the cell membrane and the nucleus. Many metabolic reactions occur in the cytoplasm, which consists mainly of water. The organelles are suspended in the cytoplasm.

The **nucleus** is the control center of the cell. It is surrounded by a nuclear membrane and contains the genetic material, which is found in the **chromosomes**. The chromosomes are made of **DNA** (**deoxyribonucleic acid**) and protein.

The **ribosomes** are tiny organelles that are suspended in the cytoplasm and attached to the membranes of the endoplasmic reticulum. Protein synthesis takes place at the ribosomes.

The **mitochondria** are the sites of most reactions of aerobic **cellular respiration**, the process by which energy is released from nutrient molecules (such as glucose). Most of the ATP produced by aerobic respiration is synthesized in the mitochondria.

The **vacuoles** are fluid-filled organelles surrounded by membranes. In single-celled organisms, digestion occurs in food vacuoles and excess water collects in contractile vacuoles, which pump it out of the cell. Plant cells contain very large vacuoles that may fill much of the cell's interior. In animal cells, there are relatively few vacuoles, and they are small.

The **chloroplasts** are small, pigment-containing organelles found in the cytoplasm of plants, algae, and some protists. Photosynthesis takes place in the chloroplasts.

The *cell wall* is a nonliving structure found outside the cell membrane of plant, algal, and fungal cells. It provides strength and rigidity, but does not interfere with the passage of materials into or out of the cell.

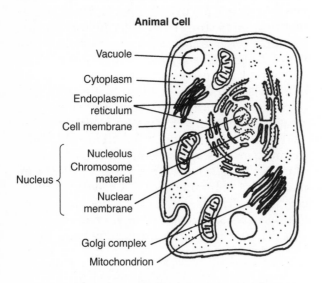

Animal Cell

Vacuole
Cytoplasm
Endoplasmic reticulum
Cell membrane
Nucleolus
Chromosome material
Nuclear membrane
Nucleus
Golgi complex
Mitochondrion

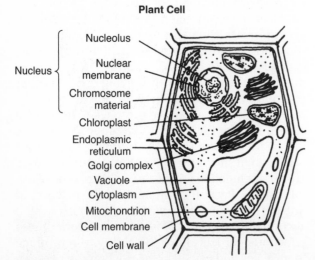

Plant Cell

Nucleolus
Nuclear membrane
Chromosome material
Nucleus
Chloroplast
Endoplasmic reticulum
Golgi complex
Vacuole
Cytoplasm
Mitochondrion
Cell membrane
Cell wall

Figure 2-1. Typical cell organelles.

QUESTIONS

12. The unit of structure and function of all living things is (1) an organ (2) an atom (3) a cell (4) a nucleolus

13. According to the cell theory, which statement is correct? (1) Viruses are true living cells. (2) All cells are basically different in structure. (3) Mitochondria are found only in plant cells. (4) All cells come from preexisting cells.

14. Chloroplasts and mitochondria are examples of (1) cells (2) tissue (3) organelles (4) organs

15. Which statement best describes the term "theory" as used, for example, in the "cell theory"? (1) A theory is never revised as new scientific evidence is presented. (2) A theory is an assumption made by scientists and implies a lack of certainty. (3) A theory refers to a scientific explanation that is strongly supported by a variety of experimental data. (4) A theory is a hypothesis that has been supported by one experiment performed by two or more scientists.

16. The term "selectively permeable" is used in reference to the (1) nucleus (2) cell wall (3) cytoplasm (4) cell membrane

17. The part of a cell that is in most direct contact with the environment is the (1) nucleus (2) cell membrane (3) mitochondrion (4) vacuole

18. Plant cell organelles that contain photosynthetic pigments are (1) chloroplasts (2) ribosomes (3) chromosomes (4) cell walls

19. An observable difference between an onion's epidermal (skin) cells and a human's cheek cells is that the onion's cells have a (1) cell membrane (2) nucleus (3) vacuole (4) cell wall

20. The sites of protein synthesis in the cytoplasm are the (1) ribosomes (2) chromosomes (3) nuclei (4) vacuoles

21. The watery environment in which most life activities of a cell take place is the (1) cell membrane (2) chloroplast (3) cytoplasm (4) vacuole

22. Transport of materials into and out of a cell is most closely associated with the (1) nucleus (2) cell wall (3) ribosome (4) cell membrane

23. Which organelle contains genetic material and controls most cell activities? (1) nucleus (2) cell membrane (3) vacuole (4) endoplasmic reticulum

24. The cell organelles that are the sites of aerobic cellular respiration in both plant and animal cells are the (1) mitochondria (2) vacuoles (3) chloroplasts (4) nuclei

25. An increase in the concentration of ATP in a muscle cell is a direct result of which life function? (1) cellular respiration (2) reproduction (3) digestion (4) excretion

26. A nonliving cell structure is a (1) cell membrane (2) nucleus (3) cell wall (4) mitochondrion

27. In a cell, information that controls the production of proteins must pass from the nucleus to the (1) cell membrane (2) chloroplasts (3) mitochondria (4) ribosomes

28. The arrows in the diagram below indicate the movement of materials into and out of a single-celled organism. The movements indicated by these arrows are directly involved in (1) the maintenance of homeostasis (2) photosynthesis only (3) excretion only (4) the digestion of minerals

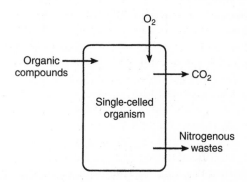

29. In the diagram of a single-celled organism shown below, the arrows indicate various activities taking place. Which systems perform these same activities in humans? (1) digestive, circulatory, and immune (2) excretory, respiratory, and reproductive (3) respiratory, excretory, and digestive (4) respiratory, nervous, and endocrine

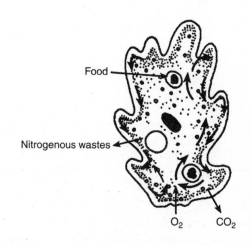

30. Explain why scientists have difficulty including viruses in any category of living organisms.

Chapter 2: Cellular Processes of Living Things   **21**

TOOLS AND METHODS OF CELL STUDY

There are various scientific tools and methods that enable the up-close study of cell structures and functions. These different techniques and types of equipment are used to study cells and cell parts at varying levels of magnification and in different conditions. For example, some tools are used for the study of live cells, while others can be used only for the examination of preserved (dead) cells. Some of these tools and techniques are described below.

Compound Light Microscope
A microscope that uses two lenses or sets of lenses to form an enlarged image is called a *compound light microscope.* Light passes through the specimen, the objective lens, and the ocular lens, or *eyepiece*, before reaching the eye. The objective lens produces a magnified image that is further enlarged by the ocular lens. The main parts of a compound light microscope are shown in Figure 2-2. The functions of these parts are listed in Table 2-1. The amount of enlargement of an image produced by the lenses of a microscope is its *magnifying power.* For a compound microscope, magnifying power is found by multiplying the magnifying power of the objective lens by the magnifying power of the ocular lens. For example, if the magnifying power of the objective is 40 × (40 times) and that of the ocular is 10 × (10 times), the total magnification is $40 \times 10 = 400 \times$ (400 times). The greater the magnification of a specimen, the smaller the field of vision, or observable area. The *resolution*, or resolving power, is the capacity of the microscope to show, as separate, two points that are close together.

Other Types of Microscopes
A microscope that has an ocular lens and an objective lens for each eye is called a binocular or *dissecting microscope.* Dissecting microscopes, which produce a three-dimensional image, have relatively low magnifying power and are used for viewing fairly large, opaque specimens. For more magnifying power and the ability to observe unstained, living cells, a *phase-contrast microscope* is used; it makes visible parts that cannot be seen with an ordinary light microscope. The most powerful kind of microscope is the *electron microscope*, which can magnify an object more than 400,000×. Unlike other microscopes, the electron microscope uses an electron beam focused by electromagnets, instead of light and lenses. One disadvantage of the electron microscope is that only dead specimens can be viewed with it.

Techniques of Cell Study
Tiny instruments that can be used, with the aid of a microscope, to remove or transfer the parts of a cell are *microdissection* instruments. For example, with the use of microdissection instruments, a nucleus can be transferred from one cell to another. The laboratory instrument that is used to separate small particles or materials on the basis of density is the *ultracentrifuge*. In fact,

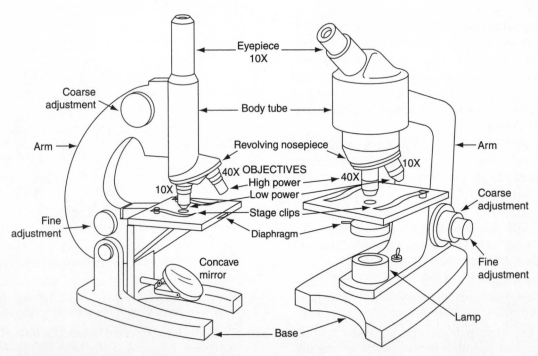

Figure 2-2. Main parts of the compound light microscope.

Table 2-1

Parts of the Compound Light Microscope and Their Functions

Part	Function
Base	Supports the microscope
Arm	Used to carry microscope; attaches to the base, stage, and body tube
Body tube	Holds the objective lens and eyepiece
Stage	Platform on which the glass slide with the specimen is placed (over the hole in the stage through which light passes)
Clips	Hold the slide in position on the stage
Nosepiece	Holds the objective lenses; rotates so that the different objective lenses can be moved in line with the specimen and eyepiece
Coarse adjustment	Larger knob used for rough-focusing with the low-power objective
Fine adjustment	Smaller knob used for focusing with the high-power objective and for final focusing with the low-power objective
Mirror or Lamp	Directs light to the specimen (on the stage)
Diaphragm	Controls the amount of light reaching the specimen
Objective lenses	Lenses mounted on the nosepiece
Ocular lens	Lens at the top of the body tube; commonly called the *eyepiece*

various cell organelles can be isolated by the process of ultracentrifugation. The ultracentrifuge spins the sample in a test tube at very high speeds so that particles of different densities settle to the bottom of the test tube in layers. In addition, cell structures can be made clearly visible by the use of various *staining* techniques. Depending on its specific chemical makeup, a particular stain will be absorbed only by certain parts of the cell. For example, methylene blue and iodine are stains that are absorbed by the nucleus. Other parts of the cell can be made visible with other stains.

The unit used in measuring structures that can be viewed with a compound light microscope is the *micrometer* (μm). One micrometer equals 0.001 millimeter (mm); 1000 micrometers equal 1 millimeter. The diameter of the low-power field of a compound light microscope is commonly about 1500 μm. A paramecium is about 250 μm (0.25 mm) long. (Measurement with a microscope is discussed in greater detail in Chapter 10.)

QUESTIONS

PART A

31. Which of the following plant cell structures could *not* be seen when using the 10× objective of a compound microscope? (1) nucleus (2) cell wall (3) cytoplasm (4) endoplasmic reticulum

32. A microscope reveals one hundred similar cells arranged end-to-end in a space of 1 millimeter. The average length of each cell must be (1) 0.1 micrometer (2) 10 micrometers (3) 100 micrometers (4) 1000 micrometers

33. Which instrument would provide the most detailed information about the internal structure of a chloroplast? (1) a compound light microscope (2) a phase-contrast microscope (3) an electron microscope (4) an ultracentrifuge

34. If the low-power objective and the eyepiece both have a magnifying power of 10×, the total magnifying power of the microscope is (1) 10× (2) 100× (3) 1× (4) 20×

35. To separate the parts of a cell by differences in density, a biologist would probably use (1) a microdissection instrument (2) an ultracentrifuge (3) a phase-contrast microscope (4) an electron microscope

36. Which microscope magnification should be used to observe the largest field of view of an insect wing? (1) 20× (2) 100× (3) 400× (4) 900×

37. The diameter of the field of vision of a compound light microscope is 1.5 millimeters. This may also be expressed as (1) 15 micrometers (2) 150 micrometers (3) 1500 micrometers (4) 15,000 micrometers

38. To transplant a nucleus from one cell to another cell, a scientist would use (1) an electron microscope (2) an ultracentrifuge (3) microdissection instruments (4) staining techniques

39. A student used a compound microscope to measure the diameters of several red blood cells and found that the average length was 0.008 millimeter. What was the average length of a single red blood cell in micrometers? (1) 0.8 (2) 8 (3) 80 (4) 800

40. A student using a compound microscope estimated the diameter of a cheek cell to be about 50 micrometers. What is the diameter of this cheek cell in millimeters? (1) 0.050 (2) 0.500 (3) 5.00 (4) 50.0

41. A student has a microscope with a 10× eyepiece and 10× and 40× objectives. She observed 40 onion epidermal cells across the diameter of the low-power field. How many cells would she observe under high power? (1) 1 (2) 40 (3) 10 (4) 4

42. After examining cells from an onion root tip under high power, a student switches to the low-power objective without moving the slide. He would most likely see (1) more cells and less detail (2) more cells and more detail (3) fewer cells and less detail (4) fewer cells and more detail

PART B-1

43. The diagram below represents the field of vision of a microscope. What is the approximate diameter in micrometers of the cell shown in the field? (1) 50 (2) 500 (3) 1000 (4) 2000

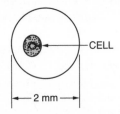

CELL

← 2 mm →

PART B-2

44. Select any three parts that are labeled in the diagram below and, for each part selected:
- identify the part;
- state the function of that part.

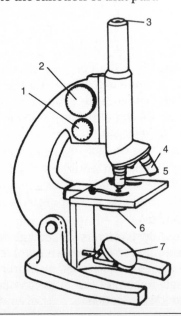

45. Explain how a biology student can calculate the magnification of a specimen when the powers of the eyepiece lens and the objective lens are known.

PART C

Answer question 46 based on the following information and data table.

46. A lab microscope has two interchangeable eyepieces and four objective lenses. The table below shows various combinations of the eyepiece and objective lenses and the apparent magnification of the specimen image produced. Use the information provided in the table to complete the missing data.

Eyepiece Lens	Objective Lens	Magnification of Image
10×		100×
	40×	400×
15×	90×	
10×		150×
		900×

47. Briefly describe how the development of the compound microscope and other magnifying instruments greatly advanced the science of biology.

BIOCHEMISTRY

The chemical reactions necessary to sustain life take place in the cells. The study of the chemical reactions of living things is called *biochemistry*.

Elements
A substance that cannot be broken down into simpler substances is called an *element*. Examples of elements include hydrogen, oxygen, sodium, and potassium. The most abundant elements in living things are **carbon, hydrogen, oxygen,** and **nitrogen**. Elements found in lesser amounts in living things include sulfur, phosphorus, magnesium, iodine, iron, calcium, chlorine, potassium, and others.

Atoms
All elements are made up of particles called **atoms**. Each element has a different kind of atom. The atoms of different elements differ in the numbers

of protons, neutrons, and electrons they contain. A *compound* is formed when two or more elements combine chemically. For example, water (H_2O) is formed by the chemical combination of two hydrogen atoms and one oxygen atom.

Chemical Bonding

The formation of compounds involves either the transfer or the sharing of electrons between atoms, resulting in the formation of chemical bonds. When atoms lose or gain electrons, they become electrically charged particles called ions, and an *ionic bond* is formed. When atoms share electrons, a *covalent bond* is formed. When a compound forms, it has properties that are different from those of the elements that make it up.

Inorganic and Organic Compounds

There are two basic classes of chemical compounds: inorganic compounds and organic compounds. Both types are found in living things.

Compounds that do not contain both carbon and hydrogen atoms are **inorganic** compounds. Inorganic compounds found in cells include water, salts, carbon dioxide, and inorganic acids, such as hydrochloric acid (HCl).

Compounds that contain both carbon and hydrogen atoms are **organic** compounds. Because carbon atoms can form four covalent bonds with other atoms, organic compounds are often large and complex. The major categories of organic compounds are carbohydrates, proteins, lipids, and nucleic acids.

Carbohydrates

Sugars and starches, which are used primarily as sources of energy and as food-storage compounds, are *carbohydrates*. These substances are made up of carbon, hydrogen, and oxygen, and the ratio of hydrogen to oxygen is always 2 to 1. The simplest carbohydrates are the *monosaccharides*, or **simple sugars**. Glucose, galactose, and fructose, each with the formula $C_6H_{12}O_6$, are simple sugars.

Some carbohydrates, such as maltose and sucrose (both $C_{12}H_{22}O_{11}$) are known as *disaccharides*, sugars whose molecules are made up of two monosaccharide molecules bonded together. For example, a maltose molecule is formed from two glucose molecules that are bonded together.

The complex carbohydrates that are made up of chains of monosaccharides are called *polysaccharides*. Starch, cellulose, and glycogen are polysaccharides that are made up of chains of glucose molecules. In plants, **starch** is a food storage compound and cellulose makes up the cell walls. In animals, glycogen is the food-storage compound.

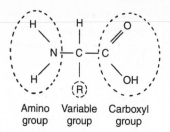

Amino group   Variable group   Carboxyl group

Figure 2-3. Generalized structure of an amino acid.

Proteins

Enzymes, hormones, and various structural parts of organisms are **proteins**. Proteins are made up of smaller **subunits** called **amino acids.**

Structure of Amino Acids. Amino acids contain the elements carbon, hydrogen, oxygen, and nitrogen. Some also contain sulfur. Figure 2-3 shows the generalized structure of an amino acid.

The $-NH_2$ is an amino group; the $-COOH$ is a carboxyl, or acid, group; and the *R* represents a variable group. The *R* group is the part of the amino acid structure that differs from one amino acid to another. Twenty different amino acids are found in the cells of living things.

Like carbohydrates, amino acids combine chemically to form more complex molecules. When two amino acids combine, they form a *dipeptide*. The bond that holds the amino acids together is called a *peptide bond*. More amino acids may combine with a dipeptide to form a *polypeptide*. A protein is made up of one or more polypeptide chains. There are a great many types of protein molecules in living things. These molecules differ in the number, kinds, and sequences of amino acids they contain.

Lipids

Fats, oils, and waxes belong to a class of organic compounds called **lipids**. They serve mainly as sources of energy and as components of structures such as cell membranes. Lipids that are solid at room temperature are *fats*, while those that are liquid are *oils*. Lipids contain carbon, hydrogen, and oxygen. The ratio of hydrogen atoms to oxygen atoms is greater than 2 to 1 and varies from one lipid to another. The building blocks of lipids are fatty acids and glycerol.

Nucleic Acids

DNA and RNA are two very important classes of organic molecules; they are both involved in the passing on of genetic information from one generation to the next. **DNA**, or *deoxyribonucleic acid*, is located in the nucleus, mitochondria, and chloroplasts of cells, where it stores vital information for the cell. DNA molecules are capable of duplicating, or *replicating*, themselves. In addition to transmitting traits, DNA is responsible for the synthesis (production) of proteins, which all cells need to survive. **RNA (ribonucleic acid)** is similar to DNA in that both molecules are *polymers*; this means they are composed of thousands of smaller chemical units that are repeated over and over again, thus forming giant molecules. RNA assists in carrying out the vital information (such as the instructions for protein synthesis) stored in the DNA molecules. In some viruses, RNA, rather than DNA, is the genetic material that allows the virus to function. For example, HIV—the human immunodeficiency virus that causes AIDS—is a type of RNA virus. (Nucleic acids are discussed in greater detail in Chapter 7: Genetics and Heredity.)

QUESTIONS

PART A

48. What is the principal inorganic solvent in cells? (1) salt (2) water (3) alcohol (4) carbon dioxide

49. Fats that are stored in human tissue contain molecules of (1) glycerol and fatty acids (2) amino acids (3) monosaccharides and disaccharides (4) nucleotides

50. One of the carbon compounds found in a cell has twice as many hydrogen atoms as oxygen atoms. This compound most likely belongs to the group of substances known as (1) nucleic acids (2) lipids (3) proteins (4) carbohydrates

51. Which formula represents an organic compound? (1) NH_3 (2) H_2O (3) $NaCl$ (4) $C_{12}H_{22}O_{11}$

52. Starch is classified as a (1) disaccharide (2) polypeptide (3) nucleotide (4) polysaccharide

53. Which organic compound is correctly matched with the subunit that composes it? (1) maltose—amino acid (2) starch—glucose (3) protein—fatty acid (4) lipid—sucrose

PART B-2

54. There are only 20 different amino acids found in living things, yet there are thousands of different proteins. Explain why this is possible.

55. Explain why starch molecules and protein molecules are both called polymers.

56. Examine each of the four molecular structures shown below. Identify each molecule as organic or inorganic and explain why it is classified as organic or inorganic.

PART C

Base your answer to question 57 on the following information and data table.

57. A lab was set up for students to analyze three unknown samples of organic molecules—a lipid, a carbohydrate, and a protein. The results of their lab tests are shown in the table below. Based on these results, identify each sample as a protein, carbohydrate, or lipid, and then state the reason for your identification of each molecule.

Unknown Sample	Elements Contained	Molecular Characteristics
A	C, H, O, and N	Polymer, high molecular mass
B	C, H, and O	Very little oxygen, much hydrogen
C	C, H, and O	Twice as much hydrogen as oxygen

58. There are four major types of organic molecules that are important in living things: carbohydrates, lipids, proteins, and nucleic acids. Select any two molecules and, for each one chosen:
   • describe the structure of the molecule;
   • state two ways that the molecule is useful to living organisms.

ENZYMES

Role of Enzymes

Chemical reactions occur continuously in living things. Each reaction requires the presence of a special protein called an **enzyme**, which regulates the rate of the reaction. In general, enzymes speed up the rate of a reaction. Enzymes are **catalysts**, substances that change the rate of a chemical reaction but are themselves unchanged by the reaction.

Enzymes are named after their *substrates*, the substances they act on. The name of an enzyme generally ends in *ase*. For example, a lipase acts on lipids, a protease acts on proteins, and maltase acts on the sugar maltose.

Enzyme Structure

An enzyme is a large, complex protein that consists of one or more polypeptide chains. In addition to the protein, some enzymes contain a nonprotein component called a *coenzyme*. If the coenzyme part is missing, the enzyme will not function. *Vitamins* often function as coenzymes. Although most enzymes are made up of proteins, some enzymes are composed of another type of organic molecule, that is, RNA. This type of enzyme is sometimes called a *ribozyme*.

The polypeptide chains that make up an enzyme are folded in a highly specific way, forming pockets on the enzyme surface into which the substrate molecule or molecules fit. The specific part of the enzyme where the substrate fits is called the *active site*.

Models of Enzyme Action

Different models can be used to describe the mechanism of enzyme action. According to the *lock-and-key model*, the active site on an enzyme has a unique three-dimensional shape that can form a complex with only one type of substrate. The substrate fits an active site just as a key fits a lock (Figure 2-4). However, it is important to note that the enzyme molecule is not rigid. Rather, according to the *induced fit model* of enzyme action, when it binds to a substrate at the active site, the enzyme molecule bends somewhat to cause, or induce, a closer fit between itself and

the substrate. This enhanced fit allows the enzyme to function in a more effective manner.

For an enzyme to affect the rate of a chemical reaction, the substrate must become attached to the active site of the enzyme, forming an *enzyme-substrate complex*. The enzyme's action occurs while the enzyme and substrate are bound together. At this time, bonds of the substrate may be weakened, causing it to break apart, or bonds may form between substrate molecules, joining them together. After the reaction is complete, the enzyme and product(s) separate, and the enzyme molecule becomes available to act on other substrate molecules.

Factors Influencing Enzyme Action

The rate of enzyme action is affected by temperature, concentrations of enzyme and substrate, and pH.

Temperature. The rate of enzyme action varies with temperature. Up to a point, the rate increases with increasing temperature (Figure 2-5). The temperature at which the enzyme functions most efficiently is called the optimum temperature. If the temperature is raised above the optimum, the rate of enzyme action begins to decrease. The decrease in enzyme action occurs because the higher temperature destroys the three-dimensional shape of the enzyme protein. In this process, known as *denaturation*, the shape of the enzyme's active site is altered so that it no longer fits the substrate. In humans, the normal body temperature of about 37°C is also the optimum temperature for most human enzymes. Denaturation of these enzymes begins at about 40°C, upsetting the body's homeostasis.

Enzyme and Substrate Concentrations. The rate of enzyme action varies with the amount of available substrate. With a high concentration of enzyme and a low concentration of substrate,

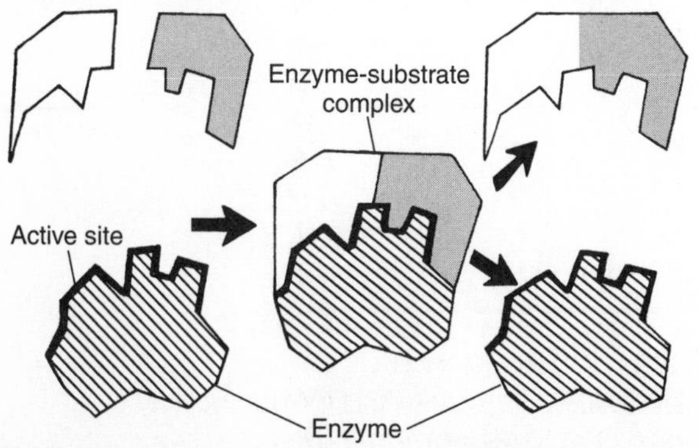

Figure 2-4. The lock-and-key model of enzyme action.

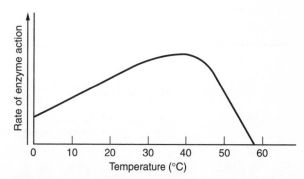

Figure 2-5. The effect of changing temperature on the rate of enzyme action.

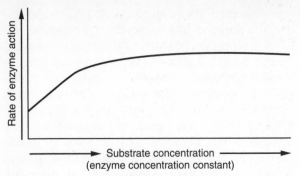

Figure 2-6. The effect of changing substrate concentration on the rate of enzyme action.

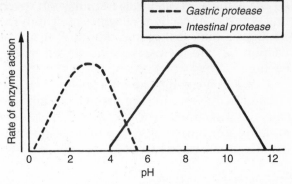

Figure 2-8. The effect of pH on the rate of enzyme action.

the rate of enzyme action increases as the substrate concentration increases (Figure 2-6). At the point where all enzyme molecules are reacting, the rate levels off, and addition of more substrate has no further effect.

pH. The rate of enzyme action varies with the pH of the environment. The **pH** scale is a measure of the hydrogen ion (H^+) concentration of a solution. Solutions with a pH of 7 are neutral. Those with a pH below 7 are acids, while those with a pH above 7 are bases (Figure 2-7).

Each enzyme has a particular pH at which it functions most efficiently. For example, most enzymes in human blood function best in neutral solutions. However, pepsin, an enzyme in the stomach, works best at a pH of 2 (a very high **acidity** level), and trypsin, an enzyme in the small intestine, works best at a pH of 8 (Figure 2-8).

QUESTIONS

PART A

59. Which of the following is characteristic of an enzyme? (1) It is an inorganic catalyst. (2) It is destroyed after each chemical reaction. (3) It provides energy for any chemical reaction. (4) It regulates the rate of a specific chemical reaction.

60. The "lock-and-key" model of enzyme action illustrates that a particular enzyme molecule will (1) form a permanent enzyme-substrate complex (2) be destroyed and resynthesized several times (3) interact with a specific type of substrate molecule (4) react at identical rates under all conditions

61. An enzyme-substrate complex may result from the interaction of molecules of (1) glucose and lipase (2) fat and amylase (3) sucrose and maltase (4) protein and protease

62. The part of the enzyme molecule into which the substrate fits is called the (1) active site (2) coenzyme (3) polypeptide (4) protease

63. A nonprotein molecule necessary for the functioning of a particular enzyme is called a (1) catalyst (2) polypeptide (3) coenzyme (4) substrate

64. Which of the following variables has the *least* direct effect on the rate of an enzyme-regulated reaction? (1) temperature (2) pH (3) carbon dioxide concentration (4) enzyme concentration

65. The diagram below represents a beaker containing a solution of various molecules involved in digestion. Which structures represent

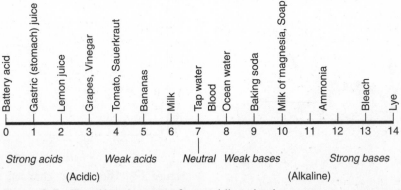

Figure 2-7. The pH scale ranges from acidic to basic.

products of digestion? (1) *A* and *D* (2) *B* and *C* (3) *B* and *E* (4) *D* and *E*

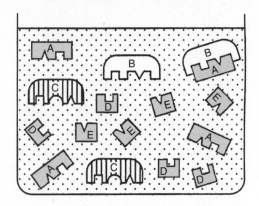

66. Enzymes have an optimum temperature at which they work best. Temperatures above and below this optimum will decrease enzyme activity. Which graph best illustrates the effect of temperature on enzyme activity?

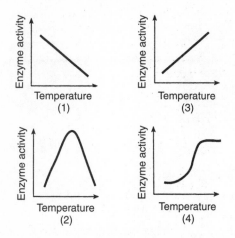

67. A word equation is shown below. This reaction is most directly involved in the process of (1) reproduction (2) protein synthesis (3) replication (4) heterotrophic nutrition

starch molecules ———biological catalyst———▸ simple sugars

68. The change in shape of enzyme molecules that occurs at high temperatures is known as (1) synthesis (2) specificity (3) replication (4) denaturation

PART B-1

Base your answers to questions 69 through 71 on the following graph and on your knowledge of biology. The graph represents the rate of enzyme action when different concentrations of enzyme are added to a system with a fixed amount of substrate.

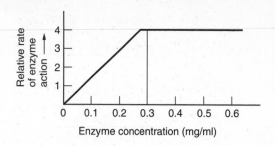

69. At which enzyme concentration does all of the available substrate react with the enzyme? (1) 0.1 mg/mL (2) 0.2 mg/mL (3) 0.3 mg/mL (4) 0.5 mg/mL

70. When the enzyme concentration is increased from 0.5 mg/mL to 0.6 mg/mL, the rate of enzyme action (1) decreases (2) increases (3) remains the same

71. If more substrate is added to the system at an enzyme concentration of 0.4 mg/mL, the rate of the reaction would most likely (1) decrease (2) increase (3) remain the same

Base your answers to questions 72 and 73 on the following graphs. Graph I shows the relationship between temperature and the relative rates of activity of enzymes A and B. Graph II shows the relationship between pH and the relative rates of activity of enzymes A and B.

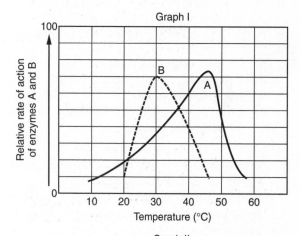

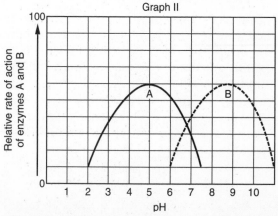

72. Under which conditions is enzyme A most effective? (1) at 40°C and a pH of 5 (2) at 45°C and a pH of 5 (3) at 45°C and a pH of 9 (4) at 50°C and a pH of 9

73. The optimum environment for enzyme B is (1) a basic medium (2) an acidic medium (3) either an acidic or a basic medium (4) a neutral medium

PART B-2

Use your knowledge of enzymes and biology to answer questions 74 and 75.

74. Fresh pineapple contains an enzyme that digests proteins. Adding fresh pineapple to gelatin (a protein) prevents it from setting or jelling. Adding cooked or canned pineapple does not have this effect and the gelatin can set normally. Explain why these different effects occur.

75. When an apple is cut open, the inside soon turns brown. This is because enzymes that are released from the cut cells react with certain molecules in the apple. Rubbing lemon juice (which contains citric acid) on the cut apple prevents it from browning. Explain why this is so.

PART C

76. The human stomach contains an enzyme called pepsin, which actively breaks down (digests) protein molecules found in food. Based on your knowledge of biology, answer the following:

• At what temperature would you expect pepsin to work best? Give a reason for your answer.

• Why would drinking very cold beverages have a negative effect on digestion of food in the stomach?

77. Draw a diagram in which you show how the enzyme maltase combines with two glucose molecules to form maltose. Label the enzyme, substrate, enzyme-substrate complex, and end product.

78. An incomplete graph is shown below. What label could appropriately be used to replace letter Z on the horizontal axis?

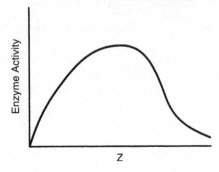

Effect of Z on Enzyme Activity

Base your answers to questions 79 through 81 on the two different cells shown below. Only cell A produces substance X. Both cells A and B use substance X.

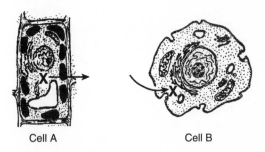

Cell A   Cell B

79. Identify substance X.

80. Identify the type of organelle in cell A that produces substance X.

81. Identify the type of organelle found in both cell A and cell B that uses substance X.

82. The enzyme catalase is found in almost all living tissues. This enzyme catalyzes the breakdown of harmful hydrogen peroxide in the body. Liver tissue is particularly rich in catalase content. Like all enzymes, catalase is affected by temperature fluctuations. Design and describe an experiment in which a person can study the activity of catalase over a range of temperatures, from 0°C to 80°C. Be sure to include an appropriate control and a data table in your experimental design.

Base your answers to questions 83 through 86 on the information below and on your knowledge of biology. Source: Living Environment Regents Exam—June 2004, p. 7. (Adapted from: Paul Recer, The Daily Gazette, April 26, 1988.)

Researchers Find New Means to Disrupt Attack by Microbes

Some of the most common and deadly bacteria do their mischief by forming a sticky scum called biofilm. Individually, the microbes are easy to control, but when they organize themselves into biofilms they can become deadly, said Dr. Barbara Iglewski of the University of Rochester.

Biofilms are actually intricately organized colonies of billions of microbes, all working in a coordinated way to defend against attack and to pump out a toxin that can be deadly.

Once they are organized, the bacteria are highly resistant to antibiotics and even strong detergents often cannot wash them away or kill them.

Iglewski and colleagues from Montana State University and the University of Iowa report in *Science* that they discovered how the microbes in the colonies communicate and found that once this conversation is interrupted, the deadly bugs can be easily washed away.

Using *Pseudomonas aeruginosa*, a common bacteria that is a major infection hazard in hospitals and among cystic fibrosis patients, the researchers isolated a gene that the bacteria uses to make a communications molecule. The molecule helps the microbes organize themselves into a biofilm—a complex structure that includes tubes to carry in nutrients and carry out wastes, including deadly toxins.

In their study, the researchers showed that if the gene that makes the communications molecule was blocked, the *Pseudomonas aerginosa* could form only wimpy [weak], unorganized colonies that could be washed away with just a soap that has no effect on a healthy colony.

83. What is one characteristic of a biofilm?

(1) presence of tubes to transport material into and out of the colony

(2) presence of a nervous system for communication within the colony

(3) ease with which healthy colonies can be broken down by detergents

(4) lack of resistance of the bacterial colony to antibiotics

84. Which statement best describes *Pseudomonas aeruginosa* bacteria?

(1) They cause mutations in humans.

(2) They are easy to control.

(3) They cause major infection problems in hospitals.

(4) They are deadly only to people with cystic fibrosis.

85. The tubes in biofilms function much like the human

(1) muscular and nervous systems

(2) circulatory and excretory systems

(3) digestive and endocrine systems

(4) reproductive and respiratory system

86. Bacteria that form biofilms may be controlled most effectively by

(1) antibiotics

(2) detergents

(3) cutting the tubes through which the bacteria communicate

(4) blocking the expression of a gene that helps the colonies to organize

CHAPTER 3 Maintenance in Living Things

Most living organisms perform all the same life functions. They obtain and process food, and distribute nutrients and other essential materials to their cells. They get rid of wastes produced as a result of cell metabolism. All of these life functions must be regulated.

Different kinds of organisms have specific structures and behavioral patterns that enable them to perform the life functions efficiently within their physical surroundings, or *environment*. These structures and behavioral patterns are called **adaptations**.

NUTRITION

Nutrition includes those activities by which organisms obtain and process food for use by the cells. The cells use **nutrients** from foods for energy, growth, repair, and regulation. Nutrition may be autotrophic or heterotrophic. In **autotrophic** nutrition, the organism can synthesize organic substances (nutrients) from inorganic substances obtained from the environment. In **heterotrophic** nutrition, the organism must ingest needed organic substances from other organisms in the environment.

Photosynthesis

The most common type of autotrophic nutrition is **photosynthesis**. Organisms that carry out photosynthesis are called *autotrophs*—a group that includes all plants, some **bacteria** (monerans), and all **algae** (types of protists). In photosynthesis, the organism uses carbon dioxide and water taken from the environment along with energy from sunlight to synthesize the organic compound glucose. Most of the chemical energy available to living organisms comes either directly or indirectly from photosynthesis. Also, most of the oxygen in the air comes from photosynthesis.

Photosynthetic Pigments. Photosynthesis requires the presence of certain colored substances called *pigments*, which "trap" light energy and convert it to a form of chemical energy that can be used by living things. *Chlorophylls* are the green pigments found in photosynthetic organisms. In most of these organisms, the chlorophyll is found in organelles called *chloroplasts* (Figure 3-1). In addition to the chlorophylls, chloroplasts may contain a variety of other pigments. These additional pigments, called *accessory pigments*, trap those wavelengths of light that chlorophyll cannot trap. For example, chlorophyll cannot trap green light. That is why plants appear green; their chlorophyll reflects the green light. The orange pigment carotene, however, can trap green light and thus makes its energy available for photosynthesis.

Chemistry of Photosynthesis. The process of photosynthesis is complex, involving several series of reactions. However, it can be summarized by the following equation:

$$\text{carbon dioxide} + \text{water} \xrightarrow[\substack{\text{chlorophyll} \\ \text{enzymes}}]{\text{light energy}} \text{glucose} + \text{water} + \text{oxygen}$$

Carbon dioxide and water are the raw materials of photosynthesis. Light energy absorbed by

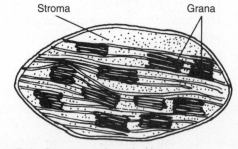

Figure 3-1. Structure of a chloroplast.

the chlorophyll is converted to chemical energy, which is used to synthesize glucose from the raw materials. Water and oxygen are released as waste products of photosynthesis.

The glucose produced by photosynthesis is used, when needed, as an energy source in cellular respiration. It can also be converted to starch, an insoluble food storage compound. Before starch can be used in any cellular process, it must be broken down to glucose by enzymes within the cell. The glucose can be used in the synthesis of other organic compounds, such as lipids and proteins.

QUESTIONS

PART A

1. By which process are carbon dioxide and water converted to carbohydrates? (1) transpiration (2) respiration (3) fermentation (4) photosynthesis

2. The conversion of light energy into chemical bond energy occurs within the cells of (1) molds (2) yeasts (3) algae (4) grasshoppers

3. Glucose molecules may be stored in plants in the form of (1) oxygen (2) starch (3) nucleic acids (4) amino acids

4. Organisms capable of manufacturing organic molecules from inorganic raw materials are classified as (1) autotrophs (2) heterotrophs (3) aerobes (4) anaerobes

5. The basic raw materials for photosynthesis are (1) water and carbon dioxide (2) oxygen and water (3) sugar and carbon dioxide (4) carbon dioxide and oxygen

6. Which word equation represents the process of photosynthesis?
    (1) carbon dioxide + water → glucose + oxygen + water
    (2) glucose → alchol + carbon dioxide
    (3) maltose + water → glucose + glucose
    (4) glucose + oxygen → carbon dioxide + water

7. Autotrophic activity in plant cells is most closely associated with the organelles called (1) mitochondria (2) ribosomes (3) vacuoles (4) chloroplasts

8. In terms of nutrition, the functional difference between animals and plants is that green plants are able to (1) synthesize glucose (2) break down carbohydrates (3) carry on aerobic respiration (4) form ATP molecules

PART B-2

Base your answers to questions 9 and 10 on the following statement and on your knowledge of biology.

Carbon exists in a simple organic molecule in a leaf and in an inorganic molecule in the air that humans exhale.

9. Identify the simple organic molecule formed in the leaf and the process that produces it.

10. Identify one molecule that humans exhale and the process that produces it.

PART C

11. State one function of each of the following in the process of photosynthesis:
    • light;
    • chlorophyll;
    • carbon dioxide;
    • water.

12. Why is photosynthesis called one of the most important processes on Earth? Give at least one example to support your answer.

13. Bromthymol blue turns to bromthymol yellow in the presence of carbon dioxide. When the carbon dioxide is removed, the solution returns to a blue color. Two green water plants were placed in separate test tubes, each containing water and bromthymol yellow. Both test tubes were corked. One tube was placed in the light, the other in the dark. After several days, the liquid in the tube exposed to the light turned blue. Based on these results, answer the following:
    • Why did the bromthymol solution turn blue in one of the tubes?
    • What does it illustrate about the activity of plants during photosynthesis?
    • What do you think occurred in the tube that was placed in the dark?

Use the information below and your knowledge of biology to answer the following question.

14. A suspension (of chloroplasts in water) from spinach leaves was kept under a bright light at a temperature of 25°C. Another suspension was kept in a dark corner of the same room. Each container had attached to it a small pipette by which the amount of oxygen released by the chloroplasts could be measured. The data table (page 34) shows the volume of oxygen produced by each suspension over a 24-hour period.

Total Volume Oxygen Produced by Chloroplast Suspension (mL)		
Time (hours)	Incubated in Light	Incubated in Dark
0	0.00	0.00
6	0.42	0.01
12	0.96	0.01
18	1.78	0.01
24	2.36	0.01

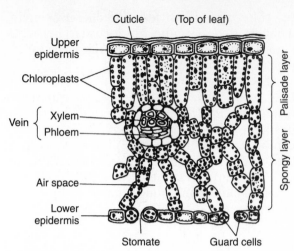

Figure 3-2. Cross section of a typical leaf.

- Describe the difference recorded in the amount of oxygen produced by the two chloroplast suspensions.
- Explain why there was a difference in the volume of oxygen produced by the two chloroplast suspensions.
- Give a scientific reason for why the suspension incubated in the dark produced only 0.01 mL of oxygen.
- Make a line graph showing the results of the experiment. Use different colors to plot the data for each of the two suspensions.
- State one way the researcher could modify the experiment to show that the results are reliable.

Adaptations for Photosynthesis

Algae and green plants are autotrophic organisms that carry on photosynthesis. A large percentage of Earth's photosynthesis occurs in unicellular algae present in the oceans. The raw materials necessary for photosynthesis are absorbed directly from the water into the cells of the algae. Most photosynthesis in terrestrial (land-dwelling) plants occurs in leaves.

Structure of Leaves. Most leaves are thin and flat, providing the maximum surface area for the absorption of light. The outermost cell layer of the leaf is the *epidermis*, which protects the internal tissues from water loss, mechanical injury, and attack by fungi (Figure 3-2). In some plants, the epidermis is covered by a waxy coating, called the *cuticle*, which provides additional protection against water loss and infection.

There are many tiny openings in the epidermis and cuticle, mainly on the undersurface of the leaf. These openings, called *stomates*, allow the exchange of carbon dioxide, oxygen, and water vapor between the environment and the moist, inner tissues of the leaf. Each stomate is surrounded by a pair of chloroplast-containing guard cells. By changing shape, the guard cells open or close the stomate opening.

Beneath the upper epidermis is the *palisade layer*, which is made up of tall, tightly packed cells filled with chloroplasts. Most of the photosynthetic activity of the leaf occurs in this layer. The cells of the epidermis are clear, so that light striking the leaf passes through to the chloroplasts in the palisade layer.

Between the palisade layer and the lower epidermis of the leaf is the *spongy layer*, which is made up of loosely arranged cells separated by interconnecting air spaces. The air spaces are continuous with the stomates. Gases from the environment enter the leaf through the stomates and diffuse from the air spaces into the cells. Other gases diffuse out of the cells into the air spaces and then out of the leaf through the stomates. The cells of the spongy layer contain chloroplasts and carry on some photosynthesis.

The conducting tissues of the leaf are found in bundles called *veins*. The conducting tissues carry water and dissolved minerals from the roots through the stems to the leaves, and they carry food from the leaves to the rest of the plant.

QUESTIONS

PART A

15. Water is lost from the leaves of a plant through (1) spongy cells (2) root hairs (3) veins (4) stomates

16. The waxy covering over the surface of a leaf is the (1) cuticle (2) epidermis (3) palisade layer (4) spongy layer

PART B-1

Base your answers to questions 17 through 20 on the following diagram, which shows a leaf cross section, and on your knowledge of biology.

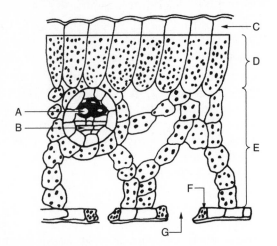

17. Which letter indicates the principal region of food manufacture? (1) *E* (2) *B* (3) *C* (4) *D*

18. Which letter indicates the area where carbon dioxide passes out of the leaf? (1) *A* (2) *G* (3) *C* (4) *D*

19. Which letter indicates a structure that regulates the size of a stomate? (1) *A* (2) *B* (3) *F* (4) *G*

20. Water and dissolved nutrients are carried by the tissues labeled (1) *D* and *E* (2) *C* and *D* (3) *A* and *B* (4) *E* and *F*

PART B-2

Refer to the following diagrams of three different leaf types to answer question 21.

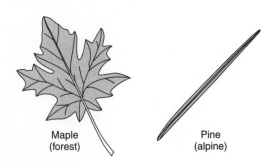

Maple (forest)

Pine (alpine)

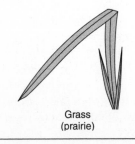

Grass (prairie)

21. How is each leaf adapted to carry out photosynthesis in the particular habitat in which the plant lives? (See diagrams for typical habitat of each leaf type.) State one adaptation for each leaf.

22. A student placed one of her tropical houseplants outside on her porch during the summer to receive some natural sunshine. A few days later, a rubbish fire broke out in a nearby vacant lot, spreading soot all over the neighborhood. Within two weeks, the plant's leaves started to turn yellow and drop off the stem. Give two scientific explanations of how the soot may have negatively affected the plant.

PART C

23. Explain why a cactus's leaves are not broad and flat but are reduced in size to the form of spines. How is this an adaptation that helps the cactus survive?

24. How is the typical leaf adapted for carrying out photosynthesis? Include the functions of the following leaf parts:

- cuticle;
- chloroplasts;
- stomates;
- guard cells.

Heterotrophic Nutrition

Organisms that cannot synthesize organic molecules from inorganic raw materials are *heterotrophs* and must obtain preformed organic molecules from the environment. Heterotrophic organisms include most bacteria, some protists, and all **fungi** and animals. Heterotrophic nutrition involves the processes of ingestion, digestion, and egestion. It generally begins with the mechanical breakdown of food, during which large pieces of food are broken down into smaller pieces by cutting, grinding, and tearing. The smaller pieces provide greater surface area for the action of enzymes during chemical digestion.

Digestion. In some heterotrophs, chemical digestion is *intracellular*—it occurs within the cell (or cells) of the organism. In most heterotrophs, however, digestion is *extracellular*—it occurs in a sac or a tube outside the cells. The end products of digestion are then absorbed into the cells.

Adaptations for Heterotrophic Nutrition

Heterotrophs obtain nutrients in a variety of ways.

Chapter 3: Maintenance in Living Things   **35**

Protists. In protists, such as the ameba and paramecium, digestion is intracellular. In the ameba, food particles are surrounded and engulfed by extensions of the cell called *pseudopods*. This process is known as *phagocytosis*. Within the cell, the food particle is enclosed in a food vacuole. In the paramecium, food particles are ingested through a fixed opening called the *oral groove*. They are moved into this opening by the beating of tiny "hairs" called *cilia*. The food particles are then enclosed in a food vacuole, which circulates in the cytoplasm (Figure 3-3).

In both the ameba and paramecium, the food vacuole merges with a *lysosome*, which is an organelle that contains digestive enzymes. The food within the vacuole is digested by these enzymes, and the end products of digestion are then absorbed into the cytoplasm. In the ameba, wastes are expelled from the cell through the cell membrane. In the paramecium, wastes are expelled through a fixed opening called the *anal pore*.

Humans. The human digestive system is essentially like that of most other **multicellular** (many-celled) animals. Food moves in one direction through a tube, and specialized organs carry out its mechanical breakdown and chemical digestion.

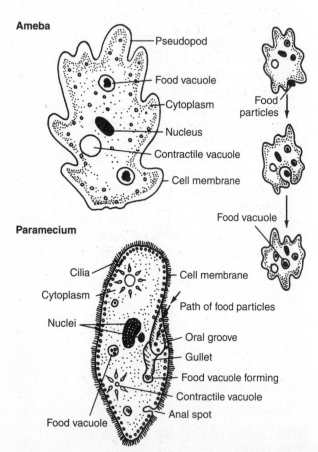

Figure 3-3. Nutrition in the ameba and paramecium.

QUESTIONS

PART A

25. Based on their pattern of nutrition, all animals are classified as (1) autotrophic (2) heterotrophic (3) photosynthetic (4) phagocytic

26. Digestion that occurs in a sac or a tube is referred to as (1) phagocytic (2) intracellular (3) extracellular (4) heterotrophic

27. A fruit fly is classified as a heterotroph, rather than as an autotroph, because it is unable to (1) transport needed materials throughout its body (2) release energy from organic molecules (3) manufacture its own food (4) divide its cells mitotically

28. The principal function of mechanical digestion is the (1) storage of food molecules in the liver (2) production of more surface area for enzyme action (3) synthesis of enzymes necessary for food absorption (4) breakdown of large molecules to smaller ones by the addition of water

29. In the paramecium, most intracellular digestion occurs within structures known as (1) ribosomes (2) endoplasmic reticula (3) mitochondria (4) food vacuoles

30. Which organism ingests food by engulfing it with pseudopods? (1) grasshopper (2) paramecium (3) ameba (4) earthworm

PART B-2

31. How does mechanical digestion aid the process of chemical digestion?

32. Briefly compare intracellular digestion and extracellular digestion.

PART C

33. Use the information below and your knowledge of biology and experimental procedures to answer the following questions.

A biology student performed an experiment to determine the rate of digestion by protease (a protein-digesting enzyme) on cooked egg white. He set up three sets of six test tubes each. Into the first set, he placed the same amount of water and pepsin plus two grams of cooked egg white into each test tube. The egg white was left in one piece in each tube. To the second set, he added the same amounts of water, pepsin, and egg white, but this time he cut the two grams of egg white into eight small pieces before placing it into each test tube.

The third set of test tubes also received the same amounts of water, pepsin, and egg white, but the egg white was finely chopped up before being placed into each test tube.

- What hypothesis was the student most likely testing?
- Predict what should occur in each setup and give a scientific explanation for your prediction.
- The student omitted a control in his experiment. Describe an appropriate control that could be used in this investigation.

TRANSPORT

Transport involves the absorption of materials through an organism's cell membranes and into its body fluids, and the circulation of materials throughout its body.

The Cell Membrane

The cell membrane surrounds the cell and regulates the passage of materials into and out of the cell.

Structure of the Cell Membrane. The currently accepted model of the structure of the cell membrane is called the *fluid mosaic model*. According to this model, the cell membrane consists of a double layer of lipids in which large protein molecules float (Figure 3-4).

Function of the Cell Membrane. The cell membrane selectively regulates the passage of substances into and out of the cell. Small molecules, including water, carbon dioxide, oxygen, and the soluble end products of digestion, pass easily through the cell membrane. Most larger molecules, such as proteins and starch, cannot pass through the cell membrane. However, molecular size is not the only factor that affects the movement of substances into or out of a cell.

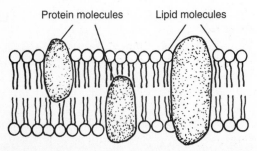

Figure 3-4. The fluid mosaic model of cell membrane structure.

The cell membrane may contain a number of special *receptor sites* to which molecules bind as they enter or leave the cell. Binding to the receptor site is often the only way these molecules can enter or leave a cell. The shape of these receptor site molecules is highly specific to the shape of the molecule being transported. For example, a common receptor site on many cell membranes regulates the passage of sodium ions and potassium ions. The maintenance of homeostasis is largely dependent on the proper functioning of receptor sites on cells. The failure of a receptor site can lead to a serious illness. The disease cystic fibrosis is linked to the failure of chloride ion receptor sites on the cell membranes of afflicted people. This results in the accumulation of mucus in a person's lungs and, eventually, death.

Diffusion and Passive Transport

All ions and molecules are in constant, random motion. When such particles collide, they bounce off each other and travel in new directions. As a result of their motion and collisions, the particles tend to spread out from an area of high concentration to an area of low concentration, a process known as **diffusion**. The difference in concentration between two such areas is known as the *concentration gradient*.

Molecules and ions that can pass through a cell membrane tend to move into or out of the cell by diffusion. The direction of diffusion depends on the relative concentration of the substance inside and outside the cell and usually results in a balance, or **equilibrium**, in the substance's concentration. Diffusion is a type of **passive transport**; it occurs because of the kinetic energy of the molecules and ions and does not require the use of additional energy by the cell.

The diffusion of water through a membrane is called *osmosis*. In osmosis, water molecules move from a region of higher concentration of water to a region of lower concentration of water until they reach an equilibrium.

Active Transport

Processes that require **active transport** involve the movement of particles through a membrane with the use of energy by the cell. In some cases, substances are moved by active transport from a region of lower concentration to a region of higher concentration (against the concentration gradient). In active transport, protein molecules embedded in the cell membrane act as carriers that aid in the transport of materials across the membrane.

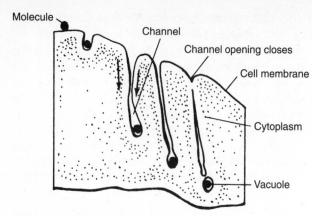

Figure 3-5. Pinocytosis.

Pinocytosis and Phagocytosis

Large, dissolved molecules can pass through a cell membrane by the process of *pinocytosis* (Figure 3-5). In pinocytosis, the cell membrane folds inward. The outer surface of the cell membrane then closes over, and the large molecule is enclosed in a vacuole inside the cell. In contrast, *phagocytosis* is the process by which a cell engulfs large, undissolved particles by flowing around them and enclosing them in a vacuole. For example, amebas use their pseudopods to engulf food particles during phagocytosis (refer to Figure 3-3).

Circulation

Circulation involves the movement of materials both within cells and throughout multicellular organisms. The movement of materials within a cell, *intracellular circulation*, takes place by diffusion and by *cyclosis*. Cyclosis is the natural streaming of cytoplasm that occurs within all cells. Intracellular circulation may also involve the movement of materials through the channels of the endoplasmic reticulum. The transport of materials throughout multicellular organisms is called *intercellular circulation*. Depending on the complexity of the organism, intercellular circulation may occur by diffusion or it may involve a specialized circulatory system with conducting, or vascular, tissues.

QUESTIONS

PART A

34. Which process would describe the movement of sugar molecules through a membrane from a region of higher concentration to a region of lower concentration? (1) osmosis (2) cyclosis (3) passive transport (4) active transport

35. In the human body, the potassium ion can pass easily through cell membranes, yet the potassium ion concentration is higher inside many cells than it is outside these cells. This condition is mainly the result of (1) passive transport (2) active transport (3) osmosis (4) pinocytosis

36. Chemical analysis indicates that the cell membrane is composed mainly of (1) proteins and starch (2) proteins and cellulose (3) lipids and starch (4) lipids and proteins

37. The flow of materials through the membrane of a cell against the concentration gradient is known as (1) passive transport (2) active transport (3) osmosis (4) pinocytosis

38. A biologist observed a plant cell in a drop of water and illustrated it as in diagram *A*. He added a 10 percent salt solution to the slide, observed the cell, and illustrated it as in diagram *B*. The change in appearance of the cell resulted from more (1) salt flowing out of the cell than into the cell (2) salt flowing into the cell than out of the cell (3) water flowing into the cell than out of the cell (4) water flowing out of the cell than into the cell

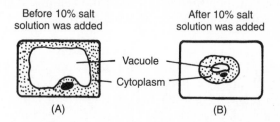

39. The natural streaming of the cytoplasm that occurs within all cells is called (1) pinocytosis (2) phagocytosis (3) osmosis (4) cyclosis

40. When a cell uses energy to move materials across its membrane, the process is known as (1) osmosis (2) active transport (3) diffusion (4) passive transport

41. The diffusion of water molecules into and out of cells is called (1) cyclosis (2) pinocytosis (3) osmosis (4) active transport

42. The net movement of molecules into cells is most dependent on the (1) selectivity of the cell membrane (2) selectivity of the cell wall (3) number of vacuoles (4) number of chromosomes

43. The process by which amebas ingest food particles is called (1) pinocytosis (2) osmosis (3) phagocytosis (4) cyclosis

PART B-1

Base your answers to questions 44 and 45 on the information and diagram on page 39 and on your knowledge of biology.

An investigation was set up to study the movement of water through a membrane. The results are shown in the diagram below.

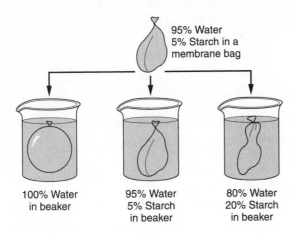

44. Based on these results, which statement correctly predicts what will happen to red blood cells when they are placed in a beaker containing a water solution in which the salt concentration is much higher than the salt concentration in the red blood cells? (1) The red blood cells will absorb water and increase in size. (2) The red blood cells will lose water and decrease in size. (3) The red blood cells will first absorb water, then lose water and maintain their normal size. (4) The red blood cells will first lose water, then absorb water, and finally double in size.

45. A red blood cell placed in distilled water will swell and burst due to the diffusion of (1) salt from the red blood cell into the water (2) water into the red blood cell (3) water from the blood cell into its environment (4) salt from the water into the red blood cell

Base your answers to questions 46 and 47 on the diagram below, which illustrates a process by which protein molecules may enter a cell, and on your knowledge of biology.

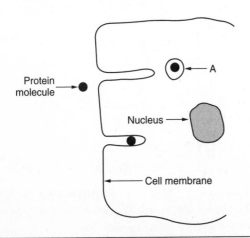

46. Which process is illustrated in this diagram? (1) pinocytosis (2) osmosis (3) diffusion (4) passive transport

47. Structure *A* is most likely a (1) ribosome (2) mitochondrion (3) nucleolus (4) vacuole

PART B-2

48. Describe the differences between active transport and passive transport. Give one example of each type of transport.

49. Freshwater protozoa (single-celled organisms) live in an environment that is very close to 100 percent water. The inside of the cell (cytoplasm) is about 90 percent water. Explain the problem these protozoa face in their environment with respect to maintaining homeostasis. Briefly describe how they are adapted to deal with the problem.

PART C

Use the information in the paragraph and table below to answer the following questions.

50. A biology student was attempting to determine the percent of water present in the cells of elodea (an aquatic plant). She placed leaves of elodea in varying concentrations of saltwater solutions and observed when plasmolysis (cell shrinking) occurred. The table summarizes the results of her experiment.

Solution Concentration	Observed Plasmolysis
0.5% NaCl	None
1.0% NaCl	None
1.5% NaCl	None
2.0% NaCl	Very slight
2.5% NaCl	Pronounced
3.0% NaCl	Pronounced

• According to the data in the table, what percent of elodea cells is water? Explain how you arrived at this conclusion.

• Give a scientific explanation for what caused the cells to shrink at a certain concentration of salt water. What process causes plasmolysis of the cells?

Transport in Plants

The transport of materials in plants involves cyclosis, osmosis, diffusion, and active transport.

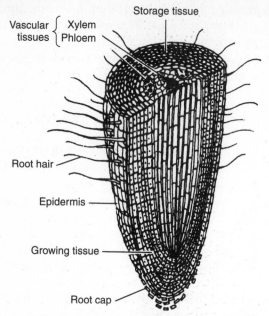

Figure 3-6. Structure of a root tip.

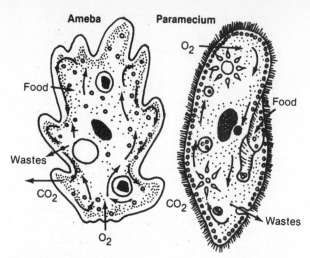

Figure 3-7. Transport in ameba and paramecium.

Some plants contain specialized transport, or *vascular*, tissues while others do not.

Roots. Roots are structures that are specialized for the absorption of water and minerals from the soil and the conduction of these materials to the stem. Roots also anchor the plant in the soil and may contain stored nutrients in the form of starch.

The surface area of the root is increased (for greater absorption) by the presence of *root hairs* just behind the growing tip (Figure 3-6). Water and minerals from the soil are absorbed through the membranes of the root hairs by osmosis, diffusion, and active transport. Materials are transported throughout the plant by two kinds of vascular tissues, *xylem* and *phloem*.

Stems. Although the structure of stems is more complex than that of roots, the xylem and phloem of the stem are continuous with the xylem and phloem of the roots.

Leaves. The xylem and phloem of the leaves, which are in bundles called *veins*, are also continuous with the xylem and phloem of the roots and stem.

Transport in Protists

Protists and other unicellular organisms have no specialized transport system. Materials enter and leave the cell by diffusion and active transport, and are circulated within the cell by diffusion and cyclosis (Figure 3-7).

Transport in Animals

Simple multicellular animals, whose cells are in direct contact with the surrounding water, have no specialized transport system. All other (that is, more complex) multicellular animals do have a specialized system for the transport of materials.

Humans. The human circulatory system is a closed system. Blood is moved through vessels by the pumping action of the heart (Figure 3-8). Human blood contains the pigment *hemoglobin*, which carries oxygen to the body tissues.

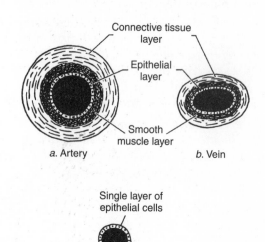

Figure 3-8. The three types of blood vessels.

QUESTIONS

51. The primary function of the root hairs in a plant is to (1) prevent excessive loss of water

(2) provide increased surface area for absorption (3) conduct water and minerals upward (4) conduct organic food materials upward and downward

52. A circulatory system in which the blood remains within vessels is called (1) a closed circulatory system (2) an open circulatory system (3) an internal circulatory system (4) an external circulatory system

53. The epidermal (outermost) cells of a plant's roots can continue to absorb water even when the concentration of water in the soil is very low, even lower than that in its cells. Explain the biological process that enables the root epidermal cells to do this.

54. Use your knowledge of biology to answer the following questions comparing a single-celled organism (such as an ameba) and a multi-celled organism (such as a human):

- What are two similarities in their transport systems?
- What are two differences in their transport systems?

RESPIRATION

The life processes of all organisms require energy. There is potential energy in the chemical bonds of organic molecules such as glucose. However, this energy cannot be used directly in cell metabolism. During *cellular respiration*, these bonds are broken; the energy that is released is temporarily stored in the bonds of the energy-transfer compound called *ATP* (adenosine triphosphate). This process occurs continuously in the cells of all organisms.

Cellular Respiration

Cellular respiration involves a series of enzyme-controlled reactions in which the energy released by the breakdown of the chemical bonds in glucose is transferred to the high-energy bonds of ATP. When ATP is broken down by hydrolysis (the addition of water), ADP (adenosine diphosphate) and phosphate (P) are produced, and energy is released for use by the cell.

The conversion of ATP to ADP is a reversible reaction catalyzed by the enzyme ATP-ase. In living organisms, ATP is constantly being converted to ADP, and the energy released is used for the reactions of cell metabolism. The ADP is then converted back to ATP by the reactions of cellular respiration.

Anaerobic Respiration. In most organisms, cellular respiration requires the presence of free oxygen, and the process is known as *aerobic respiration*. In a few kinds of organisms, free oxygen is not used, and the process is known as *anaerobic respiration*, or *fermentation*. Some cells, such as muscle cells, which normally carry on aerobic respiration, can carry on anaerobic respiration in the absence of oxygen. Other cells, such as yeast and some bacteria, which carry on anaerobic respiration, lack the enzymes necessary for aerobic respiration. There is a net gain of only two ATP molecules for each molecule of glucose used in anaerobic respiration.

$$H_2O + ATP \;\underset{\longleftarrow}{\overset{ATP\text{-}ase}{\longrightarrow}}\; ADP + P + energy$$

Aerobic Respiration. In aerobic respiration, glucose is broken down completely to carbon dioxide and water by a series of enzyme-controlled reactions. Under ideal circumstances, these reactions, which take place mainly in the mitochondria, produce a net gain of 36 ATP molecules. However, the actual number of ATP molecules produced can vary from as few as 20 to as many as 38 per glucose molecule broken down, depending on various factors in the cell's environment.

$$glucose + oxygen \xrightarrow{enzymes} water + carbon\ dioxide + ATP$$

$$C_6H_{12}O_6 + 6O_2 \xrightarrow{enzymes} 6H_2O + 6CO_2 + 36\ ATP$$

Adaptations for Respiration

The oxygen used in aerobic cellular respiration comes from the environment, and the carbon dioxide produced must be excreted into the environment. Although the chemical processes of respiration are similar in most organisms, living things show a variety of adaptations for the exchange of these respiratory gases.

Protists. In simple organisms, such as protists, all or most of the cells are in direct contact with the environment. So the exchange of respiratory gases takes place by diffusion through the thin, moist cell membranes (Figure 3-9, page 42).

Plants. In plants, respiratory gases are exchanged through the leaves, stems, and roots. The exchange of respiratory gases occurs by diffusion through the cell membranes of internal cells, which are surrounded by intercellular spaces. The intercellular spaces open to the environment through

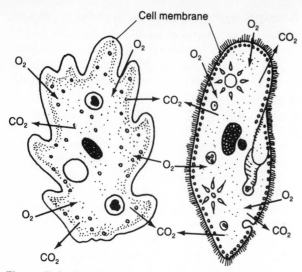

Figure 3-9. Respiration in ameba and paramecium.

the stomates, openings on the undersurface of the leaf.

Humans. In humans, the exchange of respiratory gases takes places at thin, moist membranes within the lungs. Hemoglobin aids in the transport of oxygen in the blood. Carbon dioxide and oxygen are carried between the respiratory surface in the lungs and the environment by a system of air tubes.

QUESTIONS
PART A

55. Most animals make energy available for cell activity by transferring the potential energy of glucose to ATP. This process occurs during (1) aerobic respiration only (2) anaerobic respiration only (3) both aerobic and anaerobic respiration (4) neither aerobic nor anaerobic respiration

56. In animal cells, the energy to convert ADP to ATP comes directly from (1) hormones (2) sunlight (3) organic molecules (4) inorganic molecules

57. The organelles in which most of the reactions of aerobic cellular respiration take place are the (1) ribosomes (2) chloroplasts (3) lysosomes (4) mitochondria

58. The substances that most directly control the rate of reaction during cellular respiration are known as (1) enzymes (2) phosphates (3) monosaccharides (4) disaccharides

59. Which end product is of the greatest benefit to the organism in which respiration occurs? (1) glucose (2) carbon dioxide (3) ATP molecules (4) water molecules

60. Protists obtain oxygen from their environment through (1) stomates (2) cell membranes (3) vacuoles (4) mitochondria

61. Which process usually uses carbon dioxide molecules? (1) cellular respiration (2) asexual reproduction (3) active transport (4) autotrophic nutrition

62. In humans, respiratory gases are exchanged between the lungs and the environment through (1) air tubes (2) hemoglobin (3) vacuoles (4) stomates

63. Arrows A, B, and C in the diagram below represent the processes necessary to make the energy stored in food available for muscle activity. The correct sequence of processes represented by A, B, and C is (1) diffusion → synthesis → active transport (2) digestion → diffusion → cellular respiration (3) digestion → excretion → cellular respiration (4) synthesis → active transport → excretion

Food $\xrightarrow{A}$ Simpler molecules $\xrightarrow{B}$ Mitochondria $\xrightarrow{C}$ ATP in muscle cells

PART B-1

Base your answers to questions 64 through 68 on the diagram below, which represents a cellular process in animals, and on your knowledge of biology.

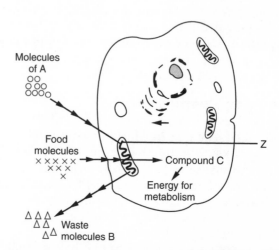

64. The items labeled as food molecules most likely represent (1) starch (2) glucose (3) phosphate (4) chlorophyll

65. Compound C most likely represents some molecules of (1) oxygen (2) glucose (3) ATP (4) DNA

66. If this cell is carrying on aerobic respiration, *B* represents molecules of a waste product known as (1) carbon dioxide (2) ATP (3) ethyl alcohol (4) phosphate

67. If this represents a kidney cell from the human body, the molecules of *A* are most probably (1) carbon dioxide (2) enzymes (3) lipids (4) oxygen

68. The cell organelle labeled Z is called a (1) chloroplast (2) mitochondrion (3) nucleolus (4) vacuole

PART B-2

69. Why must all organisms must carry out cellular respiration?

70. Briefly compare the processes of cellular respiration in aerobic and anaerobic organisms. Include the following:
    - the function of oxygen;
    - the net gain of ATP molecules.

PART C

Base your answers to question 71 on the following information and on your knowledge of biology.

71. A biologist was culturing some muscle cells from a mouse (an aerobic organism) in a petri dish. He was interested in measuring the amount of ATP produced by the muscle cells when the cells were supplied with glucose. At the beginning of the experiment, the cells were producing large quantities of ATP. He then added a substance called malonic acid to the cell culture, and the amount of ATP produced fell to near zero.
    - Which organelles in the muscle cells were most likely affected by the malonic acid?
    - Propose a testable hypothesis concerning respiration in cells treated with malonic acid.
    - Predict the effect of malonic acid on an anaerobic organism and explain your prediction.

EXCRETION

The metabolic activities of living cells produce waste materials. The life process by which the wastes of metabolism are removed from the body is called *excretion*.

Table 3-1

The Waste Products of Metabolism

Metabolic Activity	Wastes
Respiration	Carbon dioxide and water
Dehydration synthesis	Water
Protein metabolism	Nitrogenous wastes
Certain metabolic processes	Mineral salts

Wastes of Metabolism

The wastes of various metabolic processes are shown in Table 3-1. Some wastes are **toxins** (compounds that are poisonous to body tissues), while other wastes are nontoxic. In animals, toxic wastes are excreted from the body. In plants, toxic wastes are sealed off and stored, sometimes in vacuoles. Some nontoxic wastes are excreted, while others are recycled and used in metabolic activities.

Nitrogenous, or nitrogen-containing, wastes are produced by the breakdown of amino acids. Different kinds of organisms produce different kinds of nitrogenous wastes, including *uric acid*, which is nontoxic; *urea*, which is moderately toxic; and *ammonia*, which is highly toxic.

Adaptations for Excretion

In the simplest organisms, wastes pass from the cells directly into the environment. More complex organisms have a specialized excretory system.

Protists. In general, the excretion of wastes in protists is accomplished by diffusion through the cell membrane (Figure 3-10). In freshwater protozoans, such as the ameba and paramecium, water continuously enters the cell by osmosis. In these

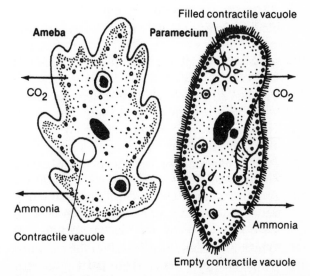

Figure 3-10. Excretion in ameba and paramecium.

organisms, the excess water collects in organelles called *contractile vacuoles*. The contractile vacuoles burst at the surface of the cell, expelling the water back into the environment. This process involves active transport.

In freshwater protozoans, the nitrogenous waste product is ammonia. Although it is very toxic, ammonia is also very soluble in water, and thus it can be easily excreted from the cells of these organisms.

In photosynthetic protists, such as algae, some of the carbon dioxide produced by cellular respiration can be recycled and used in photosynthesis. Some of the oxygen produced by photosynthesis can be used in cellular respiration.

Plants. In plants, as in algae, some of the waste gases produced by photosynthesis and cellular respiration are recycled. Excess gases diffuse out of the plant through stomates on the leaves, tiny openings on the stems, and epidermal cells on the roots.

Humans. In humans, carbon dioxide is excreted by the lungs. Water, salts, and urea are excreted by the kidneys. These waste products are passed out of the body through various specialized tubes, or passageways.

QUESTIONS

PART A

72. Metabolic wastes of animals most likely include (1) water, carbon dioxide, oxygen, and salts (2) carbon dioxide, nitrogenous compounds, water, and salts (3) hormones, water, salts, and oxygen (4) glucose, carbon dioxide, nitrogenous compounds, and water

73. Which activity would most likely produce nitrogenous waste products? (1) protein metabolism (2) glucose metabolism (3) lipid metabolism (4) starch metabolism

74. The leaf structures that are closely associated with both respiration and excretion are the (1) root hairs (2) stomates (3) waxy surfaces (4) epidermal cells

75. Protists can function without an organized excretory system because their cells (1) do not produce wastes (2) change all wastes into useful substances (3) remove only solid wastes (4) are in direct contact with a water environment

76. Which statement best describes the excretion of nitrogenous wastes from paramecia? (1) Urea is excreted by nephrons. (2) Uric acid is excreted by nephrons. (3) Urea is excreted through tiny tubules. (4) Ammonia is excreted through cell membranes.

77. Most toxic products of plant metabolism are stored in the (1) stomates (2) vacuoles (3) root cells (4) chloroplasts

78. In freshwater protozoans, the organelles involved in the maintenance of water balance are (1) food vacuoles (2) mitochondria (3) contractile vacuoles (4) pseudopods

PART B-2

79. Identify two functions the kidneys perform that help maintain homeostasis in a human.

PART C

Base your answer to the following questions on the table below and on your knowledge of biology.

80. The table compares three nitrogenous waste products, their toxicity levels, their solubility in water, and the habitats in which the organisms that produce them typically live.

Type of Waste	Toxicity	Solubility	Habitat of Organism
Ammonia	Very high	Very good	Aquatic (in water)
Urea	Moderate	Good	Land, most generally
Uric acid	Low	None	Land, often desert

• What connection exists between the habitat of an organism and the toxicity of its nitrogenous waste?

• What connection exists between the solubility of each nitrogenous waste and its toxicity?

• State a possible biological benefit of the connections among waste toxicity, waste solubility, and an organism's habitat.

81. Organisms produce waste products as a result of their metabolic activities. These wastes include carbon dioxide; water; mineral salts, and nitrogenous wastes such as ammonia, urea, and uric acid. Select any three of these metabolic wastes and answer the following:

• How are these wastes produced in an organism?

• Why must these wastes be removed from an organism?

REGULATION

Regulation involves the control and *coordination* of life activities. In all organisms, there are chemicals that regulate life activities. In multicellular animals, there is nerve control in addition to chemical control. Both nerve control and chemical control aid organisms in their maintenance of homeostasis.

Nerve Control

Nerve control depends mainly on the functioning of **nerve cells**, or *neurons*, which are specialized for the transmission of impulses from one part of the body to another.

Structure of a Nerve Cell. The three parts of a nerve cell are the *dendrites*; the cell body, or *cyton*; and the *axon* (Figure 3-11). Dendrites are composed of many branches, but the axon has branches mainly at the end that is farthest from the cell body. Impulses are received by the dendrites and passed to the cell body, which contains the nucleus and other organelles. From the cell body, impulses pass along the axon to its terminal (end) branches.

Impulses. An *impulse* is a region of electrical and chemical, or *electrochemical*, change that travels over the membrane of a nerve cell. When electrochemical impulses reach the terminal branches (end brush) of an axon, they stimulate the release of chemicals called *neurotransmitters*.

Neurotransmitters and Synapses. The junction between adjacent nerve cells is called a *synapse*. At the synapse, the nerve cells do not touch; there is a small gap between them. When impulses reach the terminal branches of the axon of one nerve cell, they stimulate the release of neurotransmitters, such as acetylcholine, which diffuse across the gap of the synapse. The neurotransmitter stimulates impulses in the dendrites of the second nerve cell. In this way, impulses pass from one nerve cell to another (Figure 3-12).

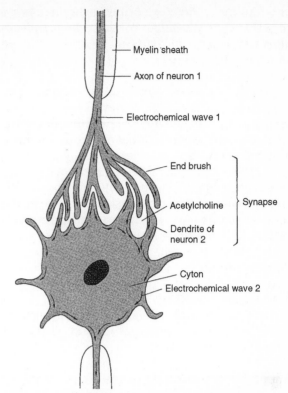

Figure 3-12. Nerve impulses are transmitted from one nerve cell to the next by chemicals that diffuse across the gap (at the synapse).

The axons of some nerve cells have junctions with a muscle or a gland. In such cases, the chemicals released by the terminal branches of the axon stimulate contraction of the muscle or secretion by the gland.

Stimulus and Receptors. Any change in the external or internal environment that initiates impulses is called a **stimulus** (plural, *stimuli*). Stimuli are detected by specialized structures called **receptors**. Each kind of receptor is sensitive to a particular kind of stimulus; for example, eyes are sensitive to light, ears to sound, and so on.

Responses and Effectors. The reaction of an organism to a stimulus is called a **response**. The response to a stimulus is carried out by *effectors*, generally the muscles or glands.

Adaptations for Nerve Control

Even the simplest animals have some type of nerve cells that transmit impulses.

Simple Animals. In some animals, there is no brain, but there is a structure called a *nerve net* that transmits messages throughout the body. In

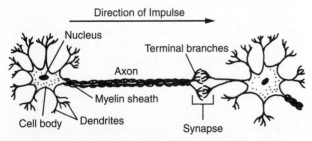

Figure 3-11. Structure of a typical nerve cell.

other animals, there is a primitive brain, a *nerve cord* that runs the length of the body, *peripheral nerves* that serve all parts of the body, and, in some, sense receptors.

Humans. Humans have a central nervous system that consists of a highly developed brain and a dorsal (spinal) nerve cord that runs down the back. The central nervous system permits impulses to travel in one direction along definite pathways. There is also a peripheral nervous system that consists of an elaborate network of nerves. The peripheral nervous system carries signals between the central nervous system and all parts of the body. In addition, there are many highly developed sense organs, such as smell, sight, hearing, taste, and skin receptors.

Chemical Control
In both plants and animals, various aspects of their life activities are controlled by chemicals called **hormones**.

Plant Hormones. In plants, there are no organs specialized for the production of hormones. Plant hormones are produced in greatest abundance in the cells of actively growing regions, such as the tips of roots and stems and in buds and seeds. The hormones produced in these regions affect the growth and development of cells in other parts of the plant. The effects of hormones vary with their concentration and with the type of tissue they act on.

Animal Hormones. Unlike plants, many animals do have organs specialized for the synthesis and secretion of hormones. These organs, called *endocrine glands,* or ductless glands, release their secretions directly into the bloodstream. Hormones are found in a wide variety of animals, both vertebrates and invertebrates. The various hormones control the animals' metabolic activities, as well as their metamorphosis and reproduction.

QUESTIONS
PART A

82. Animal cells that are specialized for conducting electrochemical impulses are known as (1) nerve cells (2) synapses (3) nephrons (4) neurotransmitters

83. A hawk gliding over a field suddenly dives toward a moving rabbit. The hawk's reaction to the rabbit is known as a (1) stimulus (2) synapse (3) response (4) impulse

84. The transmission of nerve impulses at synapses involves chemicals called (1) hormones (2) neurotransmitters (3) enzymes (4) nucleic acids

85. Neurotransmitters, such as acetylcholine, are initially detected by which part of a nerve cell? (1) dendrites (2) nucleus (3) terminal branches (4) mitochondrion

86. The nucleus of a nerve cell is found in the (1) dendrite (2) axon (3) synapse (4) cell body

87. Structures that detect stimuli are called (1) effectors (2) receptors (3) synapses (4) cell bodies

88. The secretions of endocrine glands are known as (1) enzymes (2) hormones (3) pigments (4) neurotransmitters

89. A chemical injected into a tadpole caused the tadpole to undergo rapid metamorphosis into a frog. This chemical was most probably a(n) (1) enzyme (2) neurotransmitter (3) hormone (4) blood protein

90. The two systems that directly control homeostasis in most animals are the (1) nervous and endocrine (2) endocrine and excretory (3) nervous and circulatory (4) excretory and circulatory

PART B-2

91. Compare the central nervous system and the peripheral nervous system. What are the main structures and functions of each system?

92. Explain why the endocrine glands are also referred to as ductless glands? How is this related to their function?

LOCOMOTION

Locomotion is the ability to move from place to place. Among many protists and animals, locomotion improves the organism's ability to survive. It increases chances of finding food and shelter, avoiding predators and other dangers, and finding a mate.

Adaptations for Locomotion
Many protists and almost all animals are capable of some form of locomotion, or **movement**. Such organisms are said to be *motile*. The hydra is generally a *sessile* organism; it tends to remain in one place, fastened to another structure. However, it does have fibers that permit some limited movements.

Protists. There are three basic forms of locomotion among protists. In the ameba, locomotion

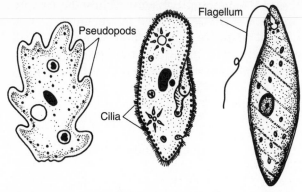

Pseudopods

Flagellum

Cilia

Figure 3-13. Locomotion in ameba, paramecium, and euglena (an alga). (Not drawn to scale.)

Humans. Humans have an internal skeleton, or *endoskeleton*. Locomotion is accomplished by the interaction of muscles and jointed appendages (bones and cartilage).

is by ameboid motion, in which the cell cytoplasm flows into the pseudopods. This causes the organism to move in the direction of the newly formed extension of its cytoplasm. In the paramecium, locomotion involves cilia, which are short, hairlike organelles that cover the outer surface of the cell. The cilia wave back and forth in a coordinated way, moving the cell through the water. Some algae and other protozoans move by means of *flagella*, long, hairlike organelles that can pull the cell through the water (Figure 3-13).

QUESTIONS

PART A

93. Locomotion increases an animal's opportunity to do all of the following *except* (1) obtain food (2) find a mate and reproduce (3) escape from predators (4) transmit impulses

94. Which structures are *not* associated with locomotion among protists? (1) flagella (2) cilia (3) pseudopods (4) tentacles

95. Which organism is able to move due to the interaction of its muscles and skeleton? (1) ameba (2) paramecium (3) human (4) hydra

PART B-2

96. Describe four survival advantages of locomotion.

97. Some organisms are sessile, or incapable of movement. Explain how their survival might be aided in other ways.

READING COMPREHENSION

Base your answers to questions 98 through 101 on the information below and on your knowledge of biology. Source: Science News *(April 30, 2005): vol. 167, no. 18, p. 275.*

Gene Therapy Slows Down Alzheimer's Disease

Putting extra copies of the gene for a cellular growth factor into the brains of people with Alzheimer's disease slows the degenerative condition, a new study suggests.

Alzheimer's disease kills neurons, the brain cells that orchestrate message signaling throughout the nervous system. The gene added in this study encodes nerve growth factor (NGF), a protein that keeps these cells alive and so facilitates signaling among them.

The vehicle for the human gene was the patients' own skin cells. Researchers took a bit of skin tissue from each of eight people diagnosed with early Alzheimer's disease and used a non-replicating virus to transfer genes for human NGF into the skin cells.

The scientists then injected these genetically modified cells into each patient's brain. However, two of the patients were excluded from the study soon after that surgery because of bleeding in their brains.

Over the next two years, positron-emission tomography scans of the other patients revealed increased metabolic activity in their brains, a sign of neuron rejuvenation. An autopsy on one of the excluded patients, who died of a heart attack during the study, revealed that the implanted cells were making NGF. Nearby neurons appeared healthy.

These biological findings paralleled changes in the patients' behavior. Although standard testing indicated that the patients, on average, continued their mental declines during the 2 years after surgery, the pace of cognitive loss was only half as great as the patients had been experiencing before undergoing the gene therapy, says study coauthor Mark H. Tuszynski of the University of California, San Diego in La Jolla. The findings will appear in an upcoming *Nature Medicine*.

The patients scored best on tests administered more than 6 months after the surgery, suggesting that the transplanted cells took several months to rev up their production of NGF, he says.

This marks the first time that researchers have surgically intervened in Alzheimer's disease to the benefit of patients, Tuszynski says.

Although the gene therapy slows the pace of Alzheimer's decline more than drugs currently prescribed for the disease do, he adds, the surgery is unlikely to represent a cure because it doesn't address the fundamental symptom of Alzheimer's disease: the accumulation of waxy plaques in a person's brain.

"The magnitude of effect shown here is not terribly great, but any positive benefits for Alzheimer's patients would be good," say Curt R. Freed of the University of Colorado School of Medicine in Denver. "This could signal a new phase of treatment for this disorder."

However, Freed notes that skin cells used as a gene-delivery vehicle "may not be as natural to the brain as are the brain's own cells."

With that in mind, Tuszynski is teaming with researchers to inject an innocuous [harmless] virus, loaded with the gene for NGF, directly into the brains of Alzheimer's patients. The scientists expect this virus to install the gene into neurons, which would then crank out NGF, says David A. Bennett of Rush University Medical Center in Chicago, who is collaborating on the project.

98. Explain the function of human NGF (nerve growth factor).

99. How was the gene for NGF introduced into the brains of Alzheimer's patients?

100. What two facts in the article show that this gene therapy does not actually cure Alzheimer's disease?

101. Describe how researchers might modify the technique to improve results in patients.

Human Biology

NUTRITION

Humans are heterotrophs—they must ingest the nutrients they need, including carbohydrates, proteins, lipids, vitamins, minerals, and water. Carbohydrates, lipids, and proteins are made up of large molecules that must be digested before they can be absorbed and used by the cells. Vitamins, minerals, and water are made up of small molecules that can be absorbed without being digested. Specific nutritional requirements of humans depend on the age, gender, and activity of the individual.

Human Digestive System

The human digestive system is an **organ system** that consists of a one-way digestive tube called the *gastrointestinal*, or GI, tract and accessory **organs** (Figure 4-1). Food is moved through the GI tract by rhythmic, muscular contractions called *peristalsis*. As food moves through the tract, it is broken down mechanically and chemically. The accessory organs (liver, gallbladder, and pancreas) secrete enzymes and other substances into the digestive tract that aid in digestion.

Oral Cavity. The mouth, or oral cavity, contains the teeth, tongue, and openings from the salivary glands. Food is ingested through the mouth, and digestion begins there. The teeth function in the mechanical breakdown of food into smaller pieces, which provides a larger surface area for the chemical action of digestive enzymes.

The *salivary glands* secrete saliva, a fluid that passes into the mouth through ducts. Saliva contains an enzyme, amylase, which begins the chemical digestion of starch. The tongue aids in chewing and in mixing saliva with the food by moving the food around in the mouth. The tongue also moves the food mass to the back of the mouth for swallowing.

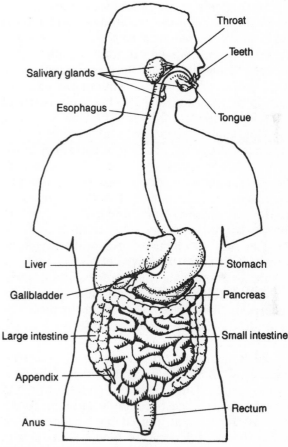

Figure 4-1. The human digestive system.

Esophagus. When the food is swallowed, it passes into the esophagus, and peristalsis of the esophagus wall moves it downward to the stomach. Digestion of starch continues while the food is in the esophagus.

Stomach. Food reaching the lower end of the esophagus enters the stomach, a muscular sac in which it is mixed and liquefied (mechanical digestion). Gastric glands in the stomach lining secrete hydrochloric acid and the enzyme gastric protease. Hydrochloric acid provides the proper pH (acidic) required for effective functioning of gastric protease, which begins the chemical digestion of proteins.

Small Intestine. Partially digested food moves from the stomach into the small intestine, a long, convoluted tube in which most digestion occurs. The walls of the small intestine are lined with intestinal glands that secrete several different enzymes. These enzymes digest proteins, lipids, and disaccharides. The liver, gallbladder, and pancreas secrete substances into the small intestine.

The *liver* produces *bile*, which passes into the *gallbladder*, where it is stored temporarily. From the gallbladder, bile passes through ducts into the small intestine, where it acts on fats, breaking them down mechanically into tiny droplets. This process, known as *emulsification*, increases the surface area of fats for subsequent chemical digestion by enzymes. Bile also helps to neutralize the acidic food mass from the stomach.

The **pancreas** produces and secretes a juice that passes through ducts into the small intestine. Proteases, lipases, and amylase in pancreatic juice, together with the enzymes secreted by the intestinal glands, complete the chemical digestion of proteins, lipids, and carbohydrates in the small intestine.

The end products of digestion, including amino acids, fatty acids, glycerol, and glucose, are absorbed through the lining of the small intestine. The intestinal lining is specially adapted for absorption. Its surface area is greatly increased by many folds and by fingerlike projections called *villi* (singular, *villus*).

Each villus contains a lacteal and capillaries (Figure 4-2). A *lacteal* is a small vessel of the lymphatic system. Fatty acids and glycerol (the end products of fat digestion) are absorbed into the lacteals; they are transported in the lymph, which is eventually added to the blood. Glucose and amino acids are absorbed into the blood of the capillaries and transported to the liver for temporary storage. From the liver, glucose and amino acids are distributed by the blood to all the cells, as they are needed.

When excess glucose is removed from the blood in the liver, it is converted to, and stored as, glycogen, an insoluble polysaccharide. When the concentration of glucose in the blood drops below a certain level, the glycogen is broken down to glucose, which is then returned to the blood. The storage of excess glucose as glycogen is an adaptation for the maintenance of a con-

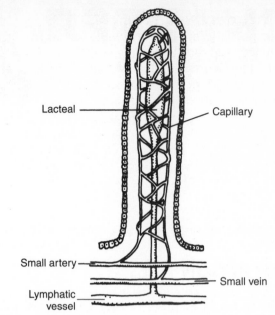

Figure 4-2. Structure of a villus.

stant blood glucose level and is an example of homeostasis.

Large Intestine. Undigested and indigestible foods and water move from the small intestine into the *large intestine,* which is shorter and wider than the small intestine. Water is reabsorbed from the undigested food into the capillaries in the wall of the large intestine. This reabsorption helps the body to conserve water. The remaining wastes, called *feces,* are moved through the large intestine by strong peristaltic action to the rectum, where they are stored temporarily. The feces are periodically egested from the body through the anus.

Mechanism of Chemical Digestion

In digestion, large, insoluble molecules are broken down into small, soluble molecules by the process of *hydrolysis.* Each of the many hydrolytic reactions of digestion is regulated by a specific hydrolytic enzyme.

Chemically, hydrolysis is the opposite of dehydration synthesis; large molecules are split with the addition of water. In a series of reactions, polysaccharides, such as starch, are also broken down by hydrolysis into monosaccharides (simple sugars).

In the presence of water and protein-digesting enzymes (proteases), proteins are broken down into their constituent amino acids. In the hydrolysis of proteins, peptide bonds are broken.

In the presence of water and lipid-digesting enzymes (lipases), lipid molecules are broken down by hydrolysis into fatty acids and glycerol.

Nutritional Requirements

A balanced diet must contain carbohydrates, proteins, and fats, as well as vitamins, minerals, and water. Ingredients labels that are printed on all packaged food items indicate how much of a particular nutrient is supplied per serving, as well as the serving's calorie content. In addition to a balanced diet, exercise is recommended to maintain proper health. (*Note:* You can see the current recommended daily allowances for all the major food groups by checking the Web site www.mypyramid.gov.)

Carbohydrates. Carbohydrates serve as the major source of energy in the body. Excess carbohydrates are converted to glycogen or fat and stored in the body as an energy reserve. *Cellulose*, a complex carbohydrate found in the cell walls of fruits, vegetables, and whole grains, provides indigestible material that serves as *roughage*. Roughage, also called *fiber*, helps to move the food mass through the intestines.

Proteins. Proteins in food are broken down into their constituent amino acids, which are then used to synthesize human proteins. Twenty different amino acids are needed for the synthesis of human proteins. Twelve of these can be synthesized in the body from other amino acids; but the other eight, called the *essential amino acids,* must be obtained from the food.

All necessary amino acids must be present at the same time for protein synthesis to occur. An inadequate supply of any essential amino acid limits protein synthesis. Meat proteins generally contain all of the essential amino acids. Such foods are called complete protein foods. Vegetable proteins are generally incomplete protein foods—they lack one or more essential amino acids. However, a variety of vegetable proteins, if eaten together, can complement each other and provide all the essential amino acids. For example, a meal of rice and beans supplies complete protein.

Fats. Fats contain relatively large amounts of potential energy and serve as an energy-storage compound in organisms. Fats are also a structural component of cell membranes.

Fats are classified as saturated and unsaturated. *Saturated fats*, which are found in meats, butter, and other animal products, are solid at room temperature. Chemically, saturated fats contain the maximum number of hydrogen atoms and have no double bonds. *Unsaturated fats* contain one or more double bonds and, thus, could hold additional hydrogen atoms. An excess of saturated fats, as well as another type of fat called a *trans fat*, in the diet is thought to contribute to cardiovascular disease. Fortunately, some forms of unsaturated fats are thought to protect against cardiovascular disease. However, it is generally considered wise to limit one's intake of all kinds of fats in order to maintain good health.

Disorders of the Digestive System

An *ulcer* is an open sore in the lining of the stomach or intestines. Ulcers may be caused by the presence of excess amounts of hydrochloric acid, which breaks down the lining of the digestive tract, or by bacterial infection. Ulcers are painful and sometimes cause bleeding.

Constipation is a condition marked by difficulty in eliminating feces from the large intestine. Constipation occurs when too much water is removed from the feces in the large intestine or when there is a reduction in peristaltic activity, slowing down the movement of waste through the large intestine. Insufficient roughage in the diet may also be a cause of constipation.

Diarrhea is a gastrointestinal disturbance characterized by frequent elimination of watery feces. This condition may result from decreased water absorption in the large intestine and increased peristaltic activity. Prolonged diarrhea may result in severe dehydration.

Appendicitis is an inflammation of the appendix, a small pouch located at the beginning of the large intestine. *Gallstones* are small, hardened cholesterol deposits that sometimes form in the gallbladder. When gallstones enter the bile duct and block the flow of bile, they cause severe pain.

QUESTIONS

PART A

1. Into which parts of the human digestive system are digestive enzymes secreted? (1) mouth, esophagus, stomach (2) stomach, small intestine, large intestine (3) mouth, stomach, small intestine (4) esophagus, stomach, large intestine

2. In humans, excess glucose is stored as the polysaccharide known as (1) glycogen (2) glycerol (3) maltose (4) cellulose

3. After a person's stomach was surgically removed, the chemical digestion of ingested protein would probably begin in the (1) mouth (2) small intestine (3) large intestine (4) liver

4. Which organ forms part of the human gastrointestinal tract? (1) trachea (2) esophagus (3) diaphragm (4) aorta

5. The intestinal folds and villi of the human small intestine function primarily to (1) increase the surface area for absorption of digested nutrients (2) excrete metabolic wastes (3) circulate blood (4) force the movement of food in one direction through the digestive tract

6. Lipase aids in the chemical digestion of (1) fats (2) proteins (3) enzymes (4) salts

7. In humans, which of the following is true of carbohydrate digestion? (1) It begins in the oral cavity and ends in the esophagus. (2) It begins in the oral cavity and ends in the small intestine. (3) It begins in the small intestine and ends in the large intestine. (4) It begins and ends in the small intestine.

8. Organisms are classified as heterotrophs if they derive their metabolic energy from (1) photosynthesis (2) inorganic raw materials (3) lightning (4) preformed organic compounds

9. Glands located within the digestive tube include (1) gastric glands and thyroid glands (2) gastric glands and intestinal glands (3) thyroid glands and intestinal glands (4) adrenal glands and intestinal glands

10. The small lymphatic vessels that extend into the villi are called the (1) veins (2) lacteals (3) glands (4) capillaries

11. The principal function of mechanical digestion is the (1) hydrolysis of food molecules for storage in the liver (2) production of more surface area for enzyme action (3) synthesis of enzymes necessary for food absorption (4) breakdown of large molecules to smaller ones by the addition of water

12. In which organ's walls does peristalsis occur? (1) liver (2) pancreas (3) oral cavity (4) esophagus

13. A person who consumes large amounts of saturated fats may increase his or her chances of developing (1) meningitis (2) hemophilia (3) pneumonia (4) cardiovascular disease

PART B-1

Base your answers to questions 14 through 18 on your knowledge of biology and on the following graph, which shows how much carbohydrates, proteins, and fats are chemically digested as food passes through the human digestive tract. The letters represent body structures, in sequence, that make up the digestive tract.

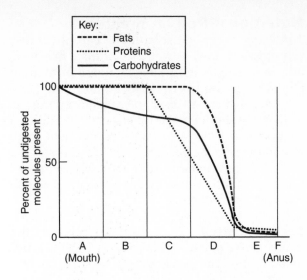

14. Proteins are digested in both (1) A and B (2) B and C (3) C and D (4) A and C

15. The organ represented by letter C is most probably the (1) esophagus (2) stomach (3) small intestine (4) large intestine

16. Enzymes secreted by the pancreas enter the system at (1) A (2) B (3) C (4) D

17. The final products of digestion are absorbed almost entirely in (1) F (2) B (3) C (4) D

18. Water is removed from the undigested material in (1) A (2) B (3) E (4) D

PART B-2

19. Use your knowledge of biology to complete the following table.

Nutrient	Digestive End Products	Where Chemical Digestion Begins	End Products Absorbed by
Starches	Simple sugars		Villi capillaries
Lipids		Small intestine	
Proteins	Amino acids		

PART C

Base your answers to questions 20 and 21 on the following food pyramid, which shows suggested daily servings for several types of food.

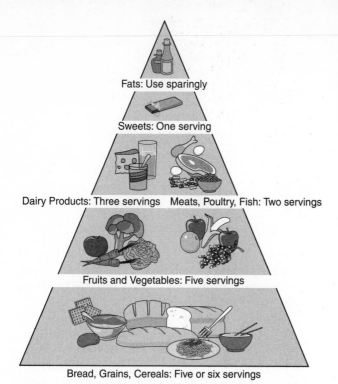

Fats: Use sparingly

Sweets: One serving

Dairy Products: Three servings   Meats, Poultry, Fish: Two servings

Fruits and Vegetables: Five servings

Bread, Grains, Cereals: Five or six servings

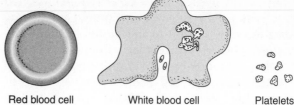

Red blood cell   White blood cell   Platelets

Figure 4-3. The three main types of blood cells.

20. Make up a balanced, daily meal plan for breakfast, lunch, and dinner for one person.

21. Explain why people are advised to ingest fats "sparingly," that is, in low amounts.

22. Based on your knowledge of human biology, answer the following:
   • Identify three organs of the human digestive system.
   • Describe the type of digestion that occurs in each organ.

23. Explain why most foods (nutrients) eaten by humans must be digested before they can be used by the body.

TRANSPORT

Transport includes the absorption and distribution of materials throughout the body. In humans, dissolved and suspended materials are transported in the blood, which is moved throughout the body by the circulatory system.

Blood

Blood consists of a fluid called *plasma* in which red blood cells, white blood cells, and platelets are suspended (Figure 4-3).

Plasma. Blood plasma consists mostly of water. It contains many dissolved materials, including inorganic ions, wastes, nutrients, and a variety of proteins. The proteins include antibodies, enzymes, hormones, and clotting factors.

Red Blood Cells. The most numerous cells in the plasma are the *red blood cells*, which are produced in the marrow of certain bones. Mature red blood cells do not have a nucleus. Within red blood cells is the red, iron-containing pigment hemoglobin, which carries oxygen between the lungs and the body tissues.

White Blood Cells. The **white blood cells** are larger than the red blood cells and contain one or more nuclei. White blood cells are produced in the bone marrow and in lymph nodes. There are several types of white blood cells, including phagocytes and lymphocytes.

Phagocytes are white blood cells that engulf and destroy bacteria at the site of an infection. The phagocytes leave the capillaries, by means of ameboid motion, and enter the body tissues. There, they engulf bacteria and other foreign matter in the same way that amebas engulf food.

Lymphocytes are white blood cells that produce special protein molecules called **antibodies**. Antibodies react chemically with foreign substances or microorganisms in the blood and inactivate them. The substances that cause antibody production are called **antigens**. Most antigens are protein in nature. An antigen–antibody reaction is referred to as an *immune response*. (The immune response is discussed in greater detail in Chapter 5.)

Platelets. The small cell fragments that are involved in the clotting of blood are called *platelets*. A platelet consists of cytoplasm surrounded by a cell membrane; it has no nucleus.

Blood Clotting. When an injury occurs, blood vessels break and blood is released. To stop the loss of blood, a blood clot forms, blocking the wound.

Clotting involves a series of enzyme-controlled reactions. All the substances required for clotting are normally present in the blood; but clot formation does not take place unless there is a break in a blood vessel. When this occurs, blood platelets are ruptured and they release an enzyme that starts the clotting reactions. The plasma protein *fibrinogen* is converted to *fibrin*, which forms a meshwork of solid fibers across the wound. Blood cells become trapped in the fibers, forming the clot.

Immunity

The ability of the body to resist a specific disease is called **immunity**. Immunity, which is provided by the **immune system** and depends on the action of antibodies in the bloodstream, can develop in two ways.

Active immunity results when antibodies are produced in response to a foreign substance (antigen) in the body. When a person develops a disease, for example chicken pox, antibodies develop against the disease-causing agent. After the illness is over, antibodies against this agent remain in the blood and protect against reinfection by the same substance or microorganism.

Active immunity is also produced by **vaccination** against a particular disease. A *vaccine* contains dead or weakened microorganisms that can stimulate antibody production but cannot cause disease.

Passive immunity develops when an individual receives antibodies from the blood of another person or from an animal. These antibodies provide temporary immunity to a particular disease. However, the "borrowed" antibodies are gradually destroyed, and the immunity they provided ends.

Allergies. In some people, exposure to certain common, foreign substances, such as dust, pollen, insect bites, foods, and medications, causes an immune response known as an *allergy*. These responses, or **allergic reactions**, are actually over-reactions of the body's immune system to a foreign substance. The antibodies produced may stimulate the release of a substance called *histamine*, which causes typical allergic responses, such as sneezing, coughing, or a rash.

Blood-typing. Knowledge of immunity has made possible the transplanting of organs and the transfusion of blood from one person to another. In both organ transplants and blood transfusions, an immune response is stimulated if the body of the recipient recognizes foreign antigens in the tissue or blood from the donor. In organ

Table 4-1

Antigens and Antibodies of the ABO Blood Group System

Blood Type	Antigens on Red Cells	Antibodies in Plasma
A	A	Anti-B
B	B	Anti-A
AB	A and B	Neither Anti-A nor Anti-B
O	Neither A nor B	Anti-A and Anti-B

transplants, an antigen–antibody reaction against the transplanted organ is called *rejection*. Donor tissue proteins must be carefully matched to those of the recipient to avoid rejection.

Blood-typing for transfusions is based on the presence or absence of antigens on the surface of red blood cells. The most important blood group system in blood-typing is the ABO blood group system. In this system, two kinds of antigens may be found on the red blood cells: A and B. In addition, the plasma of the blood may contain antibodies: anti-A and anti-B. Table 4-1 shows the antigens and antibodies for each type of blood.

Transport Vessels

Blood circulates through the human body within the blood vessels, which include *arteries, capillaries*, and *veins*. (Refer to Figure 3-8, page 40.)

Arteries. Blood is carried from the heart to all parts of the body in arteries, which are thick-walled, muscular vessels that expand and contract to accommodate the forceful flow of blood from the heart. The rhythmic expansion and contraction of the arteries produced by the heartbeat aids the flow of blood to all parts of the body; it is called the pulse.

Capillaries. With increasing distance from the heart, arteries branch into smaller and smaller vessels, finally forming capillaries, tiny blood vessels with walls only one cell layer thick. Capillaries are the site of exchange of materials between the blood and the body tissues.

Veins. Blood flows from the capillaries into the veins, thin-walled vessels that carry the blood back to the heart. Veins contain flaps of tissue that act as valves (Figure 4-4). The valves allow the blood in the veins to flow in only one direction—back toward the heart.

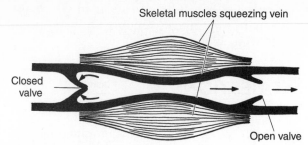

Figure 4-4. Blood flow in a vein.

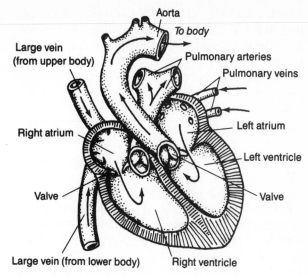

Figure 4-6. Structure of the human heart.

Intercellular Fluid and Lymph

As blood passes through the capillaries of the body, some of the plasma is forced out of the vessels and into the surrounding tissues. This fluid, which bathes all the cells of the body, is called *intercellular fluid*, or *ICF*. Materials diffusing between the cells and the blood of the capillaries are dissolved in the ICF.

Excess intercellular fluid is drained from the tissues by tiny *lymph vessels*, which are part of the *lymphatic system*. Once inside these vessels, the fluid is called *lymph*. The lymph vessels merge, forming larger vessels. Eventually, all lymph flows into two large lymph ducts, which empty into veins near the heart. In this way, the fluid lost from the blood is returned to the blood (Figure 4-5).

Major lymph vessels have enlarged regions called *lymph nodes* in which phagocytic cells filter bacteria and dead cells from the lymph. Some lymph vessels contain valves that, like those in the veins, keep the lymph flowing back toward the heart.

The Heart

Blood is pumped through the blood vessels of the body by the contractions of the heart.

Structure of the Heart. The heart has four chambers (Figure 4-6). The two upper chambers,

the *atria* (singular, *atrium*), receive blood returning to the heart from the rest of the body. The two lower chambers, the *ventricles*, pump blood out of the heart into the arteries. The walls of the ventricles are thicker and more muscular than those of the atria.

Circulation Through the Heart. The deoxygenated (oxygen-poor) blood from the body is returned to the right atrium of the heart through two large veins—one from the upper part of the body and one from the lower part of the body. This deoxygenated blood flows down from the right atrium into the right ventricle; from there it is pumped out of the heart through the pulmonary arteries to the lungs. When in the lungs, the blood gives up carbon dioxide and picks up oxygen. The oxygenated (oxygen-rich) blood is then returned through the pulmonary veins to the left atrium of the heart. The blood then passes from the left atrium into the left ventricle,

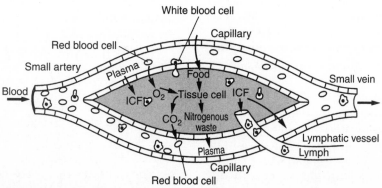

Figure 4-5. Molecules diffuse between the capillaries, intercellular fluid, and body cells.

which pumps it through the *aorta*, the largest artery in the body.

This one-way flow of blood through the heart is controlled by valves that prevent backflow of the blood. There are valves between the atria and the ventricles, between the right ventricle and the pulmonary artery, and between the left ventricle and the aorta.

Blood Pressure. The pressure exerted by the blood on the walls of the arteries during the pumping action of the heart is referred to as *blood pressure*. During the contraction phase of the heartbeat cycle, arterial blood pressure is highest. During the relaxation phase of the heartbeat cycle, blood pressure is lowest.

Pathways of Circulation

The pathway of blood between the heart and the lungs is called the *pulmonary circulation*. The circulatory pathway between the heart and all other parts of the body except the lungs is called the *systemic circulation*. The system of blood vessels that supplies the heart itself is called the *coronary circulation*.

Disorders of the Transport System

Diseases of the heart and blood vessels are called *cardiovascular* diseases. The most common form of cardiovascular disease is high blood pressure, or *hypertension*, which is characterized by elevated arterial blood pressure. This condition can be caused by a number of factors, including stress, diet, heredity, cigarette smoking, and aging. High blood pressure can damage the lining of arteries and weaken the muscle of the heart.

A blockage of the coronary artery or its branches is a *coronary thrombosis*, or heart attack. As a result of the blockage, some of the muscle tissue of the heart is deprived of oxygen and is damaged.

A narrowing of the coronary arteries may cause temporary shortages of oxygen to the heart muscle, resulting in intense pain in the chest and sometimes in the left arm and shoulder. This condition is called *angina pectoris*.

Anemia is a condition in which the blood cannot carry sufficient amounts of oxygen to the body cells. Anemia may be due to inadequate amounts of hemoglobin in the red blood cells or to too few red blood cells. One form of anemia is caused by a shortage of iron in the diet.

Leukemia is a form of cancer in which the bone marrow produces abnormally large numbers of white blood cells.

QUESTIONS
PART A

24. Which is a characteristic of lymph nodes? (1) They carry blood under great pressure. (2) They move fluids by means of a muscular pump. (3) They produce new red blood cells. (4) They contain phagocytic cells.

25. The accumulation of specific antibodies in the plasma, due to the presence of an antigen, is characteristic of (1) an immune response (2) angina pectoris (3) a coronary thrombosis (4) cerebral palsy

26. An organism develops active immunity as a result of (1) manufacturing its own antigens (2) producing antibodies in response to a vaccination (3) receiving an injection of antibodies produced by another organism (4) receiving an injection of a dilute glucose solution

27. In the human body, which blood components engulf foreign bacteria? (1) red blood cells (2) white blood cells (3) antibodies (4) platelets

28. In humans, the exchange of materials between blood and intercellular fluid directly involves blood vessels known as (1) capillaries (2) arterioles (3) venules (4) arteries

29. An injury to a blood vessel may result in the formation of a blood clot when (1) bone marrow cells decrease platelet production (2) kidney tubules synthesize clotting factors (3) ruptured platelets release enzyme molecules (4) white blood cells release antibodies

30. Oxygen carried by the blood in capillaries normally enters the body cells by (1) active transport (2) osmosis (3) diffusion (4) pinocytosis

31. Which type of vessel normally contains valves that prevent the backward flow of blood? (1) artery (2) arteriole (3) capillary (4) vein

32. The blood vessels that transport deoxygenated blood to the heart are known as (1) capillaries (2) lymph vessels (3) veins (4) arteries

33. The right ventricle is the chamber of the heart that contains (1) deoxygenated blood and

pumps this blood to the lungs (2) deoxygenated blood and pumps this blood to the brain (3) oxygenated blood and pumps this blood to the lungs (4) oxygenated blood and pumps this blood to the brain

34. Which two systems are most directly involved in providing human cells with the molecules needed for the synthesis of fats? (1) digestive and circulatory (2) excretory and digestive (3) immune and muscular (4) reproductive and circulatory

PART B-1

Base your answers to questions 35 through 39 on your knowledge of biology and on the diagram below, which represents the exchange of materials between capillaries and cells.

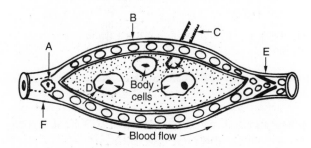

Blood flow

35. Blood vessel *B* has walls that are very thin, which enables this type of vessel to (1) transport hemoglobin to body cells (2) transport red blood cells into the tissue spaces (3) withstand the pressure of the blood coming in from veins (4) easily transport substances into and out of the blood

36. A function of cell *A* is to (1) carry oxygen (2) engulf disease-producing bacteria (3) transport digested food (4) produce hemoglobin

37. A substance that diffuses in the direction indicated by *D* is most likely (1) fibrin (2) oxygen (3) urea (4) bile

38. Which vessel most likely contains the greatest amount of carbon dioxide? (1) *F* (2) *B* (3) *C* (4) *E*

39. Excess intercellular fluid (ICF) is constantly drained off by the lymphatic vessels. Which letter represents such a vessel? (1) *E* (2) *B* (3) *C* (4) *F*

Base your answers to questions 40 through 43 on the following diagram and on your knowledge

of biology. The diagram represents the human heart; the arrows indicate the direction of blood flow.

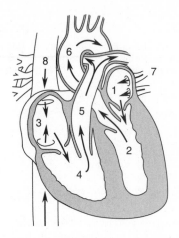

40. The aorta is represented by number (1) 1 (2) 6 (3) 8 (4) 4

41. Deoxygenated blood returns to the heart through the structure represented by number (1) 8 (2) 7 (3) 3 (4) 5

42. The chamber that pumps blood to all parts of the body except the lungs is represented by number (1) 1 (2) 2 (3) 3 (4) 4

43. Blood passes from the heart to the lungs through the structure represented by number (1) 5 (2) 6 (3) 7 (4) 8

PART B-2

44. Explain what effects faulty valves would have on a human's blood flow.

45. Describe the relationship that exists between the circulatory system and the lymphatic system.

46. The four major blood components are red blood cells, white blood cells, plasma, and platelets. Describe one major function for each blood component listed.

47. Arteries generally contain blood that has higher oxygen content than that of blood in the veins. Give a scientifically valid explanation for this observation.

48. Why are people who are anemic (have too little hemoglobin and/or too few red blood cells) often advised to take in extra iron in their diets?

49. Using appropriate information, fill in spaces *A* and *B* in the chart on page 58. In space *A*

identify an organ in the human body where molecules diffuse into the blood. In space *B* identify a specific molecule that diffuses into the blood at this organ.

An organ in the human body where molecules diffuse into the blood	A specific molecule that diffuses into the blood at this organ
A	B

50. Describe the pathway of blood flow through the heart, beginning and ending with the point at which the blood returns to the heart from the body organs.
51. Compare the structure and function of these three major types of blood vessels: arteries, veins, and capillaries.

RESPIRATION

Respiration includes cellular respiration and gas exchange. The process of cellular respiration in humans is basically the same as that in other aerobic organisms. (See Chapter 3.) Glucose is broken down completely to yield carbon dioxide and water, and ATP is formed from ADP and phosphate.

Anaerobic respiration occurs in human skeletal muscle during prolonged exercise when the amount of oxygen supplied by the circulatory system becomes inadequate for aerobic respiration. Under these circumstances, glucose is broken down in the muscle to lactic acid. The accumulation of lactic acid in skeletal muscle is thought to be responsible for muscle fatigue. When adequate oxygen is again available, the lactic acid is broken down to carbon dioxide and water.

Human Respiratory System

The human respiratory system moves respiratory gases between the external environment and the internal surfaces for gas exchange within the lungs. The respiratory system consists of a network of passageways that permit air to flow into and out of the lungs (Figure 4-7).

Nasal Cavity. Air generally enters the respiratory system through the nostrils and passes into the *nasal cavity*. This cavity is lined with a ciliated

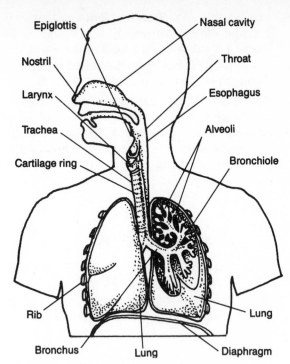

Figure 4-7. The human respiratory system.

mucous membrane that cleans, warms, and moistens the air.

Pharynx. From the nasal cavity, air passes into the *pharynx*, the area where the oral cavity and nasal cavity meet. Air passes through the pharynx on its way to the trachea.

Trachea. The *trachea*, or windpipe, is a tube through which air passes from the pharynx to the lungs. The opening (from the pharynx) to the trachea is protected by a flap of tissue called the *epiglottis*. During swallowing, the epiglottis covers the opening of the trachea so that food and liquids cannot enter the air passages. During breathing, the opening of the trachea is uncovered. (In the top of the trachea is the *larynx*, or voice box, which functions in speech.) The walls of the trachea contain rings of cartilage that keep the trachea open so that the passage of air remains unobstructed. The trachea is lined with a ciliated mucous membrane. Microscopic particles in the inhaled air are trapped by mucus, and the beating of the cilia sweeps the mucus upward toward the pharynx (to be expelled from the body by coughing or sneezing).

Bronchi and Bronchioles. The lower end of the trachea splits, forming two tubes called the *bronchi* (singular, *bronchus*). The bronchi, like the trachea, are lined with a mucous membrane and ringed with cartilage. Each bronchus extends into a lung, where it branches into smaller and smaller tubes called *bronchioles*.

The bronchioles are lined with a mucous membrane, but they lack cartilage rings. At the end of each bronchiole is a cluster of tiny, hollow air sacs called *alveoli*.

Alveoli. The lungs contain millions of alveoli (singular, *alveolus*). The walls of the alveoli are thin and moist and are surrounded by capillaries. The alveoli are the functional units for gas exchange in the human respiratory system. Oxygen diffuses from the alveoli into the surrounding capillaries, while carbon dioxide and water diffuse from the capillaries into the alveoli.

Lungs. Each bronchus with its bronchioles and alveoli make up a *lung*.

Breathing

Air moves into and out of the lungs during *breathing*. The lungs are highly elastic but contain no muscle tissue. They expand and contract in response to pressure changes in the chest cavity brought about by the actions of the rib cage and the diaphragm.

During *inhalation*, the ribs push upward and outward and the diaphragm moves down, enlarging the chest cavity. The enlargement of the chest cavity reduces the pressure around the lungs, which expand, and air flows into the lungs. In *exhalation*, the ribs move inward and downward and the diaphragm moves up. The chest cavity becomes smaller and air is forced out of the lungs (Figure 4-8).

Gas Exchange. The air that enters the alveoli is rich in oxygen. The blood in the capillaries surrounding the alveoli is oxygen-poor and contains the wastes of cellular respiration—carbon

dioxide and water. The oxygen diffuses from the alveoli into the blood, where it enters the red blood cells and becomes loosely bound to the hemoglobin.

The oxygen and hemoglobin separate in the capillaries of the body tissues. The oxygen diffuses out of the capillaries, through the intercellular fluid, and into the body cells. Carbon dioxide and water diffuse out from the cells and into the blood. When the blood returns to the lungs, these wastes diffuse into the alveoli and are expelled from the body in the exhaled air.

Breathing Rate. The rate of breathing is controlled by the breathing center in the medulla of the brain. The breathing center is sensitive to the concentration of carbon dioxide in the blood. When the carbon dioxide level is high, nerve impulses from the breathing center are sent to the rib muscles and to the diaphragm to increase the breathing rate, which speeds up the rate of excretion of carbon dioxide from the body. As the carbon dioxide level in the blood drops, the breathing rate decreases. This regulation of carbon dioxide levels is one example of the **feedback mechanisms** by which the body maintains homeostasis.

Disorders of the Respiratory System

Bronchitis is an inflammation of the linings of the bronchial tubes. As a result of such swelling, the air passages become narrowed and filled with mucus, causing breathing difficulties and coughing. *Asthma* is an allergic reaction characterized by a narrowing of the bronchial tubes, which results in difficulty in breathing. *Emphysema* is a disease in which the walls of the alveoli break down, decreasing the surface area for gas exchange. Emphysema is marked by shortness of breath, difficulty in breathing, and decreased lung capacity.

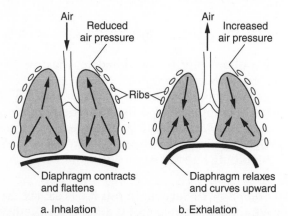

Figure 4-8. Movement of diaphragm as air goes into and out of the lungs.

QUESTIONS

PART A

52. The alveoli in humans are structures most closely associated with (1) gas exchange (2) anaerobic respiration (3) glandular secretion (4) neural transmission

53. In humans, the center that detects and regulates the amount of carbon dioxide in the blood is situated in the (1) cerebrum (2) diaphragm (3) medulla (4) rib muscles

54. The exchange of air between the human body and the environment is a result of the rhythmic contractions of the rib cage muscles and the (1) diaphragm (2) lungs (3) trachea (4) heart

55. The breathing rate of humans is regulated mainly by the concentration of (1) carbon dioxide in the blood (2) oxygen in the blood (3) platelets in the blood (4) white blood cells in the blood

Base your answers to questions 56 through 60 on the diagram below, which represents part of the human respiratory system.

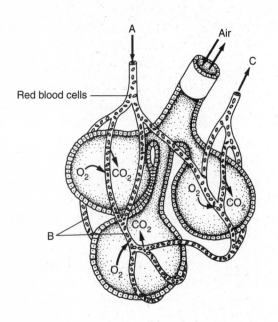

56. The blood vessels labeled *B* that are surrounding these air sacs are called (1) arteries (2) capillaries (3) veins (4) lymphatic ducts

57. These air sacs are known as (1) alveoli (2) bronchi (3) bronchioles (4) tracheae

58. The heart chamber that most directly pumps blood to the vessel network at *A* is the (1) right atrium (2) left atrium (3) right ventricle (4) left ventricle

59. The process most directly involved in the exchange of gases between these air sacs and blood vessels is called (1) active transport (2) pinocytosis (3) hydrolysis (4) diffusion

60. Compared to blood entering at *A*, blood leaving the vessel network at *C* has a lower concentration of (1) oxygen (2) hemoglobin and carbon dioxide (3) carbon dioxide (4) oxygen and hemoglobin

Base your answers to questions 61 through 63 on your knowledge of biology and on the diagram below, which represents a model of the human respiratory system.

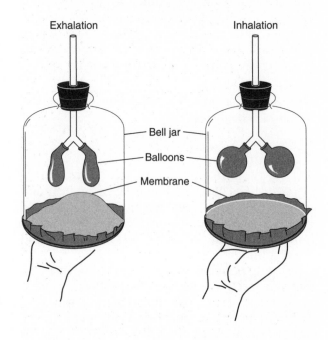

61. Explain which parts of the respiratory system the Y-tube, balloons, and rubber membrane represent.

62. Describe what happens to the balloons when the rubber membrane is pulled downward.

63. Give a scientific reason for your answer to question 62. How does this process apply to humans?

64. Breathing rate is controlled by the respiratory center in the brain, which responds to carbon dioxide levels in the blood. High levels of carbon dioxide increase the breathing rate; low levels decrease the breathing rate. Give a scientific reason for why a person's breathing rate increases during and after vigorous exercise.

65. State how the alveoli in our lungs satisfy the conditions needed to be a good respiratory surface. Include at least two conditions that allow gas exchange to take place.

66. Place the following terms in a sequence that shows the correct pathway air takes during breathing in a human (starting with taking it in from the environment): *bronchi*, *alveoli*, *nose/mouth*, *trachea*, and *bronchioles*.

EXCRETION

The metabolic wastes of humans include carbon dioxide, water, salts, and urea. Excretory wastes pass from the cells into the blood and are carried to the excretory organs that expel them from the body. The excretory organs include the lungs, liver, sweat glands, and kidneys.

Lungs. The *lungs* function in the excretion of carbon dioxide and water vapor, which are the wastes of cellular respiration.

Liver. The *liver* is a large organ that performs many functions essential to human survival. One of the excretory functions of the liver is to get rid of excess amino acids. The amino groups are removed and converted into *urea*, which is excreted by the kidneys. The remaining amino acid molecules are broken down by cellular respiration. The liver is also responsible for the breakdown of red blood cells.

Sweat Glands. The *sweat glands* of the skin excrete wastes, including water, salts, and a small amount of urea. These wastes pass by diffusion from capillaries into the sweat glands and then through ducts to pores on the surface of the skin (Figure 4-9). The mixture of wastes and water excreted by the sweat glands is called sweat, or *perspiration*.

Perspiration functions primarily in the regulation of body temperature. The evaporation of sweat from the surface of the skin occurs when heat is absorbed from skin cells, and it serves to lower the body temperature. This method of temperature regulation is another example of homeostasis.

Urinary System

The human urinary system consists of the kidneys, ureters, urinary bladder, and urethra (Figure 4-10).

Kidneys. Human *kidneys* perform two major functions: they remove urea from the blood, and they regulate the concentrations of most of the substances in the body fluids. Blood is carried to each kidney by a large artery. Within the kidney, the artery divides and subdivides into smaller and smaller arteries and then into balls of capillaries called *glomeruli* (singular, *glomerulus*). Each glomerulus is part of a *nephron*, the functional unit of the kidney (Figure 4-11, page 62). There are about one million nephrons in each kidney.

A nephron consists of a glomerulus surrounded by a cup-shaped structure called *Bowman's capsule*. Extending from the capsule is a long, coiled tubule that is surrounded by capillaries. As blood flows through the glomerulus, water, salts, urea, glucose, and some amino acids diffuse out of the blood into Bowman's capsule. This process is called *filtration*. As these substances—referred to as the *filtrate*—pass through the long, coiled tubule of the nephron, glucose, water, amino acids, and some of the salts are reabsorbed by active transport into the blood in the capillaries surrounding the tubule. The fluid that remains in the tubules consists of water, urea, and salts, and is called *urine*. Urine passes from the small tubule

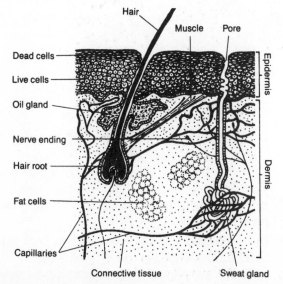

Figure 4-9. Structure of the skin.

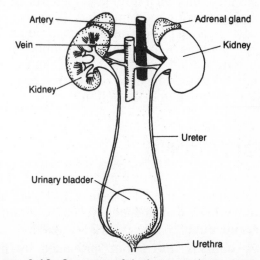

Figure 4-10. Structure of the human urinary system.

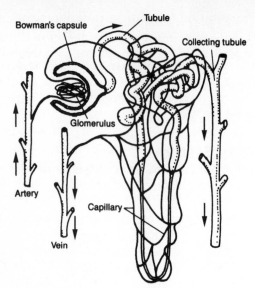

Figure 4-11. Structure of a nephron.

of the nephron into larger tubules and then to a ureter.

Ureters and Urinary Bladder. Urine flows from each kidney into a large tubule called the *ureter*. The ureters carry the urine to the urinary bladder, a muscular organ in which urine is stored temporarily.

Urethra. Urine is periodically expelled from the bladder into a tube called the *urethra*. This tube leads to the outside of the body.

Diseases of the Urinary System

Diseases of the kidneys affect the body's ability to eliminate normal amounts of metabolic wastes. *Gout* is a condition that produces symptoms similar to arthritis and is caused by deposits of uric acid in the joints. Victims of gout suffer from severe pain and stiffness in the joints. Diets that are extremely high in protein result in the production of large amounts of urea, which the kidneys must remove from the blood. The extra strain on the kidneys in eliminating these wastes may result in a kidney disorder, or **malfunction**.

QUESTIONS

PART A

67. Which human body system includes the lungs, liver, skin, and kidneys? (1) respiratory (2) digestive (3) transport (4) excretory

68. In humans, the filtrate produced by the nephrons is temporarily stored in the (1)

glomerulus (2) alveolus (3) gallbladder (4) urinary bladder

69. What is the principal waste from excess amino acids in humans? (1) salt (2) urea (3) uric acid (4) carbon dioxide

70. In humans, the organ that breaks down red blood cells and amino acids is the (1) kidney (2) liver (3) gallbladder (4) small intestine

71. The main components of urine, besides water, are (1) amino acids and fatty acids (2) urea and salts (3) ammonia and bile (4) hydrochloric acid and bases

72. In humans, urine is eliminated from the bladder through the (1) urethra (2) ureter (3) nephron (4) collecting tubule

73. The basic structural and functional excretory units of the human kidney are known as (1) nephridia (2) nephrons (3) alveoli (4) ureters

74. The excretory organ that is also associated with the storage of glycogen is the (1) stomach (2) lung (3) kidney (4) liver

PART B-1

Base your answers to questions 75 through 77 on your knowledge of biology and on the diagram below, which illustrates a nephron and its capillaries.

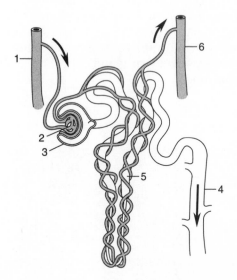

75. Into which structure does the filtrate first pass? (1) 5 (2) 6 (3) 3 (4) 4

76. In which area is water being reabsorbed? (1) 5 (2) 2 (3) 3 (4) 4

77. In which area does urine collect? (1) 1 (2) 2 (3) 6 (4) 4

78. Briefly describe how each of the following functions as an excretory organ:
- the liver;
- the skin;
- the lungs.

Base your answers to questions 79 and 80 on the diagram and information presented below. The diagram represents a nephron from which samples of fluid were extracted. The samples were recovered from the areas labeled A and B in the diagram. The concentrations of five substances in the fluid extracted from both sites were compared and the results are listed in the table.

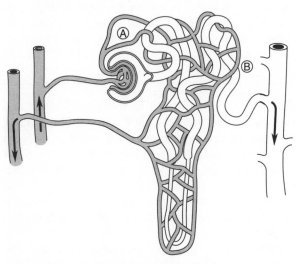

Substance	Concentration at A	Concentration at B
Water	High	Low
Urea	Moderate	High
Glucose	Moderate	Zero
Amino acids	High	Zero
Salts	Low	High

79. Explain the change in the concentrations of water and urea from area *A* to area *B*.

80. Why are there no amino acids present in the fluid extracted from area *B,* yet the concentration at area *A* was high?

NERVOUS SYSTEM

Regulation in humans involves the interaction of the nervous and endocrine systems. The two systems are similar in that they both secrete chemicals and both play a major role in the maintenance of homeostasis. In general, they differ in that the responses of the nervous system are more rapid and of shorter duration than those of the endocrine system.

Nerve Cells

The nervous system is made up of **nerve cells**, or *neurons*, which are adapted for the transmission of impulses. The nervous system contains three different types of nerve cells, which differ both in structure and function; these are the sensory neurons, motor neurons, and interneurons.

Sensory neurons transmit impulses from the sense organs, or receptors, to the brain and the spinal cord. Sense organs include the eyes, ears, tongue, nose, and skin.

Motor neurons transmit impulses from the brain and spinal cord to the *effectors*, that is, to the muscles and the glands.

Interneurons are found in the spinal cord and brain; they transmit nerve impulses from sensory neurons to motor neurons.

Nerves

The nerve cells, or parts of nerve cells, are bound together in bundles called *nerves*. There are three kinds of nerves: *sensory nerves*, which contain only sensory neurons; *motor nerves*, which contain only motor neurons; and *mixed nerves*, which contain both sensory and motor neurons.

Central Nervous System

The two main divisions of the human nervous system are the *central nervous system,* which includes the brain and spinal cord, and the *peripheral nervous system*, which includes all the nerves outside the central nervous system.

The Brain. The *brain* is a large mass of nerve cells located in the cranial cavity. It is surrounded and protected by the bones of the skull. The three major parts of the brain are the cerebrum, the cerebellum, and the medulla (Figure 4-12, page 64). Each controls different functions of the body.

In humans, the *cerebrum* is the largest part of the brain. It is the center for thought, memory, and learning; it receives and interprets messages from the sense organs; and it initiates all voluntary, or conscious, movements.

The *cerebellum* is located below and behind the cerebrum. It coordinates all motor activities and is involved in maintaining the body's balance.

The *medulla* is located at the base of the brain and connects the brain and the spinal cord. The medulla controls many important involuntary

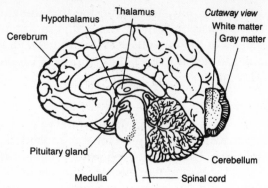

Figure 4-12. Structure of the human brain.

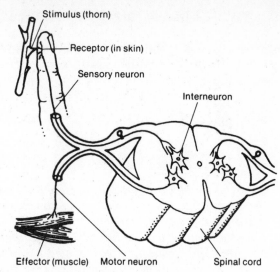

Figure 4-13. A typical reflex arc.

activities in the body, including breathing, heartbeat, blood pressure, and peristalsis.

The Spinal Cord. The medulla of the brain is continuous with the *spinal cord*, which is surrounded and protected by the vertebrae of the backbone, or spinal column. The spinal cord coordinates activities between the brain and other body structures. Impulses from sense receptors throughout the body are transmitted by sensory neurons to the spinal cord. In the spinal cord, impulses are transmitted by interneurons to the brain. Impulses from the brain are carried by motor neurons through the spinal cord and then to the appropriate effectors.

Peripheral Nervous System

The peripheral nervous system includes all neurons, both sensory and motor, outside the central nervous system. These nerve cells carry impulses between the central nervous system and the rest of the body. The two main divisions of the peripheral nervous system are the somatic nervous system and the autonomic nervous system.

The *somatic nervous system* includes all the nerves that control the movements of the voluntary muscles of the body, as well as the sensory neurons that transmit impulses from sense receptors to the central nervous system.

The *autonomic nervous system* consists of the nerves that control the activities of smooth muscle, cardiac muscle, and glands. The activities of this system, which are involuntary (not under voluntary control), include regulation of the heartbeat and circulation, respiration, and peristalsis.

Behavior and the Nervous System

All animals, including humans, have behaviors that help them maintain homeostasis and aid their survival. These behaviors are controlled by the nervous system. Some behaviors are inborn, while others are learned.

Habits. A *habit* is a kind of learned behavior that becomes automatic through repetition. The repetition establishes pathways for nerve impulse transmission that permit a rapid, automatic response to a particular stimulus.

Reflexes. An automatic, inborn response to a particular stimulus is called a *reflex*. In a reflex response, impulses follow a set pathway called a *reflex arc* (Figure 4-13). In this pathway, impulses pass from a receptor to a sensory neuron to an interneuron (in the spinal cord) to a motor neuron to an effector. Although impulses may also pass from an interneuron to the brain, the reflex response is controlled by the spinal cord and occurs without the involvement of the brain. Reflexes are generally protective in nature, allowing a rapid response to a potentially dangerous stimulus.

Disorders of the Nervous System

Cerebral palsy is a group of diseases caused by damage to the parts of the brain that control voluntary movement. This damage occurs during embryonic development. *Meningitis* is an inflammation of the membranes that surround the brain and spinal cord. Meningitis may be caused by viral or bacterial infections, and symptoms include headache, muscle stiffness, fever, and chills. A *stroke* is a disorder in which the brain is damaged as a result of a *cerebral hemorrhage* (a broken blood vessel) or a blood clot (in a blood vessel) in the brain. *Polio* is a disease that affects the central nervous system; it may result in paralysis. Polio is caused by a virus and can be prevented by immunization. *Alzheimer's* is a degenerative disease in which neurons in the brain are gradually destroyed. This fatal illness generally strikes older people and begins with such symptoms as forgetfulness, mood swings, and unusual behavior. As the disease progresses, the person becomes less and less capable of handling simple daily tasks such as dressing, bathing, and eating on his or her own.

QUESTIONS
PART A

81. The major function of a motor neuron is to (1) transmit impulses from the spinal cord to the brain (2) act as a receptor for environment stimuli (3) transmit impulses from sense organs to the central nervous system (4) transmit impulses from the central nervous system to muscles or glands

82. Nerves are composed of bundles of (1) muscle cells (2) neurons (3) phagocytes (4) bone cells

83. Which part of the human central nervous system is involved primarily with sensory interpretation and thinking? (1) spinal cord (2) medulla (3) cerebrum (4) cerebellum

84. The somatic nervous system contains nerves that run from the central nervous system to the (1) muscles of the skeleton (2) heart (3) smooth muscles of the gastrointestinal tract (4) endocrine glands

85. If the cerebellum of a human were damaged, which of the following would probably result? (1) inability to reason (2) difficulty in breathing (3) loss of sight (4) loss of balance

86. Which is the correct route of an impulse in a reflex arc?
    (1) receptor → sensory neuron → interneuron → motor neuron → effector
    (2) effector → receptor → motor neuron → sensory neuron → interneuron
    (3) sensory neuron → effector → motor neuron → receptor → interneuron
    (4) motor neuron → sensory neuron → interneuron → effector

87. The brain and the spinal cord make up the (1) autonomic nervous system (2) peripheral nervous system (3) central nervous system (4) somatic nervous system

88. Impulses are transmitted from receptors to the central nervous system by (1) receptor neurons (2) sensory neurons (3) interneurons (4) motor neurons

Base your answers to questions 89 through 92 on the following diagram of the human brain.

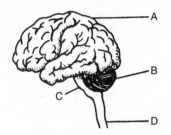

89. Injury to which part would most likely result in loss of memory? (1) A (2) B (3) C (4) D

90. Which part of the brain controls the involuntary movements of the digestive system? (1) A (2) B (3) C (4) D

91. Which part of the brain is involved with balance and the coordination of body movements? (1) A (2) B (3) C (4) D

92. Sight and hearing are functions of the structure labeled (1) A (2) B (3) C (4) D

Base your answers to questions 93 and 94 on the diagram below and on your knowledge of biology.

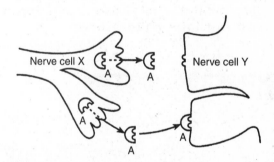

93. The process represented in the diagram best illustrates (1) cellular communication (2) muscle contraction (3) extraction of energy from nutrients (4) waste disposal

94. Which statement best describes the diagram? (1) Nerve cell X is releasing receptor molecules. (2) Nerve cell Y is signaling nerve cell X. (3) Nerve cell X is attaching to nerve cell Y. (4) Nerve cell Y contains receptor molecules for substance A.

95. Use the following terms to complete the chart on page 66, which outlines the human nervous system: somatic nervous system, brain,

cerebrum, cerebellum, autonomic nervous system, medulla, spinal cord.

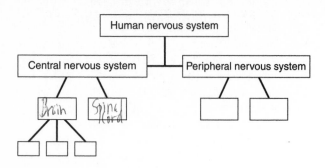

96. Briefly describe the main functions of the three major regions of the human brain:

- the cerebrum;
- the cerebellum;
- the medulla.

97. Explain how a reflex arc works to protect the human body from a potentially dangerous stimulus. Provide an example.

98. The human nervous system is one of the most complex in the world. Give one example of how our brains have led to the progress that we consider to be "advanced" compared to that of other animals.

ENDOCRINE SYSTEM

The human endocrine system is made up of the endocrine glands, which secrete **hormones** directly into the blood. The hormones are transported by the circulatory system to the organs and tissues on which they act.

Endocrine Glands
The glands of the human endocrine system are the hypothalamus, pituitary, thyroid, parathyroids, adrenals, islets of Langerhans, and gonads (ovaries and testes) (Figure 4-14).

Hypothalamus. Hormone-secreting cells are present in a small part of the brain called the *hypothalamus*. The hormones of the hypothalamus influence the activities of the pituitary gland.

Pituitary Gland. Many hormones are secreted by the *pituitary gland*, which is located at the base of the brain. Some pituitary hormones regulate the activities of other endocrine glands.

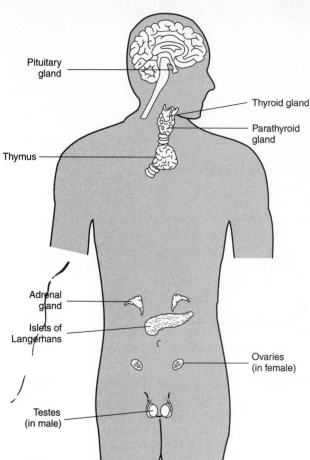

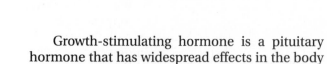

Figure 4-14. Structure of the human endocrine system.

Growth-stimulating hormone is a pituitary hormone that has widespread effects in the body in addition to stimulating the growth of long bones.

Thyroid-stimulating hormone (TSH) is a pituitary hormone that stimulates the secretion of the thyroid hormone thyroxin.

Follicle-stimulating hormone (FSH) is a pituitary hormone that stimulates the development of follicles in the ovaries of females. In males, it influences sperm production.

Thyroid Gland. The iodine-containing hormone *thyroxin* is produced by the *thyroid gland*, which is located in the neck. Thyroxin regulates the rate of metabolism in the body cells and is essential for normal physical and mental development.

Parathyroid Glands. Embedded in the back of the thyroid gland are the *parathyroid glands*, which secrete the hormone *parathormone*. Parathormone controls calcium metabolism. Calcium is required for normal nerve function, blood clotting, and the growth of teeth and bones.

Adrenal Glands. An *adrenal gland* is located on the top of each kidney. The outer layer of the adrenal glands is the adrenal cortex; the inner layer is the adrenal medulla.

The *adrenal cortex* secretes two types of steroid hormones. One type stimulates the conversion of fats and proteins to glucose, thereby increasing the level of glucose in the blood. The other type stimulates the reabsorption of sodium from the kidney tubules into the bloodstream. The concentration of sodium in the blood affects blood pressure and water balance.

The *adrenal medulla* secretes the hormone *adrenaline*, which increases the blood glucose level and accelerates the heartbeat and breathing rates. Adrenaline is released in times of stress and heavy exercise.

Islets of Langerhans. The small groups of endocrine cells that are found throughout the pancreas are called the *islets of Langerhans*. These endocrine cells secrete the hormones insulin and glucagon.

The hormone **insulin** promotes the absorption of glucose from the blood into the body cells, thereby lowering the blood glucose level. It also stimulates the conversion of glucose to glycogen in the liver and in skeletal muscle.

The hormone *glucagon* increases the blood glucose level by promoting the conversion of glycogen to glucose in the liver and skeletal muscle. The glucose then passes from the organs back into the blood. Through their opposite effects, insulin and glucagon function to help the body maintain homeostasis by keeping the blood glucose level within certain limits.

The Gonads. The male and female *gonads*—the testes and ovaries—both function as endocrine glands. The **testes** (singular, *testis*) secrete the male sex hormone **testosterone**, which stimulates the development of the male reproductive organs and secondary sex characteristics; it also stimulates the production of **sperm**. The **ovaries** secrete the female sex hormones **estrogen** and **progesterone**. Estrogen influences the development of the female reproductive organs and secondary sex characteristics; it also stimulates the production of **egg** cells. Progesterone stimulates the thickening of the uterine lining in preparation for the implantation of an **embryo** (the fertilized egg cell).

Negative Feedback

The secretion of hormones by the endocrine glands is regulated by a mechanism known as *negative feedback*. In many cases, the level of one hormone in the blood stimulates or inhibits the production of a second hormone. The blood level of the second hormone in turn stimulates or inhibits the production of the first hormone. For example, the relationship between the pituitary's secretion of thyroid-stimulating hormone (TSH) and the thyroid's secretion of the hormone thyroxin is a classic type of *negative feedback mechanism*.

When the concentration of thyroxin in the blood drops below a certain level, the pituitary is stimulated to secrete TSH. This hormone, in turn, then stimulates the secretion of thyroxin by the thyroid. When the blood thyroxin concentration reaches a certain level, the further secretion of TSH by the pituitary is inhibited. In this way, the body can regulate thyroxin levels—just as it regulates carbon dioxide levels—in order to maintain **stability**, or homeostasis.

Disorders of the Endocrine System

A *goiter* is an enlargement of the thyroid gland that is most commonly caused by a lack of iodine in the diet. *Diabetes* is a disorder in which the islets of Langerhans do not secrete adequate amounts of insulin into the bloodstream and, as a result, the blood glucose level is elevated. Disorders in the pituitary gland may affect the release of growth hormone, resulting in a negative effect on a person's growth.

Recent advances in recombinant DNA technology have allowed the synthesis of human growth hormone, as well as insulin. These hormones (made by genetically engineered bacteria) can be used to replace or supplement the inadequate amount of hormone being made by the person with the disorder.

QUESTIONS

PART A

99. Which of the following is not an endocrine gland? (1) thyroid (2) salivary gland (3) pancreas (4) testis

100. The part of the brain that is most directly related to the endocrine system is the (1) cerebrum (2) medulla (3) hypothalamus (4) cerebellum

101. Which structure secretes the substance it produces directly into the bloodstream? (1) gallbladder (2) salivary gland (3) adrenal gland (4) skin

102. The hormones insulin and glucagon are produced by the (1) thyroid (2) pituitary (3) pancreas (4) liver

103. Which hormone lowers blood sugar levels by increasing the rate of absorption of glucose by the body cells? (1) follicle-stimulating hormone (2) insulin (3) parathormone (4) adrenalin

104. A person was admitted to the hospital with abnormally high blood sugar level and a very high sugar content in his urine. Which gland most likely caused this condition by secreting lower than normal amounts of its hormone? (1) pancreas (2) parathyroid (3) salivary (4) thyroid

105. Which hormone stimulates activity in the ovaries? (1) testosterone (2) thyroid stimulating hormone (3) insulin (4) follicle stimulating hormone

106. A person's rate of metabolism is regulated by a hormone secreted by the (1) parathyroids (2) thyroid (3) pancreas (4) adrenals

107. Estrogen, which influences the development of secondary sex characteristics, is secreted by the (1) pituitary (2) adrenals (3) parathyroids (4) ovaries

108. In humans, the level of calcium in the blood is regulated by the (1) pancreas (2) thyroid (3) adrenals (4) parathyroids

109. The mechanism that regulates the secretion of hormones by endocrine glands is called (1) peristalsis (2) active transport (3) negative feedback (4) filtration

110. Insufficient iodine in the diet may cause goiter, a disorder of the (1) adrenal glands (2) pancreas (3) pituitary gland (4) thyroid gland

PART B-2

Base your answers to questions 111 through 114 on the graph below, which shows the levels of glucose and insulin present in a person's blood after eating a meal.

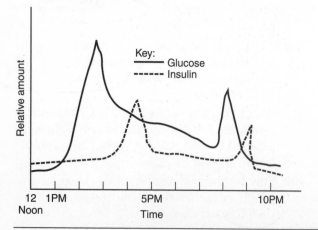

111. At approximately what times did the glucose level spike in this person?

112. What effect does insulin seem to have on the blood glucose level?

113. Describe the relationship between the levels of glucose and the levels of insulin.

114. What is the most probable reason for the time lag between the spikes in glucose level and the spikes in insulin level?

PART C

115. Use your knowledge of biology to answer the following questions about how a negative feedback mechanism works:

- How do the pituitary gland and thyroid gland affect each other?
- How does this feedback mechanism help in the maintenance of homeostasis?
- What hormones produced by each gland are part of this feedback mechanism?

116. The diagrams below represent some of the systems that make up the human body. Select one of the pairs of systems and:

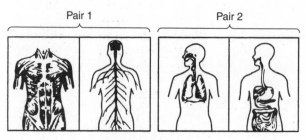

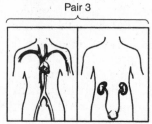

- identify each system in the pair you selected;
- state one function of each system in the pair;
- explain how the two systems work together to help maintain homeostasis.

68 Reviewing Biology: The Living Environment

LOCOMOTION

Locomotion, or movement, in humans involves the interaction of bones, cartilage, muscles, tendons, and ligaments.

Bones

The human skeleton is made up mainly of bones of various shapes and sizes. All bones are made of *bone tissue*, which is quite hard and rigid. Bones provide support and protection for the soft parts of the body; they are the sites of attachment for muscles; and, at joints, bones act as levers, enabling the body to move when the attached muscles contract. The production of new red blood cells and white blood cells occurs in the marrow of certain long bones.

Cartilage

In addition to bone, the human skeleton contains *cartilage*, a type of flexible, fibrous, elastic connective tissue. In embryos, most of the skeleton is made of cartilage. After birth, a child's cartilage is gradually replaced by bone, so that in adults, almost all of the cartilage has been replaced. In adults, cartilage is found at the ends of ribs, between vertebrae, at the ends of bones, and in the nose, ears, and trachea. Cartilage provides cushioning and flexibility at joints, and support and pliability in structures such as the nose and ears.

Joints

The places in the skeleton where the bones are connected to each other are called *joints*. Joints make movement of the skeleton possible. There are several kinds of movable joints in the human body. *Hinge joints*, which can move back and forth, are in the elbow and knee. *Ball-and-socket joints*, which are capable of circular movements, are found in the shoulder and hip. The neck has a *pivot joint*, which can move in a half circle. The bones of the skull are joined in *immovable joints*.

Muscles

Unlike other body tissues, muscle tissue has the capacity to contract, or shorten. All movement in the body involves muscle tissue. There are three types of muscle tissue in the human body: skeletal muscle, smooth muscle, and cardiac muscle (Figure 4-15).

Skeletal Muscle. The voluntary muscles attached to the bones of the skeleton are made of *skeletal muscle* tissue. Muscle tissue of this type appears striated, or striped, when viewed with a microscope, and is also known as *striated muscle*.

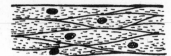
Smooth muscle

Cardiac muscle

Skeletal muscle

Figure 4-15. Smooth, cardiac, and skeletal muscle.

The contraction of skeletal muscle is controlled by the nervous system, which makes coordinated movements possible.

Skeletal muscles generally operate in antagonistic pairs; the contraction of one muscle of the pair extends the limb, while contraction of the other muscle flexes the limb. Figure 4-16 shows the muscles of the upper arm. The triceps is the extensor, while the biceps is the flexor. When the biceps contracts, the triceps relaxes, and the arm flexes, bending at the elbow. When the triceps contracts, the biceps relaxes, and the arm is extended.

Smooth Muscle. When viewed with a microscope, *smooth muscle* tissue does not appear striated. This type of muscle, which is also called

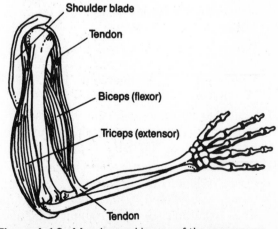

Figure 4-16. Muscles and bones of the upper arm.

visceral muscle, is found in the walls of the digestive organs and arteries, as well as in other internal organs. Smooth muscles are not under voluntary control.

Cardiac Muscle. The *cardiac muscle* tissue is found only in the heart. Although it appears striated when viewed with a microscope, cardiac muscle tissue is not under voluntary control, and its structure is different from that of skeletal muscle tissue.

Tendons and Ligaments

Muscles are attached to bones by tough, inelastic, fibrous cords of connective tissue called *tendons.* Bones are connected together at movable joints by *ligaments,* which are composed of tough, elastic connective tissue.

Disorders of Locomotion

Arthritis is an inflammation of the joints, which can be very painful and make movement difficult. *Tendonitis* is an inflammation of a tendon, usually where it is attached to a bone. This condition occurs most commonly in athletes.

QUESTIONS
PART A

117. Which type of muscle tissue found in the walls of the human stomach is most closely associated with the process of peristalsis? (1) striated (2) cardiac (3) voluntary (4) smooth

118. Bones are attached to each other at movable joints by (1) elastic ligaments (2) cartilaginous tissues (3) smooth muscles (4) skeletal muscles

119. Which is *not* a major function of cartilage tissues in a human adult? (1) giving pliable support to body structures (2) cushioning joint areas (3) adding flexibility to joints (4) providing skeletal levers

120. Which type of connective tissue makes up the greatest proportion of the skeleton of a human embryo? (1) ligaments (2) cartilage (3) tendons (4) bone

121. Which structure contains pairs of opposing skeletal muscles? (1) stomach (2) small intestine (3) heart (4) hand

122. Which statement most accurately describes human skeletal muscle tissue? (1) It is involuntary and striated. (2) It is involuntary and lacks striations. (3) It is voluntary and striated. (4) It is voluntary and lacks striations.

123. In the human elbow joint, the bone of the upper arm is connected to the bones of the lower arm by flexible connective tissue called (1) tendons (2) ligaments (3) muscles (4) neurons

For each phrase in questions 124 through 128, select the human body structure in the list below that is best described by that phrase.

Human Body Structure

A. Bones

B. Cartilage tissues

C. Ligaments

D. Smooth muscles

E. Tendons

F. Voluntary muscles

124. Cause peristalsis in the digestive tract (1) *B* (2) *C* (3) *D* (4) *F*

125. Serve as extensors and flexors (1) *A* (2) *D* (3) *E* (4) *F*

126. Serve as levers for body movements (1) *A* (2) *B* (3) *C* (4) *E*

127. Bind the ends of bones together (1) *B* (2) *C* (3) *D* (4) *E*

128. Attach the muscles to bones (1) *B* (2) *C* (3) *D* (4) *E*

PART B-2

Refer to Figure 4-16, which shows the bones and (upper arm) muscles of the human arm to answer questions 129 through 131.

129. What happens to the arm when the biceps contracts?

130. What happens to the arm when the triceps contracts?

131. Why are the biceps and triceps considered an opposing pair of muscles?

PART C

132. Briefly compare the functions of the following paired structures of the muscular and skeletal systems:
    - smooth muscle to skeletal muscle;
    - tendon to ligament;
    - bone to cartilage.

133. How do the skeletal and muscular systems work together to produce locomotion?

134. State two advantages that locomotion gives to an organism. Explain how they aid survival.

Base your answers to questions 135 through 138 on the information below and on your knowledge of biology. Source: Science News *(April 30, 2005): vol. 167, no. 18, p. 285.*

When the Stomach Gets Low on Acid

A shortage of stomach acid can lead to cancer, possibly as a result of bacterial overgrowth and chronic inflammation, a study in mice indicates.

Too much stomach acid is a well-studied problem that can cause more than simple gastritis, an inflammation of the stomach lining. Excess acid can lead to heartburn and cause chronic inflammation of the esophagus, esophageal scarring, and even cancer.

Turning the tables, scientists recently found that too little stomach acid might cause its own problems, including pneumonia.

In the new study of low stomach acid, Juanita L. Merchant, a gastroenterologist at the University of Michigan in Ann Arbor, and her colleagues studied 20 mice, half of which were genetically engineered to lack gastrin, the hormone that orchestrates stomach-acid secretion. Six of the mice lacking gastrin developed stomach tumors at 12 months of age, but none of the normal mice did, the researchers report in the March 31 *Oncogene*.

The mice lacking gastrin also had fewer stomach-lining cells die off, which is a normal, tumor-suppressing action. In this process, the body detects runaway cell growth and sends the aberrant cells into suicide mode. The gastrin-deficient mice lacked RUNX3, a protein that in normal mice can activate such programmed cell death. Merchant hypothesizes that inflammation brought on by excess bacterial growth might suppress RUNX3 production.

It's too early to draw a parallel between acid-deficient mice lacking all gastrin from birth and people who regularly take acid-blocking drugs for acid-reflux disease, Merchant says.

135. How might a lack of stomach acid lead to stomach cancer?

136. State three health problems caused by excess stomach acid.

137. Explain the connection between gastrin and stomach acid.

138. What role does RUNX3 play in preventing tumors in normal mice?

CHAPTER 5

Homeostasis and Immunity

A DYNAMIC EQUILIBRIUM

Under normal circumstances, an organism is able to maintain stability, or homeostasis, in relation to both its internal and external environments. This maintaining of a **dynamic equilibrium** means that despite the fact that environmental conditions may change, an organism responds by taking corrective actions that restore healthy conditions within its body.

For example, to maintain its normal temperature of about 37°C, the human body can make simple adjustments to keep its temperature within a safe range. If the body is too cold, small blood vessels in the skin may constrict in order to direct blood flow to the vital, internal organs. In addition, the body may shiver to generate more heat. If the body is overheated, blood vessels near the skin surface can dilate (open wider) to promote blood flow to the skin in order to lose heat to the surrounding air. The skin may also produce perspiration (sweat) as a means of lowering the body temperature.

Feedback Mechanisms

Many homeostatic adjustments in organisms involve interactions called *negative feedback mechanisms*, or feedback loops. An initial change in one part of the loop (in response to some changing condition) stimulates a reaction in another part of the system. When the condition has been corrected, the second part of the loop feeds back information to the first part, shutting it off. This, in turn, shuts off the response (that it initially caused) in the second part. Homeostasis is maintained. If conditions in the body change, the feedback system is triggered into action again.

A very common negative feedback loop involves the pituitary gland, the thyroid gland, and their hormones. If the level of thyroxin (a thyroid

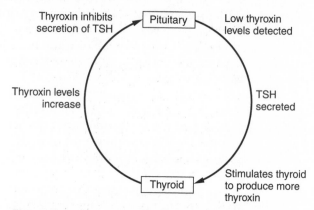

Figure 5-1. A negative feedback loop, involving the pituitary and thyroid glands.

hormone that regulates metabolism) is too low, the pituitary gland secretes thyroid-stimulating hormone (TSH). This causes the thyroid to increase its production of thyroxin. As the levels of thyroxin in the blood increase, the pituitary gland stops secreting TSH. Without the TSH, the thyroid slows down its secretion of thyroxin. In this way, the levels of thyroxin are maintained within normal limits (Figure 5-1).

WHEN HOMEOSTASIS FAILS: ILLNESS AND DISEASE

Causes of Disease

Any breakdown in an organism's ability to maintain or restore equilibrium can result in illness, disease, or even death. The causes of disease are many and varied. Diseases that are caused by factors inside the body are usually inherited and due to defective genetic traits. Diseases that are caused by factors outside the body, and that can be passed from one organism to another, are called *infectious* diseases. Factors that cause such diseases include microorganisms, or **microbes**, that are harmful. These disease-causing microbes, called **pathogens**, may include bacteria, fungi, protozoa, and worms, as well as nonliving particles of protein and nucleic acid, called *viruses*.

Another type of infectious particle is called a *prion*. Like viruses, these misshaped pieces of proteins have the unusual ability to increase in number when inside a living organism. It is thought that prions, which form clumps that kill brain cells, cause normal proteins to fold and become abnormal. A well-known disease caused by prions is *mad cow disease* (bovine spongiform encephalopathy), a fatal illness that destroys the nervous system in cattle. Sheep also may suffer from a nervous-system disease caused by prions, called *scrapie*. The degenerative nervous disorder Creutzfeldt-Jakob disease, or *kuru*, which infects some people, is also caused by prions.

Sometimes, unhealthy habits and/or risky behaviors can jeopardize health and lead to illness. Poor nutrition, cigarette smoking, and abuse of alcohol and drugs can all result in serious illness and a breakdown of homeostasis. For example, excessive consumption of alcoholic beverages can cause cirrhosis of the liver, a fatal disease.

Cancer. Disease may also occur when certain cells in the body behave abnormally due to a genetic mutation. Such cells can divide uncontrollably and result in the growth of *tumors*. Tumors may be benign (not spreading) or malignant (spreading). Uncontrolled growth, or *metastasis*, of malignant cells is known as **cancer**. When cancer cells spread throughout the body, they interfere with the functioning of normal cells. In such cases, the cancer can become life threatening. Although cancer may occur spontaneously, certain factors are known to increase the risk of developing it. Tobacco smoking, unhealthful diet, genetic factors, and exposure to **radiation** and certain chemicals called *carcinogens* are all thought to play a part in causing cancer.

Symptoms of Disease. Some diseases show their symptoms as soon as they begin to develop or soon after they are triggered by a pathogen. An example is influenza (the "flu"), which is caused by viruses. Other diseases may take several days, weeks, or even years before their symptoms appear. **AIDS** (*a*cquired *i*mmuno*d*eficiency *s*yndrome) and cancer are examples of diseases that may develop in the body for years before their symptoms appear.

QUESTIONS

PART A

1. The term that describes a body's overall ability to maintain homeostasis is (1) negative loop system (2) low maintenance (3) dynamic equilibrium (4) infectious

2. Pathogens may include all of the following *except* (1) fungi (2) protozoa (3) bacteria (4) plants

3. Viruses differ from other pathogens in that *only* viruses (1) contain a true nucleus (2) can reproduce on their own every 20 minutes (3) consist only of protein and nucleic acid (4) are able to infect healthy cells

4. Which of the following represents a correct cause-and-effect sequence? (1) cirrhosis of the liver → excessive alcohol consumption (2) low thyroxin levels → increase in TSH secretion (3) symptoms of disease → exposure to pathogen (4) dilation of blood vessels in skin → overheating of the body

5. Infectious particles known as prions, which cause nervous system diseases such as scrapie and *kuru*, consist of misshaped pieces of (1) bacteria (2) fungi (3) proteins (4) viruses

6. When a certain plant is without water for an extended period of time, guard cells close openings in the leaves of the plant. This activity conserves water and illustrates (1) cellular communication involving the action of nerve cells and receptor sites (2) an increase in rate of growth due to a low concentration of water (3) maintenance of a dynamic equilibrium through detection and response to stimuli (4) a response to one biotic factor in the environment

PART B-2

Base your answer to the following question on the diagram below and on your knowledge of biology.

7. What term or phrase does letter *X* most likely represent?

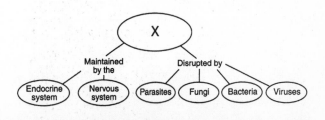

8. Select two of the following risk factors—drug abuse; poor nutrition; genetic factors; radiation; tobacco smoking—then:

- define each of the two risk factors you have selected;

- explain how each risk factor can interfere with proper functioning of the immune system.

Base your answers to questions 9 through 11 on the diagram of a negative feedback loop, shown below, and on the following data: Structure A releases a substance, X, that can stimulate structure B to release its substance, Y.

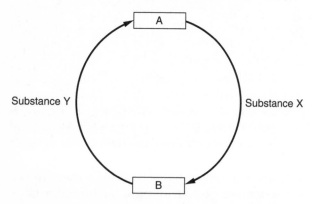

9. Under what conditions would structure *A* probably release substance *X*?

10. Explain what happens when the levels of substance *Y* get too high.

11. What type of substances are *X* and *Y* most likely to be?

THE IMMUNE SYSTEM: PROTECTION AGAINST DISEASE

The human body is well protected against invading pathogens. The first line of defense prevents harmful microorganisms from getting into the body by blocking their entry. The skin, when unbroken, provides an effective physical barrier to nearly all pathogenic organisms. Secretions such as tears, saliva, and mucus provide an effective physical and chemical barrier; they contain enzymes that destroy pathogens or help trap and flush them out of the body.

Nevertheless, some pathogens manage to elude the first line of defense and gain entry. They may do so through breaks in the skin (cuts and scrapes) or through the eyes and natural openings in the body, such as the mouth and nostrils. Once inside, these invaders are confronted by the *immune system*, the body's primary defense mechanism. Invaders may be destroyed by being engulfed by special cells or by being chemically marked for destruction and elimination.

Functions of the Immune System

How does the immune system function? All cells have very specific proteins on their plasma membrane surfaces. The immune system is able to recognize proteins on cells that are foreign and to distinguish them from its own body's proteins. These invading foreign proteins are referred to as *antigens*.

Specialized Blood Cells. The human immune system consists of specialized white blood cells and lymphatic organs such as the spleen, thymus, and tonsils. The system also has a number of *lymph nodes* that participate in defense mechanisms. Some white blood cells, called *macrophages*, engulf and digest pathogens (Figure 5-2). After destroying the pathogens, these white cells often die, too.

White blood cells called *T cells* are specialized to kill pathogens or mark them for destruction. Other white blood cells, called *B cells*, produce very specific *antibodies* against the pathogens. Antibodies have a chemical structure that precisely matches the shape of the antigen with which they react. Once the match has been made, the pathogen is destroyed. Some of the antibody-producing blood cells remain in the body's immune system as "memory cells." These specialized cells

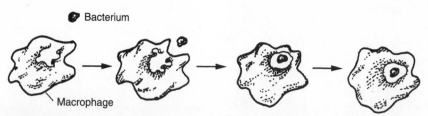

Figure 5-2. The macrophage is a type of white blood cell that engulfs and destroys invading pathogens.

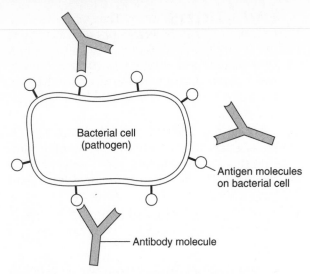

Figure 5-3. The immune system produces antibodies that are specific to the invading foreign antigens.

can quickly mount an attack if the body is invaded again by the same pathogen (Figure 5-3).

Vaccinations

A *vaccine* is a weakened pathogen or its antigen that is injected into an organism. *Vaccinations* are given to people and animals to provide *immunity* against particular pathogens. Once recognized by the immune system, the invading antigen causes antibodies that are specific to it to be made. The cells that produce the antibodies remain as memory cells in the person. Thus, vaccinations provide *active acquired immunity* (see following section). If the actual pathogen invades the body at a later time, these memory cells can launch an immediate response and attack the invaders, often before they have a chance to cause any disease symptoms. Some examples of vaccines that are given to people include those for the flu (influenza), MMR (measles, mumps, rubella), and hepatitis. In the United States, all children must be vaccinated for several diseases before they are even permitted to attend school. Some examples of vaccines that are given to animals include distemper shots and rabies shots for pet dogs.

Types of Immunity

There are several types of immunity to disease that a person may have or acquire. When a person contracts a disease (or receives a vaccination) that he/she later recovers from, his/her immune system creates antibodies that are specific to the pathogens or antigens that caused the disease. These antibodies remain in the bloodstream even after the disease symptoms have ceased. If the person becomes infected again with the same type

of pathogen or antigen, the antibodies already present attack the invaders so that he/she will either recover faster or not even become ill a second time. This type of immunity is called **acquired immunity**, and it can be either *active* or *passive*.

In **active immunity**, the person is exposed to the pathogen or antigen directly. Active immunity may result from direct exposure to a pathogen or antigen from the environment or in the form of a vaccine. In either case, the person produces his or her own antibodies against the disease.

In **passive immunity**, antibodies are transferred from another source to the person. For example, during fetal development, antibodies cross the placenta from the mother to her fetus, thus providing immunity. Additionally, if a mother breast-feeds her infant, antibodies also enter the baby through the mother's milk. Providing antibodies through an injection may also induce passive immunity. For example, if a person is bitten by a rabid animal, an injection containing antibodies to the rabies virus is given to the person to prevent the development of rabies, a fatal disease.

Innate immunity in a person is determined genetically. This type of immunity is present at birth and has no relationship to exposure to pathogenic organisms or antigens. There are certain illnesses that people are naturally immune to but that may infect other animals. For example, people do not get all of the same diseases that dogs or cats can develop.

PROBLEMS IN THE IMMUNE SYSTEM

Overreactions of the Immune System

In some individuals, the immune system overreacts to certain stimuli or antigens that are harmless to most other people. Unfortunately, these severe reactions cause, rather than prevent, suffering and illness for the person.

Allergic Reactions. An *allergic reaction* is a strong response to *allergens* in pollen, animal fur, mold, insect stings, foods, and so on. The sufferer may experience sneezing, watery and itchy eyes, a runny nose, hives, coughing, and/or swelling. These uncomfortable symptoms are triggered by the immune system's release of substances called *histamines*. Although these allergy symptoms are inconvenient, they are responses made by the body in an attempt to expel the invading antigen. In some cases, the swelling may be so severe in the sinuses or throat that it interferes with breathing. An extreme type of allergic reaction, known as *anaphylactic shock,* occurs in some people in response to bee or wasp stings (and even in response to certain medications). This condition causes severe swelling and can be truly life threatening.

Autoimmune Diseases. In very rare cases, the immune system accidentally targets some of the body's own cell proteins as antigens. Once the immune system has identified an antigen as foreign, the cells bearing that protein are attacked as if they were invading foreign pathogens. This reaction produces a condition known as an *autoimmune disease* (*auto* meaning "self"). Examples of such serious diseases include rheumatoid arthritis (which causes inflammation and pain in the joint membranes) and lupus erythematosus (which causes painful swelling of the skin and joints, fever, rash, hair loss, fatigue, and sensitivity to light).

Immune Response to Transplants. People who receive transplanted organs, such as a heart, liver, or kidney (due to a *malfunction* of their own organ), may also experience problems with their immune response. Because the organ is recognized as foreign, the immune system may launch an attack against it, causing the body to reject the new organ. Physicians attempt to match the chemistry of the organ donor as closely as possible with that of the recipient, in order to minimize the risk of organ rejection. In addition, *immunosuppressant* drugs may be used to lessen the immune response. However, use of these medications can leave the transplant recipient quite vulnerable to infection by various microbes.

A Damaged or Weakened Immune System

HIV (*h*uman *i*mmunodeficiency *v*irus), the agent that causes AIDS, damages the immune system by destroying specific T cells known as *helper T cells.* This leaves the affected person with a severely limited immune response. For that reason, AIDS is called an *immunodeficiency disease.* In fact, AIDS sufferers are prone to and frequently die from a variety of diseases that a healthy person's immune system could probably conquer (especially with the use of medicine), rather than from the virus itself.

Finally, as a person gets older, his or her immune system gradually weakens in its ability to respond to pathogens or cancerous cells. Consequently, older adults may be more prone than younger individuals to becoming ill or developing (malignant) tumors. Fatigue, stress, substance abuse, and poor nutrition can also contribute to a weakened immune response.

QUESTIONS

12. Which cells are important components of the human immune system? (1) red blood cells (2) liver cells (3) white blood cells (4) nerve cells

13. A blood test showed that a person had increased levels of antibodies. This may indicate that the person has (1) an infection (2) diabetes (3) low blood pressure (4) an enlarged thyroid

14. Antibodies are produced by the body's (1) T cells (2) lymph node cells (3) B cells (4) liver cells

15. Substances that trigger a defensive response by the immune system are called (1) antibodies (2) antigens (3) lymph nodes (4) macrophages

16. A similarity between antibodies and enzymes is that both (1) are lipids (2) are produced by liver cells (3) can make blood vessels dilate (4) have very specific shapes and functions

17. Which statement does *not* describe an example of a feedback mechanism that maintains homeostasis? (1) The guard cells close the openings in leaves, preventing excess water loss from a plant. (2) White blood cells increase the production of antigens during an allergic reaction. (3) Increased physical activity increases heart rate in humans. (4) The pancreas releases insulin, helping humans to keep blood sugar levels stable.

18. Vaccines are given to people in order to (1) disrupt their homeostasis (2) immunize them against certain diseases (3) inject T cells and B cells into them (4) test if they can destroy the pathogen

19. A person's sneezing, coughing, and watery eyes right after exposure to cat hair are all indications of (1) an autoimmune disease (2) an infection caused by the cat (3) an allergic reaction (4) early warning signs of cancer

20. The use of a vaccine to stimulate the immune system to act against a specific pathogen is valuable in maintaining homeostasis because (1) once the body produces chemicals to combat one type of virus, it can more easily make antibiotics (2) the body can digest the weakened microbes and use them as food (3) the body will be able to fight invasions by the same type of microbe in the future (4) the more the immune system is challenged, the better it performs

Answer the following question based on your knowledge of biology and on the diagram below, which represents what can happen when homeostasis in an organism is threatened.

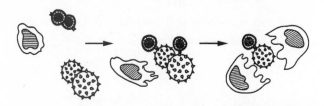

21. Which statement provides a possible explanation for these events? (1) Antibiotics break down harmful substances by the process of digestion. (2) Specialized cells tag and/or engulf invading microbes during an immune response. (3) Embryonic development of essential organs occurs during pregnancy. (4) Cloning removes abnormal cells produced during differentiation.

22. Which statement describes an example of active acquired immunity? (1) Humans generally do not get equine encephalitis, a disease of horses. (2) After having mumps as a child, an adult does not generally have a recurrence of the disease. (3) A patient receives an antibiotic to fight off a respiratory infection. (4) Breast milk provides many antibodies to a nursing infant.

23. Antibodies that cross the placenta from mother to baby during fetal development provide the child with (1) innate immunity (2) active acquired immunity (3) passive acquired immunity (4) induced passive immunity

PART B-2

Base your answers to questions 24 through 26 on the graph below, which shows the relationship between exposure to an antigen and the antibody response that followed.

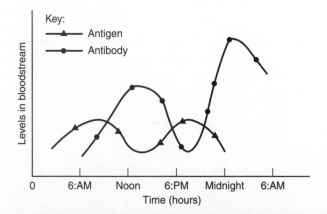

24. At what times did the antigen reach maximum levels in the bloodstream?

25. What relationship exists between the antigen levels and antibody levels in this graph?

26. The second peak of the antibody level is much greater than the first peak. Explain why.

Base your answers to question 27 through 29 on the paragraph below and on your knowledge of biology.

A boy contracted the viral disease chicken pox when he was a first grader. His doctor kept him out of school for two weeks until he recovered from the illness. Two years later, when his younger brother came down with chicken pox, the older boy did not catch it again, even though they shared a bedroom and were in close contact with one another.

27. What type of immunity to the chicken pox virus did the older boy develop? State one reason to support your answer.

28. How did the older boy's immune system protect him against chicken pox when he contracted the virus in the first grade?

29. Why didn't the older boy catch chicken pox again when his younger brother had it two years later?

PART C

30. A pharmaceutical company is proposing that its new product, *Immunoblast*, can help strengthen a person's immune system. Design an experiment in which you could test the effectiveness of this new product. Include the following steps:
   - state the problem you are investigating;
   - propose a suitable hypothesis;
   - write the experimental procedure you would follow;
   - list the data you would collect to test your hypothesis.

31. Each year, before the start of the flu season, older adults are advised to get a flu shot, or vaccination, to protect them. Answer the following questions:
   - Why is the flu shot recommended more for older adults than for younger ones?
   - How does the vaccine protect people from the flu?
   - Why is a new flu vaccination needed every year?

32. Describe two ways that the risk of organ rejection can be minimized in a transplant patient. Discuss one problem that is associated with one of the methods you have described.

33. AIDS is an infectious disease that has reached epidemic proportions. Briefly describe the nature of this disease and be sure to include:

- the type of pathogen that causes AIDS;
- the specific body system that is attacked by that pathogen;
- the effect on the body when this system is weakened by AIDS;
- *two* ways to prevent or control the spread of infectious diseases such as AIDS.

Base your answers to questions 34 through 37 on the information below and on your knowledge of biology. Source: Science News *(July 9, 2005): vol. 168, no. 2, p. 29.*

Stem Cell Shift May Lead to Infections [and] Leukemia

Researchers have long wondered why elderly people suffer more infections and have a greater chance of developing myeloid leukemia, a type of blood cancer, than younger people do. Now, research in mice suggests that the aging of blood-producing stem cells could be responsible for both conditions.

With age, the body of a person or other animal loses it capacity to sustain its tissues and organs. "Since we know the cells mediating this maintenance are stem cells, it doesn't take a great leap of faith to think that stem cells are at the heart of that failure," says Derrick Rossi of Stanford University.

To examine whether the aging of stem cells contributes to infections and leukemia, Rossi and his colleagues irradiated young and old mice to kill off their blood-making stem cells. The scientist then transplanted such stem cells from young donor mice into elderly irradiated animals and from old donors into young irradiated animals.

After several weeks, the researchers found that young animals' stem cells transplanted into the old mice produced the different types of blood cells in ratios much like those in young mice that haven't been irradiated. However, the young animals that received old animals' stem cells had significantly fewer new lymphoid blood cells—which make cells that battle infections—than normal young animals do.

After examining gene activity in the stem cells transplanted from old animals, Rossi's team found a boost in activity among genes responsible for creating myeloid cells. These create red blood cells and some other blood components. Many myeloid-production genes have been associated with myeloid leukemia in people.

The scientists conclude in the June 28 *Proceedings of the National Academy of Sciences* that a shift from lymphoid-cell production to myeloid-cell production could be responsible for the increases both in infections and in risk of leukemia that come with old age.

34. According to the article, what do researchers think is the reason that elderly people develop infections and blood cancer more often than young people do?

35. Briefly describe the experiment the research team conducted on mice to test their hypothesis.

36. Describe the change in gene activity that occurs in the blood-making stem cells of older animals.

37. What conclusions about blood cells and illness in elderly people did the scientists make as a result of their experiments on mice?

CHAPTER 6

Reproduction and Development

The survival of a species depends on reproduction, that is, the production of new individuals. There are two ways that organisms can reproduce: **asexually** and **sexually**. In *asexual reproduction*, only one parent is involved, and the new organism develops from a cell or cells of the parent organism. In *sexual reproduction*, there are usually two parents, and each one contributes a specialized **sex cell** to the new organism. The two sex cells, one from each parent, fuse to form the first cell of the new generation.

MITOSIS

All cells arise from other cells by cell division, during which the nucleus duplicates, or **replicates**, and the cytoplasm divides in two, forming two cells. The process of **mitosis** (a nuclear process) is the orderly series of changes that results in the duplication of the complete set of chromosomes and the formation of two new nuclei that are identical to each other and to the nucleus of the original parent cell. The division of the cytoplasm occurs either during or after mitosis, and it results in the formation of two new, identical daughter cells. The effect of this is that all the cells that come from a single cell are genetically identical to it and to each other; they are all *clones*.

Events of Mitosis

During the period between cell divisions, the chromosome material is dispersed in the nucleus in the form of *chromatin*. At the beginning of mitosis, before the chromosomes become visible as distinct units, the chromatin replicates. It then contracts, forming a visible set of double-stranded chromosomes. Each double-stranded chromosome consists of two identical strands, or *chromatids*, joined by a *centromere* (Figure 6-1).

During the early stages of mitosis, the nuclear membrane disintegrates and disappears, while a network of fibers called the *spindle apparatus* forms. In animal cells, two small organelles called *centrioles* move to the opposite ends, or *poles*, of the cell, where they appear to be involved in the formation of the spindle apparatus. Plant cells generally lack centrioles, but the spindle apparatus forms without them, and the movement of chromosomes is similar to that in animal cells.

The double-stranded chromosomes become attached to the spindle apparatus and line up along the cell's center, or equator. The two chromatids of each double-stranded chromosome separate and move to opposite poles of the cell. Current evidence suggests that this movement results from the shortening of the spindle fibers, which causes the chromatids to move to either one pole of the cell or the other. A nuclear membrane forms around each of the two sets of single-stranded chromosomes, thus forming two daughter nuclei—identical to each other and to the original nucleus (Figure 6-2, page 80).

Division of the Cytoplasm
In animal cells, the cytoplasm is divided when the cell membrane "pinches in" at the cell's center, separating the two nuclei and dividing the cytoplasm into approximately equal halves.

In plant cells, the cytoplasm is divided when a *cell plate* forms across the center of the cell. The cell plate then forms the new cell walls.

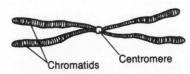

Figure 6-1. A double-stranded chromosome.

79

Plant Cell   Animal Cell

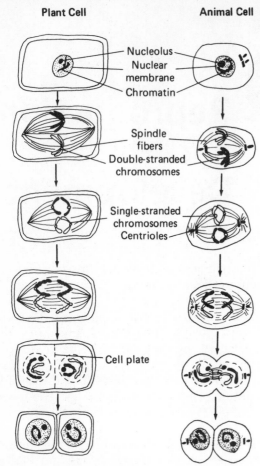

Figure 6-2. Stages of mitosis.

Uncontrolled Cell Division

In multicellular organisms, cells sometimes undergo abnormal and rapid divisions, resulting in growths called *tumors*, which invade surrounding tissues and organs and interfere with their normal activities. Such tumors are linked to a group of diseases known, collectively, as *cancer*. (For more information, refer to Chapter 5.)

QUESTIONS

PART A

1. Each of the two daughter cells that results from the normal mitotic division of the original parent cell contains (1) the same number of chromosomes but has genes different from those of the parent cell (2) the same number of chromosomes and has genes identical to those of the parent cell (3) one-half the number of chromosomes but has genes different from those of the parent cell (4) one-half the

number of chromosomes and has genes identical to those of the parent cell

2. The following list describes some of the events associated with normal cell division:

   A. nuclear membrane formation around each set of newly formed chromosomes

   B. pinching in of cell membrane to separate daughter nuclei and divide cytoplasm

   C. replication of each chromosome to form sets of double-stranded chromosomes

   D. movement of single-stranded chromosomes to opposite ends of the spindle fibers

   What is the normal sequence in which these events occur? (1) $A \rightarrow B \rightarrow C \rightarrow D$ (2) $C \rightarrow B \rightarrow D \rightarrow A$ (3) $C \rightarrow D \rightarrow A \rightarrow B$ (4) $D \rightarrow C \rightarrow B \rightarrow A$

3. What is the result of normal chromosome replication? (1) Lost or worn-out chromosomes are replaced. (2) Each daughter cell is provided with twice as many chromosomes as the parent cell. (3) The exact number of centrioles is produced for spindle fiber attachment. (4) Two identical sets of chromosomes are produced.

4. Normally, a complete set of chromosomes is passed on to each daughter cell as a result of (1) reduction division (2) mitotic cell division (3) meiotic cell division (4) nondisjunction

5. In nondividing cells, the chromosome material is in the form of (1) chromatids (2) centrioles (3) spindle fibers (4) chromatin

6. Organelles that play a role in mitotic division in animal cells but not in plant cells are (1) centrioles (2) chromatids (3) cell plates (4) chromosomes

7. In plant cells, after the cytoplasm divides, a cell plate forms across the center of the cell and forms the new (1) cell membranes (2) chromosomes (3) cell walls (4) centrioles

PART B-2

8. Colchicine is a drug that prevents chromosomes from separating during cell division. Describe how colchicine might affect daughter cells produced by a cell during mitosis.

9. Red blood cells lose their nuclei when they become fully mature. How does this explain the fact that red blood cells cannot undergo mitosis?

10. Compare the process of mitosis in a plant cell and in an animal cell.

11. Using the Internet, research how a tumor forms. How is mitosis related to the formation of a tumor? How is a benign tumor different from a malignant tumor?

TYPES OF ASEXUAL REPRODUCTION

Asexual reproduction is the production of new organisms without the joining of nuclei from two specialized sex cells. In asexual reproduction, the new organism develops by mitotic cell divisions, and the offspring are genetically identical to the parent.

Binary Fission

The form of asexual reproduction that occurs most commonly in single-celled organisms, such as the ameba and paramecium, is *binary fission* (Figure 6-3). In this type of reproduction, the nucleus divides by mitosis, and the cytoplasm divides, forming two daughter cells of equal size. These newly formed cells are smaller than the parent cell, but they contain the same number of chromosomes.

Budding

Yeasts and some other simple organisms carry on a form of asexual reproduction called *budding*,

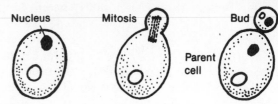

Figure 6-4. Budding in yeast.

which is basically similar to binary fission. However, in budding, the division of the cytoplasm is unequal, so that one of the daughter cells is larger than the other. The daughter cells may separate, or they may remain attached, forming a colony (Figure 6-4).

In multicellular organisms such as the hydra, budding refers to the production of a multicellular growth, or *bud*, from the body of the parent (Figure 6-5). The bud is produced by mitotic cell division, and it develops into a new organism. The new organism may detach from the parent, or it may remain attached, forming a colony.

Sporulation

In some multicellular organisms, such as bread mold, specialized cells called *spores* are produced in large numbers by mitosis. This process is called *sporulation*. Spores are generally surrounded by a tough coat, which enables them to survive harsh environmental conditions. Each spore may then develop into a new organism when environmental conditions become favorable.

Regeneration

The process of *regeneration* refers to the replacement, or regrowth, of lost or damaged body parts. For example, a lobster may regenerate a lost claw. In some cases, an entire new animal can develop from a part of the parent organism. A new sea star can develop from one arm and part of the central disk of an existing

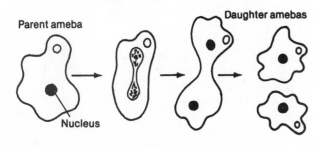

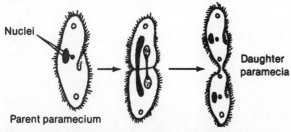

Figure 6-3. Binary fission in the ameba (top) and the paramecium (bottom).

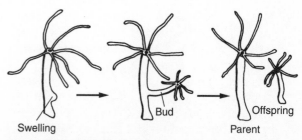

Figure 6-5. Budding in hydra.

sea star (which then regenerates the missing arm). In this case, regeneration is a type of asexual reproduction.

Invertebrates generally show a greater capacity for regeneration than vertebrates do, probably because they have many more unspecialized cells and parts than vertebrates do.

Vegetative Propagation

In plants, *vegetative propagation* involves various forms of asexual reproduction in which new plants develop from the roots, stems, or leaves of the parent plant. Examples include new plant growth from bulbs, tubers, cuttings, and runners (Figure 6-6).

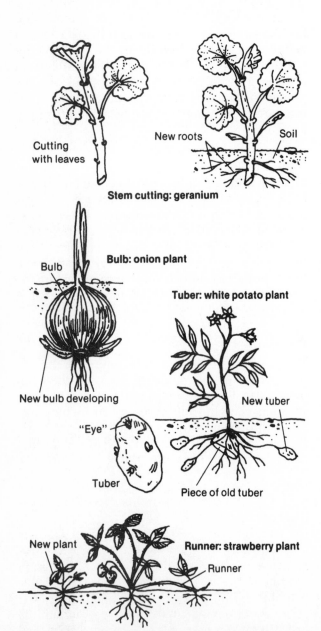

Figure 6-6. Forms of vegetative propagation.

QUESTIONS

12. Compared to the parent cell, a daughter cell produced as a result of binary fission (1) has one-half as many chromosomes (2) has twice as many chromosomes (3) is the same size, but has fewer chromosomes (4) is smaller, but has the same number of chromosomes

13. A form of asexual reproduction that occurs in yeast is (1) binary fission (2) budding (3) vegetative propagation (4) spore formation

14. What is a type of asexual reproduction that commonly occurs in many species of unicellular protists? (1) external fertilization (2) tissue regeneration (3) binary fission (4) vegetative propagation

15. A type of asexual reproduction in which new plants develop from the roots, stems, or leaves of an existing plant is called (1) binary fission (2) sporulation (3) regeneration (4) vegetative propagation

16. A form of asexual reproduction found in bread mold involves the production of large numbers of specialized cells, each surrounded by a tough coat. This process is called (1) binary fission (2) budding (3) sporulation (4) regeneration

17. Compared to vertebrates, invertebrate animals exhibit a higher degree of regenerative ability because they (1) produce larger numbers of sex cells (2) produce larger numbers of spindle fibers (3) possess more chromosomes in their nuclei (4) possess more undifferentiated cells

18. What specific type of reproduction is shown below in the diagrams of an ameba? (1) vegetative propagation (2) binary fission (3) budding (4) meiosis

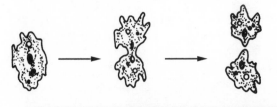

19. The chromosome content of a skin cell that is about to form two new skin cells is represented in the diagram below.

Which diagram best represents the chromosomes that would be found in the two new skin cells produced as a result of this process?
(1) 1 (2) 2 (3) 3 (4) 4

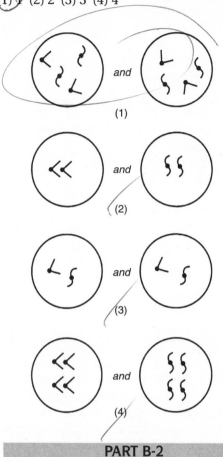

(1)

(2)

(3)

(4)

20. What role does mitosis play in asexual reproduction?

21. The ameba is a single-celled organism that reproduces asexually by mitosis. Explain why all the offspring of a single ameba can be considered clones.

22. In what ways are regeneration and vegetative propagation similar? Why are the offspring identical to the parent in both processes?

23. A scientist noted that a paramecium culture he had in his laboratory reproduced more rapidly than average when kept in a sunny corner of the room. He also observed that other paramecium cultures kept in darker parts of the room reproduced more slowly. Use your knowledge of biology to answer the following:

   • What testable question might the scientist ask based on his observations?

   • State one possible hypothesis to explain the scientist's observations.

   • State a procedure that the scientist could use to test the hypothesis.

24. The diagram below illustrates asexual reproduction in bread mold.

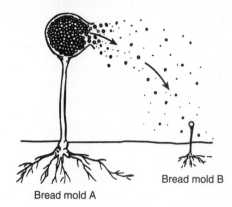

Bread mold B

Bread mold A

Reproductive structures known as spores were released from bread mold A. One of these spores developed into bread mold B. State how the genetic information in the nuclei of cells in bread mold B compares to the genetic information in the nuclei of cells in bread mold A.

MEIOSIS AND SEXUAL REPRODUCTION

In organisms that reproduce sexually, specialized sex cells, or **gametes**, are produced by *meiosis*, a special kind of cell division. One type of gamete, the *sperm cell*, is produced by the male parent, while the other type of gamete, the *egg cell*, is produced by the female parent. The fusion of the nuclei of the sperm cell and the egg cell is called **fertilization**. The resulting cell, which is called the **zygote**, undergoes repeated mitotic cell divisions to form the *embryo*.

Chromosome Number

All members of a given species have a characteristic number of chromosomes in each of their body cells. This *diploid*, or *2n, chromosome number* normally remains constant from generation to generation. For example, all human body cells have 46 chromosomes, fruit flies have 8, and garden peas have 14.

The chromosomes of a body cell are actually in the form of *homologous pairs*. The two chromosomes of each homologous pair are similar in size and shape, and control the same traits. Thus, in human body cells there are 23 pairs of homologous chromosomes (23 from the mother and 23 from the father); in fruit flies there are 4 pairs; and in garden peas there are 7 pairs).

Mature sperm and egg cells contain half the diploid number of chromosomes—they contain one member of each homologous pair. Half the diploid chromosome number is called the *monoploid*, or 1*n*, *chromosome number*. Mature sex cells (gametes) contain the monoploid (also called *haploid*) number of chromosomes; every other cell in the body contains the diploid number.

In sexually mature individuals, monoploid egg cells and sperm cells are formed in the gonads (ovaries and testes) by **meiosis**, the process of *reduction division*.

Meiosis

Meiosis occurs only in maturing sex cells and consists of two nuclear and cytoplasmic divisions but only one chromosome replication. The first meiotic division produces two cells, each containing the monoploid number of double-stranded chromosomes. The second meiotic division results in the formation of four cells, each containing the monoploid number of single-stranded chromosomes.

As a result of meiosis, a single primary sex cell with the diploid chromosome number gives rise to four cells, each with the monoploid (*n*) chromosome number. These cells mature into gametes—either sperm cells or egg cells.

Meiosis is a source of genetic variations because it provides new combinations of chromosomes for the resulting gametes. A gamete receives only one member of each pair of homologous chromosomes from the 2*n* primary sex cells. The sorting of these chromosomes during formation of the gametes is random.

Gametogenesis

The process by which sperm and eggs are produced is called *gametogenesis*. It involves meiotic cell division and cell maturation. Gametogenesis occurs in specialized paired sex organs, or *gonads*. The male gonads are the testes; the female gonads are the ovaries. In most animals, the sexes are separate; that is, each individual has either testes or ovaries. However, some animals, such as the hydra and the earthworm, have both male and female gonads; such animals are called *hermaphrodites*.

Spermatogenesis. The production of sperm is called *spermatogenesis* (Figure 6-7). The process begins with meiosis in primary sperm cells, which are diploid. As a result of meiosis, each primary sperm cell develops into four monoploid cells of equal size. As they mature, these cells lose most of their cytoplasm and develop a long, whiplike flagellum that is used for locomotion.

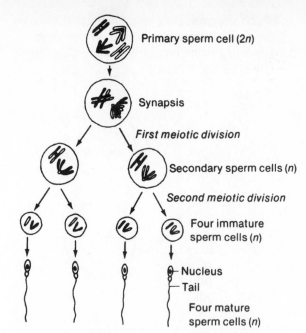

Figure 6-7. Spermatogenesis: the production of mature sperm cells.

Oogenesis. Egg cells are produced by *oogenesis* (Figure 6-8). In oogenesis, a primary egg cell undergoes meiosis. The chromosomal changes are the same as those that occur in spermatogenesis (from 2*n* to *n*). However, in oogenesis, division of the cytoplasm is unequal. The first meiotic division produces one large cell and one small one called a *polar body*. The larger cell then undergoes the second meiotic division, forming an egg cell and another polar body. The first polar body

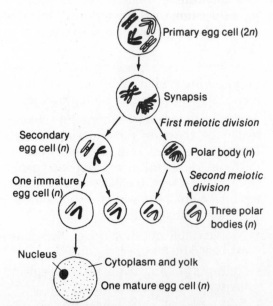

Figure 6-8. Oogenesis: the production of one mature egg cell.

Table 6-1

A Comparison of Mitosis and Meiosis

Mitosis	Meiosis
Double-stranded chromosomes line up in middle of cell in single file.	Double-stranded chromosomes line up in middle of cell in double file.
Results in diploid (2n) number of chromosomes in daughter cells.	Results in monoploid (n) number of chromosomes in daughter cells.
Occurs in all cells of the body.	Occurs only in maturing sex cells.
Results in very few genetic variations because chromosomes remain the same.	Results in many genetic variations because of random sorting and new combinations of chromosomes.

may also undergo a second meiotic division, forming two polar bodies. Oogenesis results in the production of one large, monoploid egg cell and three small polar bodies.

The polar bodies disintegrate. The advantage of the unequal cytoplasmic division is that the egg cell is provided with a large supply of stored nutrients in the form of yolk.

Comparison of Mitosis and Meiosis

The daughter cells produced by mitotic cell division have the same number and kinds of chromosomes as the original parent cell. A cell with the 2n (diploid) chromosome number produces daughter cells with the 2n (diploid) chromosome number. Mitosis produces extra body cells for growth and repair of tissues. It is also associated with asexual reproduction.

In contrast, as a result of meiotic cell division, the daughter cells have one-half the number of chromosomes of the original cell (see Table 6-1). A cell with the 2n (diploid) chromosome number produces daughter cells with the n (monoploid) chromosome number. Meiosis occurs only in the gonads during the production of gametes.

QUESTIONS

PART A

25. Monoploid gametes are produced in animals as a result of (1) meiosis (2) mitosis (3) fertilization (4) fission

26. In human males, the maximum number of functional sperm cells that is normally produced from each primary sex cell is (1) one (2) two (3) three (4) four

27. Sexually reproducing species show greater variation than asexually reproducing species due to (1) lower rates of mutation (2) higher rates of reproduction (3) environmental changes (4) sorting of chromosomes during gametogenesis

28. In animals, polar bodies are formed as a result of (1) meiotic cell division in females (2) meiotic cell division in males (3) mitotic cell division in females (4) mitotic cell division in males

29. During the normal meiotic division of a diploid cell, the change in chromosome number that occurs is represented as (1) $4n \rightarrow n$ (2) $2n \rightarrow 4n$ (3) $2n \rightarrow 1n$ (4) $1n \rightarrow \frac{1}{2}n$

30. In a species of corn, the diploid number of chromosomes is 20. What would be the number of chromosomes found in each of the normal egg cells produced by this species? (1) 5 (2) 10 (3) 20 (4) 40

31. A human zygote is normally produced from two gametes that are identical in (1) size (2) method of locomotion (3) genetic composition (4) chromosome number

32. Organisms that contain both functional male and female gonads are known as (1) hybrids (2) hermaphrodites (3) protists (4) parasites

33. In sexually reproducing species, the number of chromosomes in each body cell remains the same from one generation to the next as a direct result of (1) meiosis and fertilization (2) mitosis and mutation (3) differentiation and aging (4) homeostasis and dynamic equilibrium

PART B-1

34. The diagrams below represent the sequence of events in a cell undergoing normal meiotic cell division.

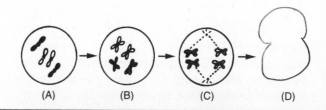

(A)　(B)　(C)　(D)

Which diagram most likely represents stage *D* of this sequence? (1) 1 (2) 2 (3) 3 (4) 4

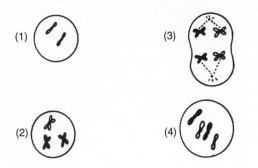

(1) (3) (2) (4)

Base your answers to questions 35 through 38 on your knowledge of biology and on the diagram below, which represents a diploid cell about to undergo meiosis. The shapes inside the cell represent homologous chromosomes.

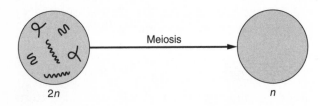

Meiosis

2n   n

35. Copy the diagram into your notebook. In the empty circle, draw one of the resulting daughter cells produced by the diploid cell at the end of meiosis.

36. What is the diploid chromosome number of the original cell?

37. What will be the chromosome number of a daughter cell produced by this cell?

38. Name an organ in which this cell might be found. Explain your answer.

---

39. Briefly compare the processes of mitosis and meiosis. What is the function of each process?

40. Compare the processes of sperm and egg production in terms of the following:
   • where each process occurs;
   • the relative numbers of gametes produced by each process.

41. Discuss why, in sexual reproduction, it is necessary for the gametes to have the monoploid number of chromosomes rather than the diploid number. How does the process of meiosis ensure that the gametes will be monoploid?

42. Explain how the daughter cells produced during meiosis may be genetically different from one another even though they result from the same original diploid cell. Why is this variation important? Why are cells produced by mitosis *not* genetically different from one another?

FERTILIZATION AND DEVELOPMENT

Fertilization is the union of a monoploid (*n*) sperm nucleus with a monoploid (*n*) egg nucleus to form a diploid (2*n*) cell, the *zygote*, which is the first cell of the new organism. Fertilization restores the diploid species number of chromosomes.

External Fertilization
The union of a sperm and an egg outside the body of the female is called *external fertilization*. External fertilization generally occurs in a watery environment and is characteristic of reproduction in frogs, most fish, and many other aquatic vertebrates.

In external fertilization, large numbers of eggs and sperm are released into the water at the same time to increase the chances that fertilization will take place and to help ensure that at least some of the fertilized eggs will develop, avoid being eaten, and survive to adulthood.

Internal Fertilization
The union of a sperm and an egg inside the moist reproductive tract of the female is called **internal fertilization**. Reproduction in most terrestrial, or land-dwelling, vertebrates, including birds and mammals, is characterized by internal fertilization.

In internal fertilization, relatively few eggs are produced at one time, since the chances that fertilization will occur are much greater with internal fertilization than with external fertilization.

Stages of Development
The early stages of embryonic development are similar in all animals. The process known as **development** begins when the zygote undergoes a rapid series of mitotic cell divisions called *cleavage*.

Cleavage. During cleavage, there is no increase in the size of the embryo—just an increase in the

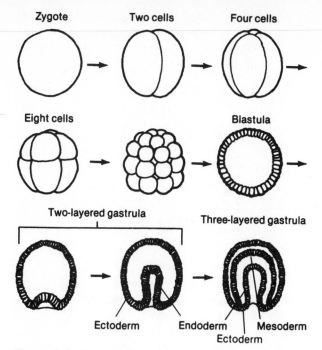

Figure 6-9. Early stages of embryonic development.

Figure labels:
Zygote — Two cells — Four cells
Eight cells — Blastula
Two-layered gastrula — Three-layered gastrula
Ectoderm — Endoderm — Mesoderm — Ectoderm

number of cells it contains (see Figure 6-9). Cell growth and specialization begin after cleavage.

Blastula Formation. The mitotic divisions of cleavage result in the formation of the *blastula*, a hollow ball made up of a single layer of cells.

Gastrulation. As mitotic divisions continue, one side of the blastula pushes inward, or indents, in a process called *gastrulation*. The resulting embryonic stage, called a *gastrula*, consists of an inner layer, or *endoderm*, and an outer layer, or *ectoderm*. A third layer, called the *mesoderm*, forms between the endoderm and ectoderm. The endoderm, mesoderm, and ectoderm are called the *germ layers*.

Differentiation and Growth. The germ layers undergo changes, or **differentiation**, to form the various tissues, organs, and organ systems of the developing animal (see Table 6-2). During differentiation, some portions of the DNA within individual cells are active, while other portions are

inactive. There are numerous methods by which sections of DNA are "switched" on or off. As a result, each cell develops into a specific type of cell, based upon which portion of its DNA is active and coding within its nucleus. For example, a stomach cell differentiates in response to signals from the active DNA in that cell, which deals with the functions of a stomach cell. Similarly, within a liver cell, different parts of the DNA—those that code for the functions of a liver cell—are active. Although each and every body cell in an animal contains the entire set of DNA for that animal, only certain parts of the DNA are active in any particular cell, depending on the type of cell it is.

Embryonic development involves growth as well as differentiation. *Growth* includes both an increase in the size of the embryonic cells and an increase in the number of cells.

External Development

Embryonic development may occur outside or inside the body of the female. Growth of the embryo outside the female's body is called *external development*.

The eggs of many fish and amphibians are fertilized externally and develop externally in an aquatic environment. The eggs of birds and many reptiles (and even a few mammals) are fertilized internally but develop externally, encased in tough, protective shells to prevent their drying out.

Internal Development

Growth of the embryo inside the female's body is called **internal development**. In most mammals, both fertilization and development are internal. The eggs of mammals have little yolk and are very small compared with the eggs of reptiles and birds. In all mammals, the young are nourished after birth by milk from the mother's mammary glands.

Placental Mammals. Most mammals are placental mammals in which the embryo develops in the **uterus**, or womb, of the female and receives food and oxygen and gets rid of wastes through the placenta.

Table 6-2

Tissues and Organs Formed from the Embryonic Germ Layers

Embryonic Layer	Organs and Organ Systems
Ectoderm	Nervous system; skin
Mesoderm	Muscles; circulatory, skeletal, excretory, and reproductive systems
Endoderm	Lining of digestive and respiratory tracts; liver; pancreas

The **placenta** is a temporary organ that forms within the uterus from embryonic and maternal tissues; it is rich in both embryonic and maternal blood vessels. Dissolved materials pass between the mother and the embryo through the blood vessels in the placenta—food and oxygen pass from the mother to the embryo, while wastes pass from the embryo to the mother. The blood of the mother and the embryo never mix.

QUESTIONS

43. In the early development of a zygote, the number of cells increases, without leading to an increase in size, in the process of (1) ovulation (2) cleavage (3) germination (4) metamorphosis

44. An embryo's three germ layers are formed during (1) gastrulation (2) fertilization (3) blastula formation (4) growth

45. In most species of fish, the female produces large numbers of eggs during the reproductive cycle. This would indicate that reproduction in fish is most probably characterized by (1) internal fertilization and internal development (2) internal fertilization and external development (3) external fertilization and internal development (4) external fertilization and external development.

46. Which type of fertilization and development do birds and most reptiles have? (1) internal fertilization and internal development (2) internal fertilization and external development (3) external fertilization and internal development (4) external fertilization and external development

47. The embryos of some mammals, such as the kangaroo and the opossum, complete their development externally. What is the source of nutrition for their last stage of development? (1) milk from maternal mammary glands (2) diffusion of nutrients through the uterine wall (3) food stored in the egg yolk (4) solid foods gathered and fed to them by the mother

48. In mammals, the placenta is essential to the embryo for (1) nutrition, reproduction, growth (2) nutrition, respiration, excretion (3) locomotion, respiration, excretion (4) nutrition, excretion, reproduction

49. Which characteristic of sexual reproduction specifically favors the survival of terrestrial animals? (1) fertilization within the body of the female (2) male gametes that may be carried by the wind (3) fusion of gametes in the outside environment (4) fertilization of eggs in the water

Base your answers to questions 50 through 53 on your knowledge of biology and on the diagram below, which represents the early stages of embryonic development.

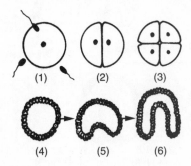

50. The structures labeled *2* and *3* are formed as a direct result of (1) meiosis (2) gastrulation (3) cleavage (4) differentiation

51. The structure in stage *4* represents a (1) zygote (2) blastula (3) gastrula (4) follicle

52. The cells of the outer layer give rise to the (1) digestive system and liver (2) excretory system and muscles (3) circulatory system and gonads (4) nervous system and skin

53. Which cells are *not* represented in any of the diagrams? (1) endoderm (2) mesoderm (3) ectoderm (4) gastrula

54. The arrows in the diagram below illustrate processes in the life of a species that reproduces sexually. Which processes result directly in the formation of cells with half the amount of genetic material that is characteristic of the species? (1) 1 and 2 (2) 2 and 3 (3) 3 and 4 (4) 4 and 5

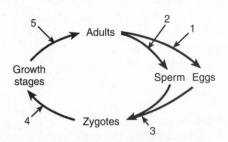

55. The development of specialized tissues and organs in a multicellular organism's embryo occurs as a result of (1) cloning (2) differentiation (3) meiosis (4) cleavage

Refer to the following four terms, which describe the early stages of embryonic development, to answer questions 56 through 58: gastrula, cleavage, zygote, blastula.

56. List these terms in the correct order, from earliest to latest stage of embryonic development.

57. For each term listed, draw a simple sketch to illustrate that stage of embryonic development.

58. Briefly describe what occurs during each of these embryonic stages.

59. Animals that are characterized by external fertilization produce many times more gametes (sperm and eggs) than do animals that have internal fertilization. Give two reasons for this observation.

60. Stem cells are cells in which the *entire* DNA is active. Explain how a stem cell is able to differentiate into any type of cell needed by the body.

HUMAN REPRODUCTION AND DEVELOPMENT

Male Reproductive System

The male reproductive system functions in the production of sperm cells, male sex hormones, and in the placement of sperm into the female reproductive system.

Sperm Production. The sperm-producing organs, the *testes*, are located in an outpocketing of the body wall called the *scrotum* (Figure 6-10). The temperature in the scrotum, which is 1 to 2°C cooler than normal body temperature, is best suited for the production and storage of sperm.

From the testes, the sperm pass through a series of ducts into which liquid is secreted by various glands. The liquid serves as a transport medium for the sperm cells and is an adaptation for life on land. The liquid and sperm together are called *semen.* Semen passes to the outside of the body through the urethra, a tube through the penis. The *penis* is used to deposit the semen in the female reproductive tract.

Hormone Production. The testes produce the male sex hormone testosterone, which regulates

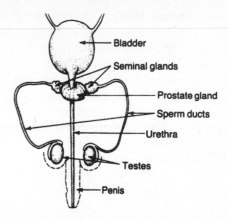

Figure 6-10. The human male reproductive system.

the maturation of sperm cells. Testosterone also regulates the development of male secondary sex characteristics, including body form, beard development, and deepening of the voice.

Female Reproductive System

The female reproductive system functions in the production of egg cells and female sex hormones.

Egg Production. The female reproductive organs, the *ovaries*, are located within the lower portion of the body cavity (Figure 6-11). In the ovaries, each egg cell is present in a tiny sac called a *follicle.* About once a month, a follicle matures and bursts, and the egg within it is released from the surface of the ovary, a process called *ovulation.* The egg cell then passes into the oviduct, or *fallopian tube,* which leads to the *uterus.* If sperm are present, fertilization may occur. If the egg is fertilized, it passes into the uterus, where embryonic development may occur. If the egg is not fertilized, it degenerates.

The lower end of the uterus, the *cervix*, opens to a muscular tube called the *vagina,* or *birth canal.* When embryonic development is complete, the baby leaves the body of the mother through the vagina.

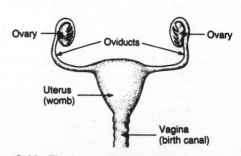

Figure 6-11. The human female reproductive system.

Hormone Production.

The ovaries produce the female sex hormones estrogen and progesterone. These hormones regulate the maturation of egg cells, as well as the development of secondary sex characteristics, including the development of the mammary glands and the broadening of the pelvis. Estrogen and progesterone are also involved in the menstrual cycle and **pregnancy**.

The Menstrual Cycle

The series of events that prepares the uterus for pregnancy is called the *menstrual cycle*. The cycle begins with the thickening of the lining of the uterine wall. The lining also becomes vascularized (filled with blood vessels). If fertilization does not occur, the thickened uterine lining breaks down and the material is expelled from the body during menstruation. The cycle then begins again.

The menstrual cycle begins at *puberty*, the stage at which the individual becomes capable of reproducing. It is temporarily interrupted by pregnancy and sometimes by illness, and ceases permanently at *menopause*. The cycle is regulated by the interaction of hormones, and lasts approximately 28 days.

The menstrual cycle consists of four stages (Figure 6-12):

(a) During the follicle stage, an egg matures and the follicle secretes estrogen, which stimulates the thickening of the uterine lining. This stage lasts about 14 days.

(b) About midway in the cycle, ovulation occurs. The egg is released from the ovary and enters the oviduct.

(c) Following ovulation, the *corpus luteum* forms from the ruptured follicle. The corpus luteum secretes progesterone, which continues the vascularization of the uterine lining started by estrogen. This stage lasts about 12 days.

(d) If fertilization does not occur, the egg cell and the thickened uterine lining break down, and the extra tissue, together with some blood and mucus, pass out of the body through the vagina. The shedding of the uterine lining is called *menstruation*. This stage lasts about four to five days.

Hormones of the Menstrual Cycle

The menstrual cycle is controlled by hormones that are released by the hypothalamus, pituitary gland, and ovaries.

During the follicle stage, the pituitary gland, under the influence of hormones from the hypothalamus, secretes FSH (follicle-stimulating

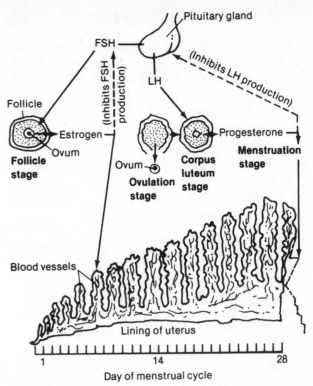

Figure 6-12. Stages of the human menstrual cycle.

hormone), which in turn stimulates the follicle to secrete estrogen. Estrogen stimulates ovulation and initiates vascularization of the uterine lining.

Increased blood estrogen levels inhibit the production of FSH by the pituitary, and the secretion of LH (luteinizing hormone) by the pituitary increases. Ovulation occurs at about this time in the cycle. After ovulation, LH stimulates the formation of the corpus luteum from the ruptured follicle. The corpus luteum secretes progesterone, which enhances the vascularization of the uterine lining.

If fertilization does not occur, the high levels of progesterone in the blood inhibit the production of LH by the pituitary. The drop in LH level causes a drop in the progesterone level. The lining of the uterus thins out, and at about the twenty-eighth day of the cycle, the shedding of the uterine lining, or menstruation, begins. The blood flow of menstruation is caused by the breakage of many small blood vessels.

The relationship between the ovarian hormones estrogen and progesterone and the pituitary hormones FSH and LH is an example of a *negative feedback mechanism*.

Fertilization and Development

If fertilization does occur in the oviduct, the zygote undergoes cleavage to form a blastula. Six to ten days later, the blastula becomes implanted in the uterine lining. Gastrulation usually occurs after implantation. The germ layers of the gastrula begin to differentiate and grow, resulting in the formation of specialized tissues and organs. The placenta and umbilical cord form, enabling the embryo to obtain nutrients and oxygen and to dispose of metabolic wastes. An amnion (membrane-enclosed sac) filled with fluid provides a watery environment that cushions the embryo, which helps to protect it from injury.

In Vitro Fertilization

Fertilization that occurs outside the body of the female (that is, by means of laboratory techniques) is known as *in vitro* (meaning "in glass") fertilization. After fertilization, the early embryo is implanted into the uterus, where development is completed.

Multiple Births

Sometimes two or more embryos develop in the uterus simultaneously. Fraternal twins develop when two eggs are released from the ovary at the same time and both are fertilized. The two eggs are fertilized by two different sperm cells. Fraternal twins may be of the same sex or opposite sexes. Identical twins develop when a zygote separates into two equal halves early in cleavage. Each half develops into an offspring. Since identical twins develop from the same zygote, they have identical genetic make-ups and are always of the same sex.

Birth and Development

The time between fertilization and birth is referred to as the *gestation period*. In humans, the gestation period is about nine months. After the first three months of gestation, the embryo is referred to as a **fetus**. At the end of the gestation period, the secretion of progesterone decreases and another hormone from the pituitary causes strong muscular contractions of the uterus. The amnion bursts, and the baby is pushed out of the mother's body through the vagina.

During *postnatal development* (development after birth), humans pass through different stages, including infancy, childhood, puberty, adulthood, and old age. Puberty begins at early adolescence. In males, puberty usually occurs between the ages of 12 and 18; in females, it usually occurs between the ages of 10 and 14.

Aging is a series of complex structural and functional changes in the body that occur naturally with the passage of time. The causes of aging are not fully understood. However, it now appears that aging may result from an interaction of both genetic and environmental factors. The aging

process ends in death, which may be described as an irreversible cessation of brain function.

QUESTIONS

61. Which of the following structures *least* affects the human female menstrual cycle? (1) pituitary (2) ovary (3) pancreas (4) corpus luteum

62. A woman gave birth to twins, one girl and one boy. The number of egg cells involved was (1) 1 (2) 2 (3) 3 (4) 4

63. A diagram of the human female reproductive structures is shown below.

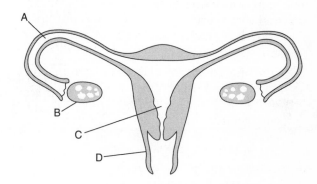

Which structure is correctly paired with its function? (1) A—releases estrogen and progesterone (2) B—produces and releases the egg (3) C—provides the usual site for fertilization (4) D—nourishes a developing embryo

64. Which structure is the membrane that serves as the protective, fluid-filled sac in which an embryo is suspended? (1) pituitary (2) placenta (3) corpus luteum (4) amnion

65. The technique of uniting a sperm cell and an egg cell outside the female's body is called (1) *in vitro* fertilization (2) internal fertilization (3) gametogenesis (4) artificial ovulation

66. Which of the following hormones is *not* involved in the regulation of the human menstrual cycle? (1) progesterone (2) estrogen (3) FSH (4) testosterone

67. Identical twins develop from (1) one egg and two sperm (2) two eggs and one sperm (3) two eggs and two sperm (4) one egg and one sperm

68. Some body structures of a human male are represented in the diagram on page 92 (top). An obstruction in the structures labeled *X* would directly interfere with the (1) transfer of sperm to a female (2) production of sperm

---

Chapter 6: Reproduction and Development   **91**

(3) production of urine (4) transfer of urine to the external environment

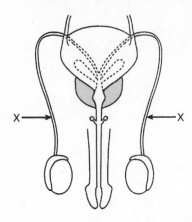

69. Fraternal twins develop from (1) one egg and two sperm (2) two eggs and one sperm (3) two eggs and two sperm (4) one egg and one sperm

70. One function of the placenta in a human is to (1) surround the embryo and protect it from shock (2) allow for mixing of maternal blood with fetal blood (3) act as the heart of the fetus, pumping blood until the fetus is born (4) permit passage of nutrients and oxygen from the mother to the fetus

---

PART B-1

Base your answers to questions 71 through 73 on the diagram below, which represents a cross section of a part of the human female reproductive system, and on your knowledge of biology.

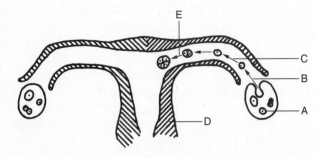

71. Which structure is prepared for the implantation of a fertilized egg as a result of the action of reproductive hormones? (1) A (2) B (3) C (4) D

72. Within which structure does fertilization normally occur? (1) A (2) B (3) C (4) D

73. Which step represents the process of ovulation? (1) A (2) B (3) C (4) D

Base your answers to questions 74 through 76 on the diagram below, which represents a stage in human embryonic development.

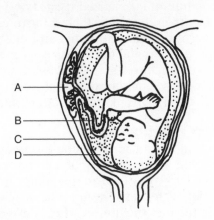

74. The exchange of oxygen, food, and wastes between the mother and the fetus occurs at structure (1) A (2) B (3) C (4) D

75. What is the function of the fluid labeled D? (1) nourishment (2) protection (3) excretion (4) respiration

76. The structure labeled C, within which embryonic development occurs, is known as the (1) oviduct (2) birth canal (3) uterus (4) placenta

For each of the processes described in questions 77 through 79, choose from the list below the correct stage of the human menstrual cycle during which that process occurs.

Human Menstrual Cycle Stage

A. Ovulation

B. Follicle stage

C. Menstruation

D. Corpus luteum stage

77. The lining of the uterus is shed: (1) A (2) B (3) C (4) D

78. An egg is released from an ovary: (1) A (2) B (3) C (4) D

79. An egg matures in an ovary: (1) A (2) B (3) C (4) D

PART B-2

Base your answers to questions 80 through 83 on the following graph, which shows a woman's changing hormone levels for FSH and estrogen over a period of 21 days, and on your knowledge of human reproductive biology.

80. Describe the relationship depicted in the graph between FSH and estrogen.

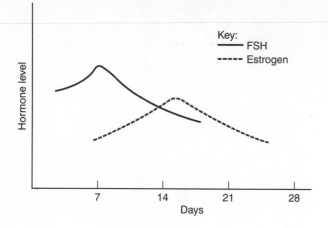

Key:
FSH
Estrogen

81. Why does the level of estrogen begin rising *after* the FSH level rises?

82. Why does the FSH level begin to fall after the seventh day?

83. On approximately what day are the woman's estrogen levels highest?

PART C

84. List the four major hormones that play a role in the menstrual cycle and discuss how they interact during the cycle.

85. Mr. and Mrs. W have been trying to conceive their first child for over one year, with no success. They decide to visit their doctors for medical tests. Blood tests on Mrs. W reveal that her FSH levels are abnormally low. Discuss how this finding might explain the couple's inability to conceive. What medical treatment might help Mrs. W to become pregnant?

86. Briefly discuss the function of the following structures in the development of the human embryo: *placenta, umbilical cord, amnion.*

87. A human is a complex organism that develops from a zygote. Briefly explain some of the steps in this developmental process. In your answer be sure to:

   • explain how a zygote is formed;

   • compare the chromosomes in a zygote with those of a parent's body cell;

   • identify one developmental process involved in the change from a zygote to an embryo;

   • identify the body structure in which fetal development usually occurs.

SEXUAL REPRODUCTION IN FLOWERING PLANTS

Flowers are the reproductive organs of the *angiosperms*, or flowering plants.

Structure of Flowers

Flowers may contain the following structures: sepals, petals, stamens, and pistils (Figure 6-13).

Sepals are leaflike structures at the base of a flower that enclose and protect the flower bud. In some species the sepals are green, while in others the sepals are white or brightly colored.

Petals are leaflike structures inside the sepals that surround the reproductive organs of the flower. Petals may be brightly colored or white and often have a sweet fragrance.

Stamens are the male reproductive organs of a flower. Each stamen consists of an oval-shaped *anther* supported by a stalk, or *filament*. *Pollen grains*, which contain monoploid sperm nuclei, are produced by meiosis by the diploid cells of the anther. The thick wall that encloses the pollen grain prevents the contents from drying out. This is an adaptation for life on land.

Pistils are the female reproductive organs of a flower. A pistil consists of a stigma, style, and ovary. The *stigma*, which is a knoblike, sticky structure, is adapted for receiving pollen grains. The stigma is supported by the *style*, a slender stalk that connects the stigma to the *ovary*, which is at the base of the pistil. In the ovary, monoploid egg cells are produced by meiosis in structures called *ovules*.

The flowers of some species contain both stamens and pistils. In other species, some flowers contain only stamens, while others contain only pistils. The flowers of some species have both sepals and petals, while the flowers of other species lack one or the other.

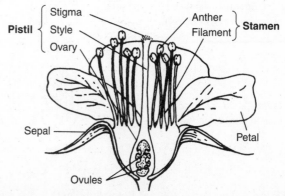

Figure 6-13. Structure of a flower.

Pollination and Fertilization

The transfer of pollen grains from an anther to a stigma is called *pollination*. The transfer of pollen from an anther to a stigma of the same flower or to a stigma of another flower on the same plant is called *self-pollination*. The transfer of pollen from an anther of one flower to the stigma of a flower on another plant is *cross-pollination*. Cross-pollination increases the chances of genetic variation in the offspring because the pollen and the egg cells are from two different plants.

Pollination may be carried out by wind, insects, or birds. Brightly colored petals and the scent of nectar attract insects and birds. Pollen grains from a flower adhere to their bodies and are carried to another flower, where they rub off on the sticky surface of a stigma.

When a pollen grain reaches a stigma, it *germinates*, or sprouts (Figure 6-14). A pollen tube grows from the pollen grain down through the stigma and style to an ovule within the ovary. The growth of the pollen tube is controlled by the tube nucleus. Two sperm nuclei and the tube nucleus pass down through the pollen tube. The sperm nuclei enter an ovule, where one sperm

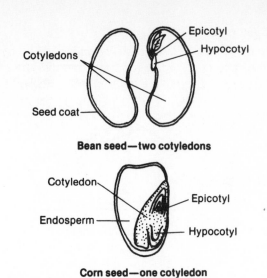

Figure 6-15. Structure of a seed.

nucleus fertilizes the egg nucleus to form a diploid (2*n*) zygote. The other sperm nucleus fuses with two *polar nuclei* in the ovule to form a triploid (3*n*) *endosperm nucleus*, which divides to form a food storage tissue. The zygote undergoes repeated mitotic divisions to form a multicellular plant embryo. After fertilization, the ovule ripens to form a *seed*, while the ovary develops into a *fruit*. The seeds of flowering plants are found inside the fruits.

Structure of a Seed

A seed consists of a seed coat and a plant embryo with one or two cotyledons (Figure 6-15). The *seed coat*, which develops from the outer coverings of the ovule, surrounds and protects the embryo. The plant embryo consists of the epicotyl, hypocotyl, and cotyledon. The *epicotyl* is the upper portion of the embryo; it develops into the leaves and upper portion of the stem. The *hypocotyl* is the lower portion of the embryo; it develops into the roots and, in some species, the lower portion of the stem. The *cotyledons* contain endosperm, the stored food that provides nutrients for the developing plant.

Fruits

The fruits of flowering plants are structures that are specialized for seed dispersal. Fruits carry the seeds away from the parent plant, which helps to prevent overcrowding. The fruits of dandelions and maples, for example, are dispersed by wind; coconuts are dispersed by water; and cockleburs are fruits that become attached to the fur of animals and are carried away as they move. Fleshy fruits are eaten by animals, and their seeds are later deposited with the animal's wastes.

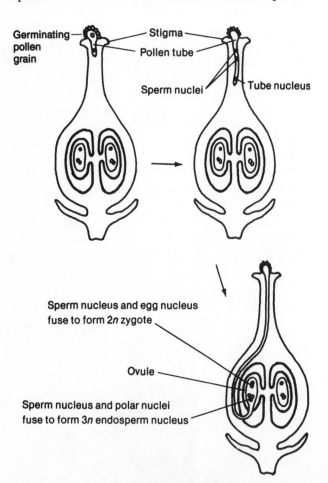

Figure 6-14. Fertilization in flowering plants.

Seed Germination

When conditions of moisture, oxygen, and temperature are favorable, seeds germinate. The embryo plant develops leaves and roots, and begins to produce its own food by photosynthesis. The development of a mature plant from an embryo involves cell division, cell differentiation, and growth.

Plant Growth

In flowering plants, only certain regions, called *meristems*, are able to undergo cell division. There are two types of meristems: cell division in the tips of roots and stems (in apical meristems) results in an increase in length; cell division between the xylem and phloem (in lateral meristems, or *cambiums*) results in an increase in the diameter of roots and stems. The cells of the meristem regions divide and then undergo elongation and differentiation, forming the different kinds of plant tissues.

QUESTIONS

PART A

88. Which reproductive structures are produced within the ovaries of plants? (1) pollen grains (2) sperm nuclei (3) egg nuclei (4) pollen tubes

89. In a flowering plant, the ovule develops within a part of the (1) style (2) anther (3) pistil (4) stigma

90. Which embryonic structure supplies nutrients to a germinating bean plant? (1) pollen tube (2) hypocotyl (3) epicotyl (4) cotyledon

91. Heavy use of insecticides in springtime may lead to a decrease in apple production in the fall, which is most probably due to interference with the process of (1) pollination (2) cleavage (3) absorption (4) transpiration

92. In a bean seed, the part of the embryo that develops into the leaves and upper portion of the stem is known as the (1) seed coat (2) epicotyl (3) hypocotyl (4) cotyledon

93. A condition necessary for the germination of most seeds is favorable (1) light (2) chlorophyll concentration (3) temperature (4) nitrate concentration

94. In flowering plants, the entire female reproductive organ is called the (1) filament (2) anther (3) style (4) pistil

95. In flowering plants, pollen grains are formed in the (1) style (2) anther (3) sepal (4) stigma

96. The seeds of a flowering plant develop from the ripened (1) fruits (2) cotyledons (3) ovules (4) endosperm

97. The endosperm of a bean seed is contained within its (1) cotyledons (2) ovules (3) stamen (4) petals

98. The fruits of a flowering plant develop from the ripened (1) seeds (2) ovules (3) ovaries (4) pollen tubes

99. Which of the following is *not* part of a plant embryo? (1) epicotyl (2) seed coat (3) hypocotyl (4) cotyledon

100. Which portion of a bean seed would contain the greatest percentage of starch? (1) seed coat (2) epicotyl (3) cotyledon (4) hypocotyl

Base your answers to questions 101 through 103 on the diagram below, which shows a cross section of a flower, and on your knowledge of biology.

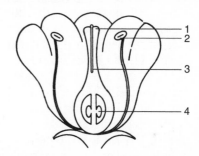

101. In this diagram, the stigma and the anther are (1) 1 and 2 (2) 1 and 4 (3) 2 and 4 (4) 2 and 3

102. Which process has occurred in this flower? (1) pollen germination (2) seed formation (3) zygote formation (4) fruit production

103. In which part would fertilization occur? (1) 1 (2) 2 (3) 3 (4) 4

Base your answers to questions 104 and 105 on the diagram below, which shows the internal structure of half of a bean seed, and on your knowledge of biology.

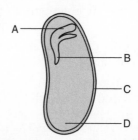

104. In which structure would most of the stored food for the embryo be found? (1) *A* (2) *B* (3) *C* (4) *D*

105. The epicotyl and the hypocotyl are represented by (1) *A* and *C* (2) *B* and *D* (3) *A* and *B* (4) *C* and *D*

106. Explain why cross-pollination increases the chances of genetic variation in the offspring of flowering plants.

107. Compare sexual reproduction in mammals and flowering plants. In what way is the process similar in these two types of organisms?

READING COMPREHENSION

Base your answers to questions 108 through 112 on the information below and on your knowledge of biology. Source: Science News *(June 4, 2005): vol. 167, no. 23, p. 366.*

Menstrual Cycle Changes the Brain

Hormone fluctuations over the course of a woman's menstrual cycle change the abundance of a particular type of receptor on the surface of nerve cells, say scientists. The finding may explain why some women with neurological disorders experience flare-ups of their conditions at the same time most months.

Previous research showed that nearly 80 percent of epileptic women have more seizures than usual during the phase of the menstrual cycle when their blood concentration of progesterone declines and that of estrogen increases. Other studies showed that women with a condition called premenstrual dysphoric disorder experience severe anxiety and depression during the same phase.

Istvan Mody and his colleagues at the University of California, Los Angeles examined brain slices from female mice. In search of changes in cells during the animals' 6-day menstrual cycles, the researchers looked at nerve-cell receptors for gamma aminobutyric acid, a chemical that inhibits neurons from firing.

The scientists found that the prevalence of a receptor subtype called delta was high when progesterone concentrations were up and estrogen concentrations were down. The same subtype was less prominent during the rest of the cycle, when the relative hormone concentrations were reversed. Nerve cells with more delta receptors were less likely to fire when stimulated with electricity than were cells with fewer delta receptors.

In a separate experiment, live mice in the high-progesterone phase of their cycle were less likely to have a seizure when given convulsion-inducing drugs than were mice in their high-estrogen phase. The high-progesterone mice also displayed less anxious behaviors than high-estrogen mice did.

Mody and his colleagues say that these results, published in the June *Nature Neuroscience*, suggest that a shortage of delta receptors may increase nerve cell activity during the low-progesterone phase of the menstrual cycle, in turn increasing anxiety and seizure susceptibility.

108. At what point in their menstrual cycle do most epileptic women suffer from more seizures?

109. What function does the chemical gamma aminobutyric acid (GABA) have in the body?

110. Using the information in this article, explain the relationship among the following factors: high progesterone levels; low estrogen levels; number of delta receptors on nerve cells; and likelihood of nerve cells to fire.

111. What conclusion did the researchers draw about the menstrual cycle and seizures, based on their studies of female mice?

112. If delta is the membrane receptor for GABA, what must be true about the three-dimensional shape of delta and GABA?

CHAPTER 7

Genetics and Heredity

FOUNDATIONS OF GENETICS

Genetics is the branch of biology that deals with patterns of **inheritance**, or heredity. *Heredity* is the biological process by which parents pass on genetic information to their offspring through their gametes. The science of genetics originated with the work of an Austrian monk, *Gregor Mendel*, who performed a series of experiments on pea plants between 1856 and 1868.

Principles of Mendelian Genetics

In his breeding experiments, Mendel (who, like everyone else at that time, had no knowledge of genes or chromosomes) made careful observations of the inheritance patterns of specific contrasting traits found in pea plants. Through a mathematical analysis of the traits found in the large numbers of offspring from his experimental crosses, Mendel developed his principles of *dominance*, *segregation*, and *independent assortment*. Mendel also concluded that the traits he observed were controlled by pairs of inherited "factors," with one member of each pair coming from each parent organism. Thus, in organisms that reproduce sexually, half of the offspring's genetic material is contributed by the female parent and half by the male parent. As a result, the offspring has traits from both parents, and is never identical to either one of them.

Gene–Chromosome Theory

The importance of Mendel's work was not recognized until the early 1900s, when the development of better microscopes enabled biologists to observe chromosome behavior during meiotic cell division. Biologists then linked the separation of homologous chromosome pairs during meiosis and their **recombination** at fertilization with the inheritance of Mendel's factors. Breeding experiments carried out by T. H. Morgan with the fruit fly, *Drosophila*, provided supporting evidence for Mendel's principles of inheritance.

Mendel's inherited, or **hereditary**, factors—now known as **genes**—are arranged in a linear fashion on the chromosomes. Each gene has a definite position, or *locus* (plural, *loci*), on the chromosome. The two alternate genes that control each trait are called *alleles*, and they are located in the same position on homologous chromosomes. This *gene–chromosome theory* explains the hereditary patterns observed by Mendel.

Gene Expression

Every organism has at least two alleles that govern every trait. As mentioned, these two genes are passed on—one from the mother and one from the father—to the offspring. The genes encode information that is expressed as the traits of the organism, a phenomenon called **gene expression**. A single gene (that is, one set of alleles) may control one or several traits. Alternatively, some traits are determined by more than one gene (that is, by more than one set of alleles).

Although all the body cells in an organism contain the same genetic instructions, the cells may differ considerably from one another in structure and function. The reason is that, in any given cell, only some of the genes are expressed, while all other genes are inactivated. For example, in liver cells, it is mainly the genes that pertain to liver functions that are active, while the other genes are inactive. The same is true of all other cells in a body. You can think of the genes on a cell's chromosomes as recipes in a cookbook: the book may contain hundreds of recipes, but if you are making a chocolate cake, you will read only the instructions for making that item. Likewise, the cell reads only the instructions for making its specific products.

Genes that are "on" are expressed, while those that are "off" are not expressed. There are many mechanisms that can switch genes on and off, including intracellular chemicals, enzymes, regulatory proteins, and the cell's environment. In addition, a particular gene may alternately be expressed or inactivated, depending on the cell's needs at the time.

SOME MAJOR CONCEPTS IN GENETICS

Dominance

In his experiments, Mendel crossed plants that were pure for contrasting traits. For example, he crossed pure tall plants with pure short plants. All the offspring of such crosses showed only one of the two contrasting traits. In the cross of tall plants and short plants, all the offspring were tall. In this type of inheritance, the allele that is expressed in the offspring is said to be *dominant*; the allele that is present but not expressed is said to be *recessive*. This pattern illustrates Mendel's principle of dominance.

By convention, the dominant allele is represented by a capital letter, while the recessive allele is represented by the lowercase form of the same letter. For example, the allele for tallness, which is dominant, is shown as T, while the allele for shortness, which is recessive, is shown as t.

If, in an organism, the two genes of a pair of alleles are the same, for example, TT or tt, the organism is said to be *homozygous*, or pure, for that trait. The genetic makeup of the organism, which is its *genotype*, is either homozygous dominant (TT) or homozygous recessive (tt). If the two genes of a pair of alleles are different, for example, Tt, the organism is said to be *heterozygous*, or *hybrid*, for that trait.

The physical appearance of an organism that results from its genetic makeup is called its *phenotype*. For example, a pea plant that is heterozygous for height has the genotype Tt and the phenotype of being tall. When an organism that is homozygous for the dominant trait is crossed with an organism that is homozygous for the recessive trait ($TT \times tt$), the phenotype of the offspring is like that of the dominant parent. Thus, the heterozygous offspring (Tt) is tall.

In studies involving genetic crosses, the organisms that are used to begin the studies are called the *parent generation*. The offspring produced by crossing members of the parent generation are called the *first filial*, or F_1, *generation*. The offspring of a cross between members of the

F_1 generation make up the *second filial*, or F_2, *generation*.

QUESTIONS

1. When a strain of fruit flies homozygous for light body color is crossed with a strain of fruit flies homozygous for dark body color, all the offspring have light body color. This illustrates Mendel's principle of (1) segregation (2) dominance (3) incomplete dominance (4) independent assortment

2. Two genes located in corresponding positions on a pair of homologous chromosomes and associated with the same characteristic are known as (1) gametes (2) zygotes (3) chromatids (4) alleles

3. For a given trait, the two genes of an allelic pair are not alike. An individual possessing this gene combination is said to be (1) homozygous for that trait (2) heterozygous for that trait (3) recessive for that trait (4) pure for that trait

4. In pea plants, flowers located along the stem (*axial*) are dominant to flowers located at the end of the stem (*terminal*). Let A represent the allele for axial flowers and a represent the allele for terminal flowers. When plants with axial flowers are crossed with plants having terminal flowers, all of the offspring have axial flowers. The genotypes of the parent plants are most likely (1) $aa \times aa$ (2) $Aa \times Aa$ (3) $aa \times Aa$ (4) $AA \times aa$

5. Curly hair in humans, white fur in guinea pigs, and needlelike spines in cacti all partly describe each organism's (1) alleles (2) autosomes (3) chromosomes (4) phenotype

6. The appearance of a recessive trait in offspring of animals most probably indicates that (1) both parents carried at least one recessive gene for that trait (2) one parent was homozygous dominant and the other parent was homozygous recessive for that trait (3) neither parent carried a recessive gene for that trait (4) one parent was homozygous dominant and the other parent was hybrid for that trait

7. Which statement describes how two organisms may show the same trait yet have different genotypes for that phenotype? (1) One is homozygous dominant and the other is heterozygous. (2) Both are heterozygous for the dominant trait. (3) One is homozygous dominant and the

other is homozygous recessive. (4) Both are homozygous for the dominant trait.

8. In cabbage butterflies, white color (W) is dominant and yellow color (w) is recessive. If a pure white cabbage butterfly mates with a yellow cabbage butterfly, all the resulting (F_1) butterflies are heterozygous white. Which cross represents the genotypes of the parent generation? (1) $Ww \times ww$ (2) $WW \times Ww$ (3) $WW \times ww$ (4) $Ww \times Ww$

9. Most of the hereditary information that determines the traits of an organism is located in (1) only those cells of an individual produced by meiosis (2) the nuclei of body cells of an individual (3) certain genes in the vacuoles of body cells (4) the numerous ribosomes in certain cells

10. The characteristics of a developing fetus are most influenced by (1) gene combinations and their expression in the embryo (2) hormone production by the father (3) circulating levels of white blood cells in the placenta (4) milk production in the mother

<hr>

PART B-2

11. Explain how two organisms can have the same phenotype but different genotypes.
12. To illustrate your answer to question 11, pick a trait and use a letter to represent it. Write the genotypes of the parents and F_1 generations for each organism.

<hr>

PART C

13. Why do the offspring of sexually reproducing organisms resemble both parents? Why are they not identical to either one of the parents?
14. Explain why the body cells of an organism can differ in structure and function, even though they all contain the same genetic information.

Segregation and Recombination

When gametes are formed during meiosis, the two chromosomes of each homologous pair separate, or *segregate*, randomly. Each gamete contains only one allele for each trait. After the gametes fuse during fertilization, the resulting (zygote) cell contains pairs of homologous chromosomes, but new combinations of alleles may be present. This process is described by Mendel's principle of segregation.

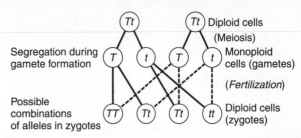

Figure 7-1. Segregation and recombination of alleles.

Figure 7-1 illustrates segregation and recombination in a cross between two individuals that are heterozygous for tallness. In a large number of such crosses, with a large number of offspring, two types of numerical ratios can be observed. In terms of genotype, the ratio is 1 homozygous dominant (TT) : 2 heterozygous (Tt) : 1 homozygous recessive (tt). In terms of phenotype, the ratio is 3 tall : 1 short. These genotype and phenotype ratios are typical for all crosses between organisms that are hybrid for one trait.

The Testcross

To determine the genotype of an organism that shows the dominant phenotype, a testcross is performed. In a *testcross*, the organism in question is crossed with a homozygous recessive organism (Figure 7-2). If the test organism is homozygous dominant, all the offspring will be heterozygous and show the dominant phenotype. If any

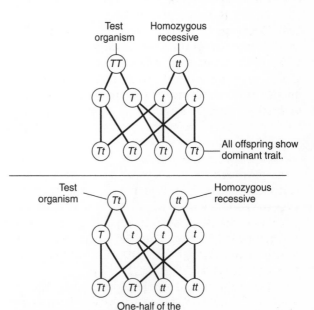

Figure 7-2. Use of a testcross to determine an organism's genotype.

offspring show the recessive phenotype, the individual being tested would have to be heterozygous.

Punnett Square

The possible offspring of a genetic cross are often shown with a diagram called a *Punnett square*. We can use a Punnett square to show the possible offspring of a cross between a heterozygous tall pea plant (*Tt*) and a homozygous short pea plant (*tt*).

The first step in using a Punnett square is to determine the possible genotypes of the gametes of each parent. In this example, the heterozygous tall plant (*Tt*) produces two types of gametes: half will contain the dominant gene for height, *T*, and half will contain the recessive gene, *t*. The gametes of the homozygous short plant (*tt*) will each contain the recessive gene for height, *t*.

As shown in Figure 7-3, the letters that represent the trait carried by the gametes of one parent are written next to the boxes on the left side of the square; the letters for the gametes of the other parent are written above the boxes on top of the square. The letters are combined to show offspring genotypes as follows: letters on top of the square are written in the boxes below them, and letters on the side are written in the boxes to the right of them. The dominant gene, when present, is written first. The pairs of letters in the four boxes represent the possible combinations of genes in the offspring of the cross. Of the possible offspring of this cross, half would be heterozygous tall (*Tt*) and half would be homozygous (recessive) short (*tt*).

Linkage

Mendel's observation of the independent inheritance of different traits was the basis for his principle of independent assortment. When the events of meiosis were discovered, it became clear that traits are inherited independently of

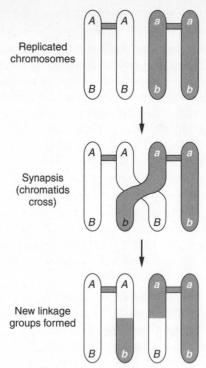

Figure 7-4. Crossing-over of chromatids.

one another only when their genes are on nonhomologous chromosomes. However, when the genes for two different traits are located on the same pair of homologous chromosomes, they tend to be inherited together. Such genes are said to be *linked*. The patterns of inheritance and phenotype ratios for linked traits are different from those of nonlinked traits (the kind observed by Mendel).

Crossing-Over

During *synapsis* in the first meiotic division, the chromatids of a pair of homologous chromosomes often twist around each other, break, exchange segments, and rejoin (Figure 7-4). This exchange of segments, called *crossing-over*, results in a rearrangement of linked genes and produces variations in offspring. Crossing-over is an important source of genetic variation in sexual reproduction.

QUESTIONS

PART A

15. Polydactyly is a characteristic in which a person has six fingers per hand. Polydactyly is dominant over the trait for five fingers. If a man who is heterozygous for this trait marries a woman with the normal number of fingers,

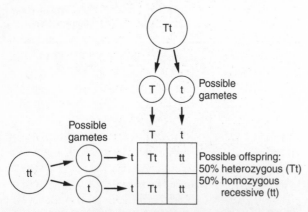

Figure 7-3. Use of a Punnett square to determine possible genotypes of offspring.

what are the chances that their child would be polydactyl? (1) 0% (2) 50% (3) 75% (4) 100%

16. A cross between two pea plants that are hybrid for a single trait produces 60 offspring. Approximately how many of the offspring would be expected to exhibit the recessive trait? (1) 15 (2) 45 (3) 30 (4) 60

17. Which principle states that during meiosis chromosomes are distributed to gametes in a random fashion? (1) dominance (2) linkage (3) segregation (4) mutation

18. In guinea pigs, black coat color is dominant over white coat color. The offspring of a mating between two heterozygous black guinea pigs would probably show a phenotype ratio of (1) two black to two white (2) one black to three white (3) three black to one white (4) four black to zero white

19. The offspring of a mating between two heterozygous black guinea pigs would probably show a genotype ratio of (1) l *BB* : 2 *Bb* : 1 *bb* (2) 3 *Bb* :1 *bb* (3) 2 *BB* :2 *bb* (4) 2 *BB* :1 *Bb* :1 *bb*

20. If a breeder wanted to discover whether a black guinea pig was homozygous (*BB*) or heterozygous (*Bb*) for coat color, the animal in question would have to be crossed with an individual that has the genotype (1) *BB* (2) *bb* (3) *Bb* (4) *BbBb*

21. Mendel's principle of independent assortment applies to traits whose genes are found on (1) homologous chromosomes (2) sex chromosomes (3) the same chromosome (4) nonhomologous chromosomes

22. The process in which the chromatids of pairs of homologous chromosomes exchange segments is called (1) linkage (2) crossing-over (3) independent assortment (4) intermediate inheritance

23. In horses, black coat color is dominant over chestnut coat color. Two black horses produce both a black-coated and a chestnut-coated offspring. If coat color is controlled by a single pair of genes, it can be assumed that (1) in horses, genes for coat color frequently mutate (2) one of the parent horses is homozygous dominant and the other is heterozygous for coat color (3) both parent horses are homozygous for coat color (4) both parent horses are heterozygous for coat color

PART B-2

24. Based on your answer to question 23, explain how two black horses could produce a chestnut-colored offspring.

Base your answers to questions 25 through 27 on the diagram below, which represents a pair of homologous chromosomes at the beginning of meiosis. The letters A, B, C, a, b, *and* c *represent pairs of alleles located on the chromosomes.*

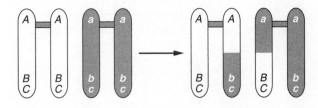

25. Compare the sets of chromosomes on the left with those on the right. Explain what has happened.

26. What process (not shown) is responsible for the observed results?

27. How does this process lead to variations among offspring?

PART C

28. When is a testcross used? Explain how it works.

29. Explain the following statement: Traits are inherited independently of one another only if their genes are on non-homologous chromosomes. You may use diagrams to support your explanation.

Sex Determination

The diploid cells of many organisms contain two types of chromosomes: *autosomes* and *sex chromosomes*. There is generally one pair of sex chromosomes, and all the other chromosomes are autosomes. In human body cells there are 22 pairs of autosomes and one pair of sex chromosomes. The sex chromosomes are called the *X* and *Y* chromosomes. Females have two *X* chromosomes, and males have one *X* and one *Y* chromosome.

During meiotic cell division, the sex chromosomes, like all other chromosome pairs, are separated (Figure 7-5, page 102). The resulting gametes contain only one sex chromosome. Since females have two *X* chromosomes, each female gamete receives an *X* chromosome. Since the genotype of males is *XY*, sperm cells may receive either an *X* or a *Y* chromosome. The sex of the offspring is determined at fertilization and depends on whether the egg is fertilized by a sperm with an *X* or a sperm with a *Y* chromosome. If the sperm has an *X* chromosome, the resulting zygote will be female (*XX*). If the sperm has a *Y* chromosome, the resulting zygote will be male (*XY*).

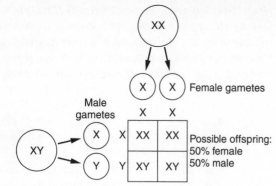

Figure 7-5. Sex determination of offspring.

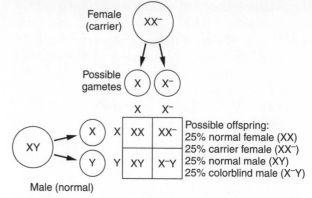

Figure 7-6. Inheritance of color blindness.

Sex-linked Traits

T. H. Morgan, in his experiments with fruit flies, found that some rare, abnormal recessive traits appear with greater frequency in males than in females. From his observations, Morgan concluded that the genes for these traits are present on the X chromosome and that there are no corresponding alleles for these traits on the Y chromosome. Genes found on the X chromosome are called *sex-linked genes*. Recessive sex-linked traits appear more frequently in males than in females because in females there is usually a normal, dominant allele on the other X chromosome, so that the phenotype is normal. In males, there is no second allele, so the presence of one recessive gene produces a recessive phenotype.

Both *hemophilia* and *color blindness* are sex-linked disorders; they occur more frequently in males than in females. Hemophilia is a condition in which the blood does not clot properly, while color blindness is an inability to distinguish certain colors. The genes for normal blood clotting and normal color vision are dominant; the genes for hemophilia and color blindness are recessive. For a female to show either of these disorders, she must have recessive genes (alleles) on both of her X chromosomes. Females with one normal, dominant gene and one recessive gene for these disorders are called *carriers*. They can pass the disorder to their offspring but do not themselves show symptoms of the disorder. Figure 7-6 shows the possible genotypes of children of a normal male and a female carrier of color blindness.

QUESTIONS

PART A

30. If a color-blind man marries a woman who is a carrier for color blindness, it is most probable that (1) all of their sons will have normal color vision (2) half of their sons will be color-blind (3) all of their sons will be color-blind (4) none of their children will have normal color vision

31. A color-blind man marries a woman with normal vision. Her mother was color-blind. They have one child. What is the chance that this child will be color-blind? (1) 0% (2) 25% (3) 50% (4) 100%

32. A color-blind woman marries a man who has normal color vision. What are their chances of having a color-blind daughter? (1) 0% (2) 25% (3) 75% (4) 100%

33. Which parental pair could produce a color-blind female? (1) homozygous normal-vision mother and color-blind father (2) color-blind mother and normal-vision father (3) heterozygous normal-vision mother and normal-vision father (4) heterozygous normal-vision mother and color-blind father

34. Which statement correctly describes the normal number and type of chromosomes present in human body cells of a particular sex? (1) Males have 22 pairs of autosomes and 1 pair of XX sex chromosomes. (2) Females have 23 pairs of autosomes. (3) Males have 22 pairs of autosomes and 1 pair of XY sex chromosomes. (4) Males have 23 pairs of autosomes.

35. Based on the pattern of inheritance known as sex linkage, if a male is a hemophiliac, how many genes for this trait are present on the sex chromosomes in each of his diploid cells? (1) 1 (2) 2 (3) 3 (4) 4

36. Traits controlled by genes on the X chromosome are said to be (1) sex-linked (2) mutagenic (3) incompletely dominant (4) homozygous

37. Use a diagram to show why, for each pregnancy, the chances of giving birth to either a boy or a girl is 50-50. Explain the results shown in your diagram.

38. Explain why hemophilia occurs more often in males than in females. Use a diagram to illustrate your answer.

GENETIC MUTATIONS

Changes in the genetic material are called **mutations**. Mutations in body cells can be passed on to new cells of the individual as a result of mitosis, but they cannot be transmitted to offspring by sexual reproduction. However, mutations in sex cells *can* be transmitted to the next generation. Mutations may involve alterations in chromosomes or alterations in the chemical makeup of genes.

Chromosomal Alterations

Chromosomal alterations involve a change in the structure or number of chromosomes. The effects of chromosomal alterations are often seen in the phenotype of an organism because each chromosome contains many genes.

Nondisjunction.   During meiosis, the two chromosomes of each homologous pair separate from each other; each gamete produced by the division receives only one member of each homologous pair. The separation of homologous chromosomes is called *disjunction*. The term *nondisjunction* refers to a type of chromosomal alteration in which one or more pairs of homologous chromosomes fails to separate normally during meiotic cell division (Figure 7-7).

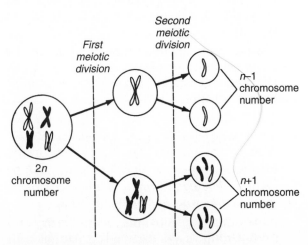

Figure 7-7. Nondisjunction of chromosomes.

As a result of nondisjunction, one of the gametes produced contains both members of the homologous pair, while another gamete contains neither chromosome. Nondisjunction results in the production of some gametes with more chromosomes than normal and some gametes with fewer chromosomes than normal. If one of these abnormal gametes is involved in fertilization, the resulting zygote will have either more than or less than the normal (2n) number of chromosomes.

Down syndrome in humans is caused by the presence of an extra chromosome number 21. Nondisjunction during gamete production in one of the parents produces a gamete with an extra chromosome 21. As a result of fertilization, this extra chromosome is transmitted to the offspring.

Polyploidy.   Occasionally during gamete formation, a complete set of chromosomes fails to undergo disjunction, and a gamete is produced that contains the diploid (2n) chromosome number. If a diploid gamete unites with a normal (n) gamete during fertilization, the resulting zygote will have a 3n chromosome number. If two 2n gametes fuse, a 4n zygote results. The inheritance of one or more complete extra sets of chromosomes is called *polyploidy*. This condition is common in plants but rare in animals. In plants, polyploid individuals are usually larger or more vigorous than the normal, diploid varieties. Certain strains of wheat, potatoes, alfalfa, apples, tobacco, and zinnias are polyploid. Some polyploid plants produce seedless fruit and are sterile.

Changes in Chromosome Structure.   Changes in the makeup of chromosomes may result from random breakage and recombination of chromosome parts. *Translocation* occurs when a segment of one chromosome breaks off and reattaches to a nonhomologous chromosome. *Addition* occurs when a segment breaks off one chromosome and reattaches to the homologous chromosome. *Inversion* occurs when a segment breaks off and reattaches in reverse on the same chromosome. *Deletion* occurs when a segment breaks off and does not reattach to any other chromosome.

Gene Mutations

A random change in the chemical makeup of the DNA (genetic material) is a *gene mutation*. The effects of some gene mutations, such as albinism, are noticeable, but other gene mutations may not produce noticeable effects.

Inheritable gene mutations tend to be harmful to the individual. For example, sickle-cell anemia and Tay-Sachs disease are caused by gene mutations. Fortunately, most gene mutations are recessive and are hidden by the normal, dominant allele. However, if both parents carry the same recessive mutant gene, there is a chance that their offspring will be homozygous recessive and show the harmful trait.

Occasionally, random gene mutations produce changes that make an individual better adapted to the environment. Over time, such helpful mutant genes tend to increase in frequency within a population.

Mutagenic Agents

Although mutations occur spontaneously, the rate of mutation can be increased by exposure to certain chemicals and forms of **radiation** that act as *mutagenic agents*. For example, forms of mutagenic radiation include x-rays, ultraviolet rays, radioactive substances, and cosmic rays. Mutagenic chemicals include formaldehyde, benzene, and asbestos fibers.

QUESTIONS

PART A

39. Which phrase best describes most mutations? (1) dominant and disadvantageous to the organism (2) recessive and disadvantageous to the organism (3) recessive and advantageous to the organism (4) dominant and advantageous to the organism

40. The failure of a pair of homologous chromosomes to separate during meiotic cell division is called (1) nondisjunction (2) translocation (3) addition (4) deletion

41. The condition in which a gamete contains the $2n$ or $3n$ number of chromosomes is called (1) translocation (2) a gene mutation (3) polydactyly (4) polyploidy

42. The presence of only one X chromosome in each body cell of a human female produces a condition known as Turner syndrome. This condition most probably results from the process called (1) polyploidy (2) crossing-over (3) nondisjunction (4) hybridization

43. A random change in the chemical structure of DNA produces (1) polyploidy (2) a translocation (3) nondisjunction (4) a gene mutation

44. Down syndrome in humans is characterized by the presence of an extra chromosome 21 in all cells of the body. The number of chromosomes present in the body cells of individuals with this condition is (1) $n + 1$ (2) $3n$ (3) $2n + 1$ (4) $4n$

45. The graph below shows the relationship between maternal age and the number of children born with Down syndrome per 1000 births.

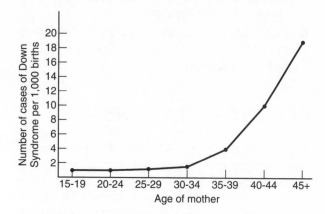

According to the graph, the incidence of Down syndrome (1) generally decreases as maternal age increases (2) is about nine times greater at age 45 than at age 30 (3) stabilized at 2 per 1000 births after age 35 (4) is greater at age 15 than at age 35

46. Ultraviolet rays, x-rays, and certain other forms of radiation can increase the rate of gene mutation. These forms of radiation are said to act as (1) mutagenic agents (2) catalysts (3) enzymes (4) indicators

47. The large size and exceptional vigor of certain varieties of wheat, apples, and zinnias are due to the possession of extra sets of chromosomes, which result from (1) incomplete dominance (2) gene mutations (3) nondisjunction of complete sets of chromosomes (4) nondisjunction of chromosome number 21 only

48. A type of chromosomal alteration in which a segment of chromosome breaks off and does not reattach to any chromosome is called (1) addition (2) inversion (3) deletion (4) translocation

49. Changes in the genetic code of a human can be transmitted to offspring if they occur in (1) cancer cells (2) gametes (3) cell membranes (4) antibodies

PART B-2

Base your answers to questions 50 through 53 on the following information about an experiment and on your knowledge of biology.

Two groups of 100 lima beans each were used. Group *A* was exposed to natural light for a period of 24 hours and then planted.

Group *B* was exposed to microwave energy for 24 hours and then planted under the same conditions as Group *A*. When the seeds germinated, the plants were observed for growth over a period of two weeks. The results are summarized in the table below.

Number of Plants

Group	Normal Growth	Stunted and/or Pale
A	83	17
B	54	46

50. What hypothesis was most likely being tested in this experiment?
51. Describe the results of the experiment.
52. Based on the data, propose a conclusion for the experiment.
53. What are some of the implications of the data?

PART C

54. Explain how it is possible for an individual to inherit an extra chromosome. List two or more human genetic disorders caused by the inheritance of an abnormal number of chromosomes.
55. Mutagens are agents that increase the rate of gene mutations in cells. Identify three types of mutagenic agents and briefly explain how each one causes mutations. Describe how people may reduce their chances of being harmed by these particular agents.

HEREDITY AND THE ENVIRONMENT

The development and expression of inherited traits can be influenced by environmental factors such as nutrients, temperature, sunlight, and so on. The relationship between gene action and environmental influence can be seen in the following examples.

Temperature affects fur color in the Himalayan rabbit. Under normal circumstances, these rabbits are white with black ears, nose, tail, and feet. (The black fur helps the rabbit absorb more heat in its extremities.) However, when some of the white fur on a Himalayan rabbit's back is shaved off and the area kept covered with an ice pack, the new hairs grow in black. The artificial change in temperature produces a change in fur color.

Experiments have shown that the production of chlorophyll requires exposure to sunlight. When parts of a leaf are covered with dark paper, chlorophyll production stops in the area that is covered. Only the exposed part produces chlorophyll, is green, and performs photosynthesis.

Stress and nutrition can affect gene expression. For example, someone who has a tall genotype may not develop a tall phenotype if his or her growth is stunted by malnutrition.

PLANT AND ANIMAL BREEDING

Using the principles of genetics, plant and animal breeders have been able to produce, improve, and maintain new varieties of plants and animals. Methods of **selective breeding** used by such people include artificial selection, inbreeding, and hybridization.

In *artificial selection*, individuals with the most desirable traits (for example, sheep with thick, soft wool) are crossed or allowed to mate in the hopes that their offspring will show the desired traits.

The offspring of selected organisms may be mated with one another to produce more individuals with the desirable traits. This technique, called *inbreeding*, involves the mating of closely related organisms. (Of course, the risk of inbreeding is that harmful recessive genes are more likely to be inherited and cause disorders in the offspring.)

Two varieties of a species may have different desirable traits. In a technique called *hybridization*, breeders cross two such varieties in the hope of producing hybrid offspring that show the desirable traits of both varieties. For example, if one variety of rose has very large petals and another variety has a very sweet scent, their hybrid might show both desirable traits.

QUESTIONS
PART A

56. If bean plant seedlings are germinated in the dark, the seedlings will lack green color. The best explanation for this condition is that (1) bean plants are heterotrophic organisms (2) bean seedlings lack nitrogen compounds in their cotyledons (3) the absence of an environmental factor limits the expression of a genotype (4) bean plants cannot break down carbon dioxide to produce oxygen in the dark
57. In many humans, exposing the skin to sunlight over prolonged periods of time results in

the production of more pigment by the skin cells (tanning). This change in skin color provides evidence that (1) ultraviolet light can cause mutations (2) gene action can be influenced by the environment (3) the inheritance of skin color is an acquired characteristic (4) albinism is a recessive characteristic

58. Identical twins were separated at birth and brought together after 13 years. They varied in height by 5 centimeters and in weight by 10 kilograms. The most probable explanation for these differences is that (1) their environments affected the expression of their traits (2) their cells did not divide by mitotic cell division (3) they developed from two different zygotes (4) they differed in their genotypes

59. A normal bean seedling that had the ability to produce chlorophyll did not produce any chlorophyll when grown in soil that was totally deficient in magnesium salts. Which statement concerning this plant's inability to produce chlorophyll is true? (1) The lack of magnesium prevented the plant's roots from absorbing water. (2) The production of chlorophyll was controlled solely by heredity. (3) The lack of magnesium caused a mutation of the gene that controlled chlorophyll production. (4) The production of chlorophyll was influenced by environmental conditions.

60. To ensure the maintenance of a desirable trait in a particular variety of plant, a farmer would use (1) binary fission (2) mutagenic agents (3) artificial selection (4) natural selection

61. The mating of very closely related organisms in order to produce the most desirable traits is known as (1) inbreeding (2) hybridization (3) karyotyping (4) crossing-over

62. Plant and animal breeders usually sell or get rid of undesirable specimens and use only the desirable ones for breeding. This practice is referred to as (1) vegetative propagation (2) artificial selection (3) natural breeding (4) random mating

63. A single gene mutation results from (1) a change in a base sequence in DNA (2) recombination of traits (3) the failure of chromosomes to separate (4) blocked nerve messages

64. The chart below shows relationships between genes, the environment, and coloration of tomato plants. Which statement best explains the final appearance of these tomato plants? (1) The expression of gene A is not affected by light. (2) The expression of gene B varies with the presence of light. (3) The expression of gene A varies with the environment. (4) Gene B is expressed only in darkness.

Inherited Gene	Environmental Condition	Final Appearance
A	Light	Green
B	Light	White
A	Dark	White
B	Dark	White

65. Some mammals have genes for fur color that produce pigment only when the outside temperature is above a certain level. This pigment production is an example of how the environment of an organism can (1) destroy certain genes (2) cause new mutations to occur (3) stop the process of evolution (4) influence the expression of certain genes

66. Identify three environmental factors that can influence phenotype. Give an example of each.

67. Describe some steps a breeder would take to produce an organism that has desirable traits.

HUMAN HEREDITY

The principles of genetics apply to all organisms. However, specific studies of human genetics are limited because humans are not suitable subjects for experimentation: human generation time is too long; there are only a small number of offspring per generation in a human family; and it is unethical to perform such experiments on humans. Knowledge of human heredity has been gathered indirectly through studies of human pedigree charts and materials obtained in the course of genetics counseling.

Human Pedigree Charts
The patterns of inheritance of certain traits can be traced in families for a number of generations. These patterns can be illustrated in *pedigree charts* that show the presence or absence of certain genetic traits in each generation. The use of a pedigree chart may also make it possible to identify carriers of recessive genes.

Human Genetic Disorders

Some diseases caused by genetic abnormalities are sickle-cell anemia, Tay-Sachs disease, and phenylketonuria. These disorders are caused by gene mutations.

Sickle-cell anemia is a blood disorder found most commonly in individuals of African descent. The disorder is caused by a gene mutation that results in the production of abnormal hemoglobin molecules and red blood cells. The abnormal hemoglobin and sickle-shaped cells do not carry oxygen efficiently, resulting in anemia. The sickle-shaped red cells also tend to obstruct blood vessels, causing severe pain. Sickle-cell anemia occurs in individuals homozygous for the trait. Both homozygous and heterozygous individuals can be detected by blood tests.

Tay-Sachs disease is a recessive genetic disorder in which nerve tissue in the brain deteriorates because of an accumulation of fatty material. The disorder is a result of the body's inability to synthesize a particular enzyme. Tay-Sachs disease, which is fatal, occurs most commonly among Jewish people of Central European descent.

Phenylketonuria (PKU) is a disorder in which the body cannot synthesize an enzyme necessary for the normal metabolism of the amino acid phenylalanine. The disease, which occurs in homozygous recessive individuals, is characterized by the development of mental retardation. Analysis of the urine of newborn infants can detect PKU. If PKU is detected, mental retardation can be prevented by maintaining a diet free of phenylalanine.

Detection of Genetic Disorders

Some human genetic disorders can be detected either before or after birth by the use of one or more of the following techniques.

Advances in genetic research have resulted in the development of simple blood and urine tests that can determine if an individual has certain genetic disorders. Carriers of sickle-cell anemia and Tay-Sachs disease can be identified by these screening techniques.

Karyotyping is a technique in which a greatly enlarged photograph of the chromosomes of a cell is prepared. The homologous pairs of chromosomes are matched together, and the chromosomes are examined to see if there are any abnormalities in number or structure.

Amniocentesis is a technique in which a small sample of amniotic fluid is withdrawn from the amniotic sac of a pregnant woman. The fluid contains fetal cells, which can be used for karyotyping or for chemical analysis. Amniocentesis is used in the identification of sickle-cell anemia, Tay-Sachs disease, and Down syndrome in fetuses.

Genetic Counseling

The various techniques described above are used by *genetics counselors* to inform concerned parents about the possible occurrence of genetic defects in their children. For couples whose families show the presence of a particular genetic disorder, a pedigree chart may be developed to predict the probability of their children's having the disorder. Amniocentesis, followed by karyotyping and chemical tests, may be performed once pregnancy is established.

QUESTIONS

PART A

68. An inherited metabolic disorder known as phenylketonuria (PKU) is characterized by severe mental retardation. This condition results from the inability to synthesize a single (1) enzyme (2) hormone (3) vitamin (4) carbohydrate

69. Which statement best describes amniocentesis? (1) Blood cells of an adult are checked for anemia. (2) Saliva of a child is analyzed for the amino acids. (3) Urine of a newborn baby is analyzed for the amino acid phenylalanine. (4) Fluid surrounding a fetus is removed for chemical and genetic analysis.

70. Which is a genetic disorder in which abnormal hemoglobin leads to fragile red blood cells and obstructed blood vessels? (1) phenylketonuria (2) sickle-cell anemia (3) leukemia (4) Down syndrome

71. Human disorders such as PKU and sickle-cell anemia, which are defects in the synthesis of individual proteins, are most likely the result of (1) gene mutations (2) nondisjunction (3) crossing-over (4) polyploidy

72. Which technique can be used to examine the chromosomes of a fetus for possible genetic defects? (1) pedigree analysis (2) analysis of fetal urine (3) karyotyping (4) blood cell tests

PART B-2

73. Give three reasons why a direct study of the inheritance of human traits is difficult to carry out.

74. Briefly describe the two ways that information about patterns of human heredity is usually obtained.

MODERN GENETICS

Biochemists have learned that the DNA of the chromosomes is the genetic material that is passed from generation to generation. Genes are sections of DNA (deoxyribonucleic acid) molecules. DNA controls cellular activities by controlling the production of enzymes.

DNA Structure

DNA molecules are very large; each is made up of thousands of repeating units called **nucleotides**. A DNA nucleotide is composed of three parts: a *phosphate group*; a molecule of the 5-carbon sugar *deoxyribose*; and a *nitrogenous base* (Figure 7-8).

There are four different nitrogenous bases found in DNA nucleotides: *adenine, cytosine, guanine*, and *thymine*. Therefore, there are four different kinds of nucleotides, depending on which nitrogenous base is present.

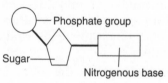

Figure 7-8. Structure of a DNA nucleotide unit.

Watson-Crick Model. In the model of DNA developed by James Watson and Francis Crick, the DNA molecule consists of two connected chains of nucleotides forming a ladderlike structure (Figure 7-9). The sides of the "ladder" are composed of alternating phosphate and deoxyribose (sugar) molecules. Each rung of the ladder consists of a pair of nitrogenous bases bonded together by hydrogen bonds. The two chains of the DNA molecule are twisted to form a spiral, or *double helix*.

The DNA model of Watson and Crick was based largely on x-ray pictures of DNA taken in a lab by molecular biologist Rosalind Franklin. The double-helix shape of the DNA molecule was apparently visible in Franklin's photograph.

The four nitrogenous bases of DNA nucleotides bond together in only one way: adenine (A) pairs with thymine (T), and cytosine (C) pairs with guanine (G). Because the bases pair together in only one way, the two strands of a DNA molecule are always *complementary*. Where there is an adenine nucleotide on one strand, there is a thymine nucleotide on the other; where there is a cytosine on one strand, there is a guanine on the other. If you know the order of bases on one strand, then you also know the order on the second strand.

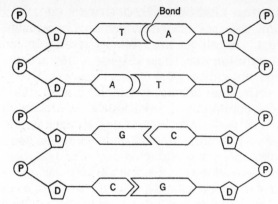

Figure 7-9. Structure of the DNA molecule.

DNA Replication

DNA, unlike any other chemical compound, can make exact copies of itself—that is, DNA can **replicate**. This process, called DNA *replication*, is a necessary part of the chromosome replication that occurs during mitosis and meiosis.

In replication, the double-stranded DNA helix unwinds; the two strands then separate, or unzip, by breaking the hydrogen bonds between the nitrogenous base pairs. Free nucleotides from the cytoplasm then enter the nucleus, where they bond to their complementary bases on the DNA strands (Figure 7-10). Replication produces two identical DNA molecules that are exact copies of the original molecule. The process of DNA replication is actually carried out by a team of several important, specific enzymes.

Gene Control of Cellular Activities

The unique qualities of an organism are determined by the DNA of its genes. The genes control

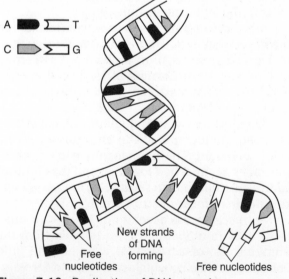

Figure 7-10. Replication of DNA strands.

enzyme synthesis, and the enzymes control cell activities. For example, a dominant gene enables people to produce the enzyme lactase, which digests milk sugar (lactose). People who lack an active copy of this gene cannot digest milk sugar and, thus, are lactose intolerant.

The hereditary information is in the sequence of the nucleotides in DNA molecules. The DNA nucleotide sequence determines the sequence of amino acids in enzymes and other proteins. The genetic control of protein synthesis involves RNA as well as DNA.

RNA

Molecules of *ribonucleic acid*, or *RNA*, are similar to DNA in that they are also made up of nucleotide units. However, in RNA nucleotides, the 5-carbon sugar *ribose* is substituted for deoxyribose, and the nitrogenous base *uracil* (U) is substituted for thymine. RNA molecules consist of one strand of nucleotides, while DNA molecules have two. There are three kinds of RNA molecules in cells: *messenger RNA* (mRNA), *transfer RNA* (tRNA), and *ribosomal RNA* (rRNA).

Messenger RNA is synthesized in the cell nucleus. Portions of a DNA molecule unwind, and the two strands separate. The RNA nucleotides pair with complementary bases on a DNA strand, thus forming a strand of messenger RNA that is complementary to the DNA strand. The DNA strand serves as a **template**, or pattern, for the synthesis of messenger RNA. In this way, the hereditary information in the nucleotide sequence of DNA is copied in complementary form into the nucleotide sequence of messenger RNA.

The sequence of nucleotides in messenger RNA contains the genetic code, which determines the amino acid sequence of proteins. The genetic code for each amino acid is a specific sequence of three nucleotides. The three-nucleotide sequence in messenger RNA that specifies a particular amino acid is called a *codon*.

Transfer RNA molecules are found in the cytoplasm. Their function is to carry amino acid molecules to the *ribosomes*, the sites of protein synthesis. Ribosomes are made up of rRNA and proteins. There are 20 different kinds of amino acids in cells, and there is a different form of transfer RNA for each amino acid. Each kind of transfer RNA has a three-nucleotide sequence, called an *anticodon*, which is complementary to a codon on the messenger RNA.

Protein Synthesis

Protein synthesis begins with the synthesis of messenger RNA molecules, which then move

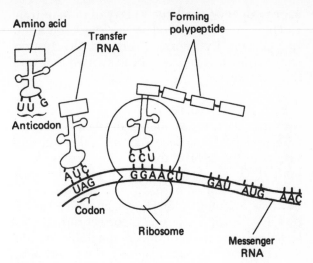

Figure 7-11. Protein synthesis (at the ribosome).

from the nucleus into the cytoplasm. In the cytoplasm, the strand of messenger RNA becomes associated with ribosomes (Figure 7-11). Amino acids are carried to the ribosomes and messenger RNA by the transfer RNAs. The anticodons of the transfer RNAs align with the codons of the messenger RNA. The amino acids carried by the transfer RNAs bond together in a sequence determined by the base sequence of the messenger RNA. The resulting chain of amino acids is a polypeptide. Some proteins consist of a single polypeptide chain, while others include two or more.

One Gene–One Polypeptide Hypothesis

According to the *one gene–one polypeptide hypothesis*, each gene controls the synthesis of a single polypeptide. A modern definition of the gene is the sequence of nucleotides in a DNA molecule necessary to synthesize one polypeptide. However, this hypothesis is now seen as overly simplified. It is now known that genes are not necessarily fixed in one place on the chromosomes. Rather, they can move to different locations, or *loci*, on the chromosomes and form new genetic codes when they are positioned next to different segments of DNA. In this way, a limited number of genes can have a much larger variety of expression than would be possible if the genes remained in one place. The geneticist Barbara McClintock first proposed the idea of movable, or "jumping," genes in the early 1940s, but other scientists largely dismissed her work at the time. Now known to exist, these movable genes, or transposable genetic elements, are called *transposons*. The recently completed Human Genome Project revealed that we have far fewer genes than had been expected. Now, the large variety of human traits can be better understood because we know about transposons. By "jumping" around the chromosomes, the same gene can be expressed in several different ways.

Gene Mutations

Any change in the sequence of nucleotides in a DNA molecule is a *gene mutation*. If the mutation occurs in the DNA of the sex cells, it may be inherited. Gene mutations may involve the *addition* or *deletion* of bases, or the *substitution* of one base for another. Sickle-cell anemia is caused by the substitution of one incorrect nitrogenous base in a gene that controls hemoglobin synthesis. The incorrect base results in the insertion of one incorrect amino acid, which in turn affects the structure and function of the hemoglobin protein.

Cloning

The term **cloning** describes the process by which a group of genetically identical offspring is produced from the cells of an organism. The cloning of plants shows great promise for agriculture, where plants with desirable qualities can be produced rapidly from the cells of a single plant. To date, the cloning of animals has been achieved in frogs, mice, sheep, goats, cows, cats, dogs, and monkeys.

Genetic Engineering

Gene splicing, or **genetic engineering**, involves the transfer of genetic material from one organism to another. This **recombining** of genes results in the formation of *recombinant DNA*. Using gene-splicing, or **biotechnological**, techniques, genes from one organism can be inserted into the DNA of another organism. Human genes that control the synthesis of insulin, interferon, and growth hormone have been introduced into bacterial cells, where they function as part of the bacterial DNA. In this way, bacterial cells are being used to synthesize certain substances needed by humans. Genetic engineering may eventually be able to correct some genetic defects and produce commercially desirable plants and animals.

Techniques of Genetic Engineering. The technique of making recombinant DNA (rDNA) molecules involves three important components.

First, a specific enzyme is needed to cut the DNA from the donor genes at a specific site. This enzyme is called a *restriction enzyme*. The enzyme is used to cut out a piece of DNA that contains one or more desired genes from the donor's DNA.

Next, a *vector* is needed to receive the donor DNA. Most frequently, a naturally occurring circular piece of bacterial DNA, called a *plasmid*, is used for this purpose.

Finally, an enzyme is used to "stitch" the donor DNA into the plasmid vector. This enzyme is called *ligase*, and it creates permanent bonds between the donor DNA and the plasmid DNA. The result is that the donor DNA is incorporated into the bacterial plasmid, forming the recombinant DNA (rDNA).

It is important that the donor and the plasmid DNA be cut with the same restriction enzyme. Since each enzyme cuts DNA only at a specific site, the two different DNAs will have matching cut ends known as "sticky ends." The nitrogenous bases exposed at these cut sites can then match up according to the base-pairing rules, A to T and G to C (Figure 7-12).

The rDNA is then inserted into bacteria. When these bacteria reproduce, they copy the rDNA plasmid along with their own DNA. The plasmid is copied thousands of times, forming a clone (a colony having identical genetic material).

In addition to copying the plasmid along with their other DNA, the bacteria *express* the genes that the plasmid carries, including the donor genes. As they reproduce, the bacteria continue to code for production of the desired protein. In this way, the bacteria can produce human proteins because they carry the genes with the instructions. This technique has made it possible to produce many chemicals that are needed by

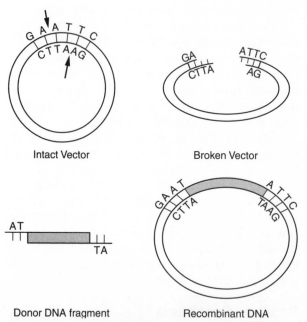

Intact Vector

Broken Vector

Donor DNA fragment

Recombinant DNA

Figure 7-12. Use of a restriction enzyme, plasmid vector, and donor DNA to form recombinant DNA.

people who cannot produce them, due to genetic disorders. Two human proteins that have been successfully synthesized by rDNA techniques are the hormone insulin and human growth hormone.

Electrophoresis

We have already learned that DNA molecules can be cut with specific enzymes known as *restriction enzymes*. These enzymes cleave DNA molecules at highly specific sites that have a certain sequence of bases, such as AAAGGG. Different restriction enzymes have specific restriction sequences that they recognize. Each time the enzyme encounters its unique restriction site it cuts the DNA molecule between a phosphate and a sugar subunit in the backbone. If DNA is incubated with a specific restriction enzyme, the molecule will be cut into many fragments of varying sizes, depending on where the restriction site is located in the DNA molecule.

DNA fragments can be separated according to size because they are electrically charged. The phosphate group at the end of a DNA fragment carries a negative charge; thus, it will be attracted to an area with a positive charge. To separate DNA fragments by means of electrophoresis, a small chamber or box is connected to an electrical source. DNA samples are loaded into small wells within a medium, called a *gel*, which is prepared so that an electric current can pass through it. When the current is switched on, the DNA fragments begin to move in response to the electrical field that is created in the gel. They move away from the wells (the negative end) toward the opposite (positive) end of the box (Figure 7-13).

The size of a DNA fragment depends on the number of base pairs it contains. The more base pairs, the larger and heavier the fragment. Heavier DNA fragments cannot move as far as smaller DNA fragments can within an electrical field. Thus, the smaller fragments will move the farthest from the wells. The result is a series of bands of DNA running from the wells to the other end of the gel. These DNA fragments are made visible to the human eye by means of a simple staining technique, a special chemical, and/or ultraviolet light.

Electrophoresis is useful in determining the sizes of DNA fragments as well as in comparing DNA from two or more different sources. If two different DNA samples (from different individuals) are cut using the same restriction enzyme and then run on a gel in separate wells, we can determine how closely matched the two samples are. If the DNA samples are very similar, they will produce similar-sized fragments that will line up next to each other on the gel. If the two samples are very different, they will produce many different-sized fragments and relatively few will match up. If DNA samples are run from identical twins, all the fragments should match up, since the DNA in these individuals is identical.

Gel electrophoresis has been extremely helpful in law enforcement and forensics. Using this technique, often referred to as "DNA fingerprinting," scientists have been able to match DNA collected at a crime scene with the DNA gathered from a suspect. Biologists also use electrophoresis to determine current and evolutionary relationships among living things.

QUESTIONS

PART A

75. Which diagram illustrates the correct structure of a segment of a DNA molecule? (1) 1 (2) 2 (3) 3 (4) 4

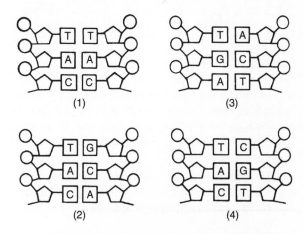

76. DNA and RNA molecules are similar in that they both contain (1) nucleotides (2) a double helix (3) deoxyribose sugars (4) thymine

77. Which series is arranged in correct order according to *decreasing* size of structures? (1) DNA, nucleus, chromosome, nucleotide, nitrogenous base (2) nucleotide, chromosome, nitrogenous base, nucleus, DNA (3) nucleus,

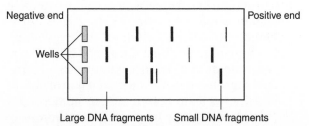

Figure 7-13. Gel electrophoresis: The smaller DNA fragments move farthest from the wells.

chromosome, DNA, nucleotide, nitrogenous base (4) chromosome, nucleus, nitrogenous base, nucleotide, DNA

78. Which substances are components of a DNA nucleotide? (1) phosphate, deoxyribose, and uracil (2) phosphate, ribose, and adenine (3) thymine, deoxyribose, and phosphate (4) ribose, phosphate, and uracil

79. Which two bases are present in equal amounts in a double-stranded DNA molecule? (1) cytosine and thymine (2) adenine and thymine (3) adenine and uracil (4) cytosine and uracil

80. Which process can be used to rapidly produce a group of genetically identical plants from the cells of a single plant? (1) screening (2) karyotyping (3) gene splicing (4) cloning

81. In humans, a gene mutation results from a change in the (1) sequence of the nitrogenous bases in DNA (2) chromosome number in a sperm cell (3) chromosome number in an egg cell (4) sequence of the sugars and phosphates in DNA.

82. Which set of statements correctly describes the relationship among the terms *chromosomes, genes,* and *nuclei*? (1) Chromosomes are found on genes. Genes are found in nuclei. (2) Chromosomes are found in nuclei. Nuclei are found in genes. (3) Genes are found on chromosomes. Chromosomes are found in nuclei. (4) Genes are found in nuclei. Nuclei are found in chromosomes.

83. The genetic code for one amino acid molecule consists of (1) five sugar molecules (2) two phosphates (3) three nucleotides (4) four hydrogen bonds

84. During the replication of a DNA molecule, separation of the DNA molecule will normally occur when hydrogen bonds are broken between (1) thymine and thymine (2) guanine and uracil (3) adenine and cytosine (4) cytosine and guanine

85. In the diagram, what substance is represented by the letter *X*? (1) ribose (2) deoxyribose (3) phosphate (4) adenine

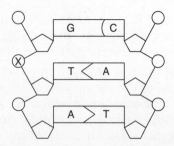

86. Which terms describe gene activities that ensure homeostasis of life processes and continuity of hereditary material? (1) oxidation and hydrolysis (2) enzyme synthesis and DNA replication (3) oxygen transport and cyclosis (4) pinocytosis and dehydration synthesis

87. The formation of recombinant DNA results from the (1) addition of messenger RNA molecules to an organism (2) transfer of genes from one organism to another (3) substitution of a ribose sugar for a deoxyribose sugar (4) production of a polyploid condition by a mutagenic agent

88. The replication of a double-stranded DNA molecule begins when the strands "unzip" at the (1) phosphate bonds (2) ribose molecules (3) deoxyribose molecules (4) hydrogen bonds

89. Cloning an individual usually produces organisms that (1) contain dangerous mutations (2) contain identical genes (3) are identical in behavior (4) produce completely different enzymes

90. The diagram below represents a section of a molecule that carries genetic information. The pattern of numbers represents (1) a sequence of paired bases (2) the order of proteins in a gene (3) folds of an amino acid (4) positions of gene mutations

91. Enzymes are used in moving sections of DNA that code for insulin from the pancreas cells of humans into a certain type of bacterial cell. This bacterial cell will reproduce, giving rise to new cells that can produce (1) human insulin (2) antibodies against insulin (3) enzymes that digest insulin (4) a new type of insulin

92. In the human pancreas, acinar cells produce digestive enzymes and beta cells produce insulin. The best explanation for this is that (1) a mutation occurs in the beta cells to produce insulin when the sugar level increases in the blood (2) different parts of an individual's DNA are used to direct the synthesis of different proteins in different types of cells (3) lowered sugar levels cause the production of insulin in acinar cells to help maintain homeostasis (4) the genes in acinar cells came from one parent, while the genes in beta cells came from the other parent

93. A gene that codes for resistance to glyphosate, a biodegradable weed killer, has been inserted into certain plants. As a result, these plants will be more likely to (1) produce chemicals that kill weeds growing near them (2) die when exposed to glyphosate (3) convert glyphosate into fertilizer (4) survive when glyphosate is applied to them

94. Gel electrophoresis is used to separate DNA fragments on the basis of their (1) size (2) color (3) functions (4) chromosomes

PART B-1

To answer questions 95 through 98, select from the list below the type of nucleic acid that is best described by the phrase. (Note: There are only three answer choices for each question.)

A. DNA

B. Messenger RNA

C. Transfer RNA

95. Genetic material responsible for the traits of an organism, that is passed from parent to offspring (1) A (2) B (3) C

96. Carries genetic information from the cell nucleus out to the ribosomes (1) A (2) B (3) C

97. Contains thymine instead of uracil (1) A (2) B (3) C

98. Carries amino acid molecules to the ribosomes in the cytoplasm (1) A (2) B (3) C

Base your answers to questions 99 through 103 on the following diagram, which represents the process of protein synthesis in a typical cell.

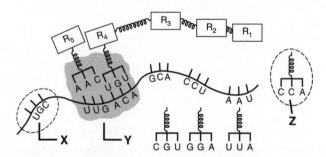

99. The original template for this process is a molecule of (1) DNA (2) messenger RNA (3) transfer RNA (4) ribosomal RNA

100. The units labeled R_1, R_2, and R_3 represent (1) nucleotides (2) RNA molecules (3) DNA molecules (4) amino acids

101. The organelle labeled Y, on which this process occurs, is the (1) nucleus (2) ribosome (3) chloroplast (4) mitochondria

102. The circled portion labeled X is known as (1) an amino acid (2) a codon (3) an anticodon (4) a single nucleotide

103. The circled portion labeled Z represents a molecule of (1) DNA (2) messenger RNA (3) transfer RNA (4) ribosomal RNA

PART B-2

104. Briefly describe two important functions of DNA.

105. Why is DNA replication critical to the survival of organisms?

Base your answers to questions 106 through 108 on the information and chart below and on your knowledge of biology.

In DNA, a sequence of three bases is a code for the placement of a certain amino acid in a protein chain. The table below shows eight amino acids with their abbreviations and DNA codes.

Amino Acid	Abbreviation	DNA Code
Phenylalanine	Phe	AAA, AAG
Tryptophan	Try	ACC
Serine	Ser	AGA, AGG, AGT, AGC, TCA, TCG
Valine	Val	CAA, CAG, CAT, CAC
Proline	Pro	GGA, GGG, GGT, GGC
Glutamine	Glu	GTT, GTC
Threonine	Thr	TGA, TGG, TGT, TGC
Asparagine	Asp	TTA, TTG

106. Which amino acid chain would be produced by the following DNA base sequence?

C-A-A-G-T-T-A-A-A-T-T-A-T-T-G-T-G-A

(1) Val—Glu—Phe—Asp—Thr—Asp

(2) Val—Pro—Phe—Asp—Asp—Thr

(3) Val—Glu—Phe—Asp—Asp—Thr

(4) Val—Glu—Phe—Thr—Asp—Asp

107. Identify one environmental factor that could cause a base sequence in DNA to be changed to a different base sequence.

108. Describe how a protein would be changed if a base sequence mutates from GGA to TGA.

109. Explain the role of each of the following items in making recombinant DNA: *restriction enzymes*, *plasmids*, and *ligase*.

110. How are the techniques of genetic engineering making it possible to treat some diseases caused by genetic disorders? Provide an example.

111. The following is a scrambled list of the techniques used in making recombinant DNA. Write these steps in the correct sequence and, for each step, explain why it is placed in that order.

Step

Cut open plasmid with restriction enzyme

Obtain synthesized protein from the bacteria

Clone bacterial cells with rDNA plasmids

Insert donor DNA into the open plasmid

Cut out donor DNA with restriction enzyme

Add ligase to bond donor DNA and plasmid

112. Animal cells utilize many different proteins. Discuss the synthesis of proteins in an animal cell. Your answer must include:

- the identity of the building blocks required to synthesize these proteins;
- the identity of the sites in the cell where the proteins are assembled;
- an explanation of the role of DNA in the process of making proteins in the cell.

Base your answers to questions 113 through 116 on the information below and on your knowledge of biology. Source: Science News *(July 9, 2005): vol. 168, no. 2, p. 19.*

Same Difference: Twins' Gene Regulation Isn't Identical

Although identical twins have identical DNA, they often harbor clear-cut differences: slight variations in appearance or stark distinctions in disease susceptibility, for example. Scientists have suggested that the interplay between nature and nurture could explain such differences, but the mechanism has been poorly understood.

A new study suggests that as identical twins go through life, environmental influences differently affect which genes are turned on and which are switched off.

Called epigenetic modification, such gene activation or silencing typically stems from two types of chemical groups that latch on to chromosomes as charms attach to a bracelet, says Manel Esteller of the Spanish National Cancer Centre in Madrid. Methyl groups that clip on to DNA tend to turn genes off. On the other hand, acetyl groups attaching to histones, the chemical core of chromosomes, usually turn genes on.

Suspecting that such epigenetic differences might account for variations between identical twins, Esteller and his team focused on the two chemical changes. The scientists recruited 80 pairs of identical twins, ranging in age from 3 to 74, from Spain, Denmark, and the United Kingdom.

After extracting DNA from blood, inner-cheek cells, and biopsied muscle, Esteller's team screened the twins' genomes for differences in epigenetic profiles between members of a pair. The researchers also had each twin or, for children, a parent answer a comprehensive questionnaire on the twins' health history and lifestyle, including diet, exercise habits, and alcohol or tobacco use.

In the youngest twins, the scientists found relatively few epigenetic differences. However, the number of differences increased with the age of the twins examined. The number of epigenetic differences in 50-year-old twins was more than triple that in 3-year-old twins. Esteller's group also saw especially large epigenetic differences between twins who had spent most of their lifetimes apart, such as those adopted by different sets of parents at birth, the team reports in an upcoming *Proceedings of the National Academy of Sciences*.

Esteller says that these results suggest that a person's environment—whether he or she is exposed to tobacco smoke, eats particular foods, or suffers an emotionally wrenching event, for example—may affect which genes are turned on or off and so how cells operate. Thus, nurture may have a heavy impact on an individual's nature.

"My belief is that people are 50 percent genetics and 50 percent environment," says Esteller. "It's important to remember that our genes give us features of who we are, but our environment can change how we are."

Arturas Petronis, who studies epigenetics at the Centre for Addiction and Mental Health in Toronto, agrees. He adds that the findings could also have wide-ranging health implications for people who aren't twins.

"About 90 percent of diseases don't follow [simple] rules for inheritance," says Petronis. "By investigating epigenetic changes, to some extent we can understand how environmental factors affect human health."

113. How can environmental influences cause genetic differences between identical twins?

114. Compare the functions of the two DNA modifiers: methyl groups and acetyl groups.

115. Describe the trend that researchers found between the age of the twins and the number of epigenetic differences in their DNA. What could account for this trend?

116. Why does the researcher say his "belief is that people are 50 percent genetics and 50 percent environment"?

CHAPTER 8 Evolution

Evolution is the process of change over time. The theory of evolution suggests that existing forms of life on Earth have evolved from earlier forms over long periods of time. These earlier forms were usually very different from the related organisms living today. Evolution accounts for the differences in structure, function, and behavior among all life forms, as well as for the changes that occur within populations over many generations.

EVIDENCE OF EVOLUTION

Observations that support the theory of evolution have been obtained from the study of the geologic record and from studies of comparative anatomy, embryology, cytology, and biochemistry. It is important to understand that the word "theory" in science has a very different meaning from the same word in everyday language. A scientific theory is a concept that is supported by much documented evidence and experimental research. Theories are based on the culmination of years of laboratory work, field observation, and extensive testing in order to validate results.

Geologic Record

Geologists estimate the age of Earth to be between 4.5 and 6 billion years old. This estimate is based on *radioactive dating* of the oldest known rocks from Earth's crust. (It is assumed that Earth is at least as old as the oldest rocks and minerals in its crust.)

In studying the geology of the planet, scientists have found many fossils, the remains or traces of organisms that no longer exist. From their studies of rocks and fossils, scientists have developed a picture of the changes that have occurred both in Earth itself and in living things on the planet.

Fossils

The earliest known fossils are traces of bacteria-like organisms that are about 3.5 billion years old. (The age of these fossils was determined by radioactive dating of the rocks in which they were found.)

Fossils of relatively intact organisms have been found preserved in ice, tar, and amber (a sticky plant resin that hardens). Mineralized bones, shells, teeth, and other hard parts of ancient organisms are sometimes found intact. (The soft parts generally decay within a short time.)

Other fossils have been formed by *petrifaction*, a process in which the tissues are gradually replaced by dissolved minerals that produce a stone replica of the original material. Imprints, casts, and molds of organisms or parts of organisms are frequently found in *sedimentary* rock. This type of rock is formed from the deposition of thick layers of soft sediments that eventually harden and turn to rock from the weight of overlying sediments and water. The fossils form when the remains of dead organisms settle to the bottom of a body of water and are quickly covered by sediment. The overlying sediment slows or halts decay. When the layers of sediment harden, traces of the buried organisms are preserved in the rock.

In studying undisturbed sedimentary rock, scientists assume that each layer is older than all the layers, or *strata*, above it. Thus, fossils in the lower strata are older than fossils in the overlying strata. Fossils in the upper strata of a sedimentary rock sample are generally more complex than fossils in the lower strata, but there is often a resemblance between them. This suggests a link between recent forms and older forms of life. The fossil record may also provide evidence of divergent evolutionary pathways of some organisms from a common ancestor (Figure 8-1).

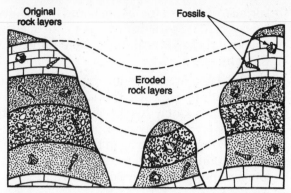

Figure 8-1. A resemblance between fossils in the upper and lower rock strata often indicates an evolutionary link between recent and older life forms.

Some fossils in older strata are unlike any organisms living today. This suggests that many previous species have died out, or gone *extinct*, over time; **extinction** means there are no longer any living members of the species. Other fossils have structures that show ancestral connections to present-day life forms. On the other hand, there are fossils that are quite similar to modern organisms, suggesting that some species have existed for a long time without much evolutionary change.

Comparative Anatomy

Another line of evidence for evolution comes from observations of basic structural, or anatomical, similarities between various organisms. *Homologous structures* are anatomical parts found in different organisms that are similar in origin and structure, although they may differ in function. For example, the flippers of whales, the wings of bats, the forelimbs of cats, and the arms of humans are homologous structures; they serve different functions, but their basic bone structures are similar (Figure 8-2). The presence of such homologous

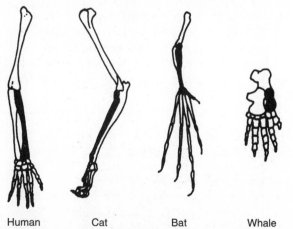

Human   Cat   Bat   Whale

Figure 8-2. The presence of homologous structures in their limb bones suggests that these mammals all evolved from a common ancestor.

structures suggests that these mammals all evolved from a common ancestor.

Comparative Embryology

Although adult organisms of different species may look very different from one another, a comparison of the early stages of their embryonic development may show similarities that suggest a common ancestry. For example, the very early embryos of such vertebrates as fish, reptiles, birds, and mammals show some similarities in structure, such as having a tail (Figure 8-3). As embryonic development continues, the characteristic traits of each species become more apparent.

Comparative Cytology

As stated in the cell theory, all living things are made up of cells. Cell organelles, including the cell membrane, ribosomes, and mitochondria, are structurally and functionally similar in most living organisms.

Comparative Biochemistry

All living things contain similar biochemical compounds. For example, the structure and function of DNA, RNA, and proteins (including enzymes) are similar in all organisms. The closer the relationship between any two organisms, the greater are their biochemical and genetic similarities. Today, biochemical relationships can be shown by means of gel electrophoresis. If two organisms

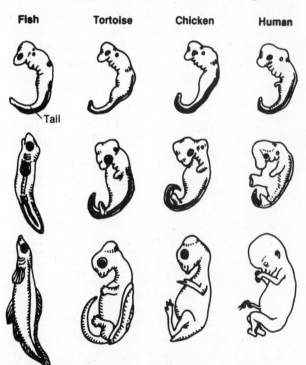

Fish   **Tortoise**   **Chicken**   **Human**

Tail

Figure 8-3. Similar features, such as a tail, in the early stages of embryonic development, point to a common ancestry for these different vertebrates.

are closely related, their DNA samples will show very similar banding patterns on the gel. Similarly, the proteins (such as enzymes) of the two organisms would be very similar, since a major function of DNA is the synthesis of proteins.

QUESTIONS

PART A

1. The front legs of a frog and the front legs of a horse are examples of structures that are (1) heterotrophic (2) homozygous (3) hermaphroditic (4) homologous

2. Which conclusion may be made when comparing fossils in undisturbed strata of sedimentary rock? (1) The fossils in the upper strata are younger than those in the lower strata. (2) The fossils in the upper strata are older than those in the lower strata. (3) The fossils in the upper strata are generally less complex than those in the lower strata. (4) There are no fossils in the upper strata that resemble those in the lower strata.

3. The similarity among the blood proteins of all mammals may be taken as evidence of evolutionary relationships based on (1) comparative anatomy (2) geographic distribution (3) comparative embryology (4) comparative biochemistry

4. The diagram below represents a section of undisturbed rock and the general location of fossils of several closely related species. According to current theory, which assumption is probably the most correct concerning species *A*, *B*, *C*, and *D*? (1) *A* is most likely the ancestor of *B*, *C*, and *D*. (2) *B* was extinct when *C* evolved. (3) *C* evolved more recently than *A*, *B*, and *D*. (4) *D* was probably the ancestor of *A*, *B*, and *C*.

Species C & D
Species C
Species A & B & C
Species A & B
Species A

5. Which assumption is a basis for the use of fossils as evidence for evolution? (1) Fossils show a complete record of the evolution of all animals. (2) In undisturbed layers of Earth's

surface, the oldest fossils are found in the lowest layers. (3) Fossils are always found deep in volcanic rocks. (4) All fossils were formed at the same time.

6. Many related organisms are found to have the same kinds of enzymes. This suggests that (1) enzymes work only on specific substrates (2) enzymes act as catalysts in biochemical reactions (3) organisms living in the same environment require identical enzymes (4) these organisms probably share a common ancestry

7. Which is an example of evidence of evolution based on comparative biochemistry? (1) Sheep insulin can be substituted for human insulin. (2) The structure of a whale's flipper is similar to that of a human hand. (3) Human embryos have a tail during an early stage of their development. (4) Both birds and bats have wings.

8. The presence of gill pouches in early-stage human embryos is considered to be evidence of the (1) likelihood that all vertebrates share a common ancestry (2) theory that the first organisms on Earth were heterotrophs (3) close relationship between fish and human reproductive patterns (4) close relationship between humans and amphibians

9. The proteins in a chimpanzee's blood are shown, through gel electrophoresis, to be similar to those in a human's blood. This is an example of which type of evidence supporting the theory of evolution? (1) comparative habitat (2) comparative anatomy (3) comparative embryology (4) comparative biochemistry

PART B-2

10. State the five main types of evidence used to support the theory of evolution.

11. The diagram below represents a cross section of undisturbed rock layers. A scientist discovers bones of a complex vertebrate species in layers *B* and *C*. In which layer would an earlier, less complex form of this vertebrate most

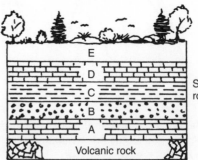

Sedimentary rock layers

Volcanic rock

likely first appear? Explain the reasons for your answer.

12. *R, S,* and *T* are three species of birds. Species *S* and *T* show similar coloration. The enzymes found in species *R* and *T* show similarities. Species *R* and *T* also exhibit many of the same behavioral patterns. Show the relationship between species *R, S,* and *T* by placing the letter representing each species at the top of the appropriate branch on the diagram below.

13. Frogs and lizards are very different animals, yet they have many similarities. Provide a brief explanation for why frogs and lizards have many characteristics in common.

THEORIES OF EVOLUTION

Any theory of evolution must attempt to explain the origin and **diversity** of life on Earth. Such a theory must also account for the wide variety of *adaptations* found among both living and extinct species. Different theories may account for different aspects of the evolutionary process. However, taken together, they can explain how life on Earth came to be and how it has progressed from the relatively simple and few to the complex and diverse.

THE HETEROTROPH HYPOTHESIS

The *heterotroph hypothesis* is one proposed explanation of how life arose and evolved on the primitive Earth. According to this hypothesis, the first life forms were heterotrophic and therefore had to obtain organic nutrients from their environment.

The Primitive Earth

It is assumed that during the period preceding the development of the first life forms, the primitive Earth was an exceptionally hot body consisting of inorganic substances in solid, liquid, and gaseous states.

The **atmosphere** of primitive Earth is thought to have had no free oxygen; instead, it consisted of hydrogen (H_2), ammonia (NH_3), methane (CH_4), and water vapor (H_2O). As Earth cooled, much of the water vapor condensed and fell as rain, which carried dissolved atmospheric gases (ammonia, methane, and hydrogen) and some minerals into the seas that formed. The seas became rich in these dissolved substances and minerals and are often described by biologists as having been a "hot, thin soup."

The primitive Earth provided an energy-rich environment. In addition to the heat, there was electrical energy in the form of lightning, **radiation** (x-rays and ultraviolet rays) from the sun, and radioactivity from rocks.

Synthesis of Organic Compounds

The large amount of available energy was the driving force for synthesis reactions on the primitive Earth. In these reactions, the inorganic raw materials in the seas became chemically bonded to form organic molecules, including simple sugars and amino acids. These organic molecules were the building blocks for the first life forms.

The scientist Stanley Miller devised an apparatus in which he simulated the conditions thought to exist in the primitive environment. His experiments showed that in the presence of heat and electrical energy, dissolved gases could combine to form simple organic compounds.

Formation of Aggregates

In time, the simple organic molecules accumulated in the seas. Eventually, they combined in synthesis reactions to form more complex organic molecules. (Such interactions between organic molecules have been demonstrated in laboratories.) Some of the large, complex molecules formed groupings, or clusters, called *aggregates*. These aggregates developed a membrane that enclosed them, thus forming a barrier between themselves and the surrounding water. This made it possible for the substances inside an aggregate to remain separate from those in the surrounding water. It is thought that aggregates absorbed simple organic molecules from the environment for "food." Thus, they carried on a form of heterotrophic nutrition. Over time, the aggregates became more complex and highly organized. Eventually, they developed the ability to reproduce. At that point, when their ability to reproduce had evolved, the aggregates are considered to have been living cells.

Heterotrophs to Autotrophs

It is thought that these early heterotrophic life forms carried on a form of anaerobic respiration, or *fermentation* (in which glucose is converted to energy and CO_2 without O_2 being present). As a result of very long periods of fermentation, carbon dioxide was added to the atmosphere. Eventually, as a result of evolution, some heterotrophic forms developed the capacity to use carbon dioxide from the atmosphere in the synthesis of organic compounds. These organisms became the first *autotrophs* (meaning "self-feeders"). Some bacteria are autotrophs, but most of the autotrophs alive today are green plants and algae.

Anaerobes to Aerobes

Autotrophic activity (photosynthesis) added oxygen molecules to the atmosphere. Over time, the capacity to use free oxygen in respiration (aerobic respiration) evolved in both autotrophs and heterotrophs.

There are both autotrophs and heterotrophs on Earth today. Some life forms still carry on anaerobic respiration; but in most life forms, respiration is aerobic. This is because aerobic respiration releases much more energy from food than does anaerobic respiration.

QUESTIONS

PART A

14. According to the heterotroph hypothesis, the first living things probably were anaerobic because their environment had no available (1) food (2) energy (3) water (4) oxygen

15. Which is one basic assumption of the heterotroph hypothesis? (1) More complex organisms appeared before less complex organisms. (2) Living organisms did not appear until there was oxygen in the atmosphere. (3) Large autotrophic organisms appeared before small photosynthesizing organisms. (4) Autotrophic activity added oxygen molecules to the environment.

16. The heterotroph hypothesis is an attempt to explain (1) how Earth was originally formed (2) why simple organisms usually evolve into complex organisms (3) why evolution occurs very slowly (4) how life originated on Earth

17. The heterotroph hypothesis states that heterotrophic life forms appeared before autotrophic forms as the first living things. A major assumption for this hypothesis is that (1) sufficient heat was not available in the beginning

for the food-making process (2) the heterotrophic organisms were able to use molecules from the sea as "food" (3) lightning and radiation energy were limited to terrestrial areas (4) moisture in liquid form was limited to aquatic areas

Base your answer to the following question on the chart below and on your knowledge of biology.

A	B	C
The diversity of multicellular organisms increases.	Simple, single-celled organisms appear.	Multicellular organisms begin to evolve.

18. According to most scientists, which sequence best represents the order of biological evolution on Earth? (1) $A \rightarrow B \rightarrow C$ (2) $B \rightarrow C \rightarrow A$ (3) $B \rightarrow A \rightarrow C$ (4) $C \rightarrow A \rightarrow B$

PART B-2

19. Identify the source of oxygen in Earth's early atmosphere; tell how this was important later to the evolution of life.

EVOLUTION BY NATURAL SELECTION

Darwin's Theory of Natural Selection

Darwin's theory was based on the presence of variations among members of a population and their interaction with the process he called natural selection. Darwin's theory includes the following main ideas:

Overpopulation: Within a population, there are more offspring produced in each generation than can possibly survive.

Competition: The natural resources, such as food, water, and space, available to a population are limited. Because there are more organisms produced in each generation than can survive, there must be *competition* among them for the resources needed for survival.

Survival of the fittest: Variations among members of a population make some of them better adapted to the environment than others. Such variability within populations means that, due to competition, the best-adapted individuals are most likely to survive.

Natural selection: The environment is the agent of natural selection, determining which adaptations or variations are helpful and which are harmful. For example, in an environment that is undergoing a particularly cold period, animals that have thicker fur than most other members of their population are more likely to survive. In this case, their variation—thicker fur—is helpful in terms of surviving the environmental pressure.

Reproduction: Individuals with useful variations tend to survive and reproduce at a higher rate than other members of their population, thus transmitting these **adaptations** to their offspring. Likewise, those individuals that do not have such favorable adaptations tend to die off within the population; so the less favorable traits are not passed on to future generations.

Speciation: The development of new species, a process called *speciation*, occurs as certain variations or adaptations accumulate in a population over many generations.

According to Darwin's theory, environmental pressures act as a force for the natural selection of the best-adapted individuals in a population—those with helpful adaptations that enable them to survive and reproduce successfully. However, Darwin's theory did not explain *how* variations arise in members of a species. (At Darwin's time—in the mid- to late-1800s—the scientific study of genes and mutations had not yet begun.) It is important to understand that *individuals* do not evolve; rather it is the *population* that evolves as the percentage of changes in its gene pool increases over time.

QUESTIONS

PART A

20. Darwin's theory of evolution did *not* contain the concept that (1) genetic variations are produced by mutations and sexual recombination (2) organisms that are best adapted to their environment survive (3) population sizes are limited due to the struggle for survival (4) favorable traits are passed from one generation to the next

21. Natural selection is best defined as (1) survival of the strongest organisms only (2) elimination of the smallest organisms by the largest organisms (3) survival of those organisms best adapted to their environment

(4) reproduction of those organisms that occupy the largest area in an environment

22. Although similar in many respects, two species of organisms exhibit differences that make each one well adapted to the environment in which it lives. The process of change that helps account for these differences is (1) evolution by natural selection (2) parthenogenesis (3) comparative embryology (4) inheritance of acquired traits

23. A key idea in Darwin's theory of evolution is that members of a population (1) are always identical (2) compete for limited resources in the environment (3) all get to reproduce and pass on their traits (4) are all equally well adapted to the environment

24. The development of new species, as variations accumulate in a population over time, is called (1) competition (2) reproduction (3) speciation (4) selection

25. Which characteristics of a population would most likely indicate the lowest potential for evolutionary change in that population? (1) sexual reproduction and few mutations (2) sexual reproduction and many mutations (3) asexual reproduction and few mutations (4) asexual reproduction and many mutations

26. The theory of biological evolution includes the concept that (1) species of organisms found on Earth today have adaptations not always found in earlier species (2) fossils are the remains of present-day species and were all formed at the same time (3) individuals may acquire physical characteristics after birth and pass these acquired characteristics on to their offspring (4) the smallest organisms are always eliminated by the larger organisms within the ecosystem

27. The graph on page 122 shows the populations of two species of ants. Ants of species 2 have a thicker outer covering than the ants of species 1. The outer covering of an insect helps prevent excessive evaporation of water. Which statement would best explain the population changes shown in the graph? (1) The food sources for species 1 increased while the food sources for species 2 decreased from January through November. (2) Disease killed off species 1 beginning in May. (3) The weather was hotter and drier than normal from April through September. (4) Mutations occurred from April through September in both species, resulting in their both becoming better adapted to the environment.

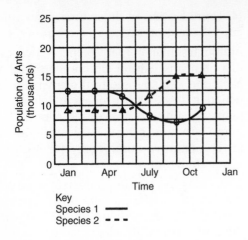

Key
Species 1 ———
Species 2 - - -

28. When a particular white moth lands on a white birch tree, its light color is beneficial, or *adaptive*, for survival. If the birch trees become covered with black soot, the white color of this particular moth in this environment would most likely (1) remain just as adaptive for survival (2) become more adaptive for survival (3) change to black to be more adaptive for survival (4) become less adaptive for survival

29. A variation causes the production of an improved variety of apple. What is the best method to use to obtain additional apple trees of this variety in the shortest period of time? (1) selective breeding (2) natural selection (3) asexual reproduction (4) hormone therapy

30. In order for new species to develop, there *must* be a change in the (1) temperature of the environment (2) migration patterns within a population (3) genetic makeup of a population (4) rate of succession in the environment

31. The diagram below shows the evolution of some different species of flowers. Which statement about the species is correct? (1) Species *A, B, C,* and *D* came from different ancestors. (2) Species *C* evolved from species *B*. (3) Species *A, B,* and *C* can interbreed successfully. (4) Species *A* became extinct.

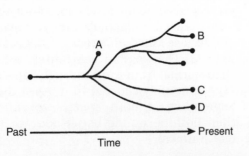

32. State the four main ideas that make up Darwin's theory of evolution by natural selection.

33. Explain what is meant by "the environment is the agent of natural selection."

MODERN EVOLUTIONARY THEORY

The modern theory of evolution includes both Darwin's ideas of variation and natural selection and the genetic basis of variations within populations.

Sources of Genetic Variations

Variations within a population result from two kinds of genetic events. First, recombination of alleles during sexual reproduction is a source of variations. Second, random and spontaneous gene and chromosome mutations produce genetic variations. Mutations may arise spontaneously in organisms, or they may be caused by exposure to *mutagenic* (mutation-causing) chemicals or radiation, such as ultraviolet rays and x-rays. These variations provide the raw material for evolution within a population.

Natural Selection and Genetic Variation

Natural selection involves the struggle of organisms to survive and reproduce in a given environment. In order for natural selection to occur, the mutations in genes or chromosomes must have an effect on the phenotype (outward appearance or behavior) of an organism. This is because natural selection works by "choosing" favorable traits that are expressed. As a result, individuals having favorable **genetic variations** are more likely to survive, reproduce, and pass those traits on to future generations. If a genetic change results in a *silent mutation*, the variation is not expressed in the phenotype; natural selection cannot work on this type of mutation.

Favorable Variations. Favorable characteristics tend to increase in (genetic) frequency within a population. Favorable variations may include physical traits, such as larger muscles and increased speed, or behavioral traits, such as better food-finding or nest-building skills.

If environmental conditions change, traits that formerly had low survival value may come to have greater survival value. Likewise, traits that were favorable may no longer be so adaptive. The survival value of traits that had been neither helpful nor harmful may also change. In all of these cases, those traits that prove to be favorable under the new environmental conditions will increase in frequency within the population.

Unfavorable Variations. Unfavorable characteristics tend to decrease in frequency from generation to generation. Individuals with non-adaptive or unfavorable traits may be so severely selected against that, over time, populations that have unfavorable traits may become extinct. Indeed, the fossil record shows that extinction is a fairly common event, having been the fate of about 99 percent of all species that have ever existed on Earth.

Geographic Isolation

Changes in gene frequencies that lead to the development of a new species are more likely to occur in small populations than in large ones. Small groups may be segregated from the main population by a geographic barrier, such as a body of water or a mountain range. As a result of this *geographic isolation,* the small population cannot interbreed with the larger, main population. In time, the isolated population may evolve into a new species.

The following factors may be involved in the evolution of a new species: (a) the gene frequencies in the isolated population may already have been different from the gene frequencies in the main population, a difference known as the *founder effect*; (b) different mutations occur in the isolated population and the main population; and (c) different environmental factors exert different selection pressures on each population. Since there is no interbreeding between the two populations, any mutations that occur in one population cannot be transmitted to the other. Over long periods of time, the two populations may become so different that they will no longer be able to interbreed even if direct contact is made. In such a case, two new species have evolved from the one. An example of this is seen in the two populations of Grand Canyon squirrels: the Kaibab and the Abert. These two squirrel populations were originally members of one species that became separated, over time, by the formation of the canyon. As a result of natural selection, the divided populations evolved to be the two different species that exist today.

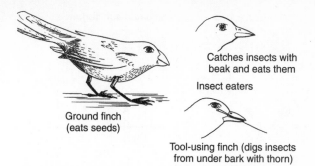

Figure 8-4. These Galápagos finches show a variety of adaptations for getting food in their particular environments.

Darwin observed the effect of geographic isolation among the finches he collected on the Galápagos Islands. Darwin hypothesized that the 14 different species he observed had evolved from a single species that had originally migrated to the islands from the mainland of South America. Over time, the different environment of each island had gradually resulted in the evolution of a new separate species (Figure 8-4).

Reproductive Isolation

Geographic isolation may eventually lead to *reproductive isolation*. The isolated population becomes so different from the main population that members of the two groups cannot interbreed, even if the geographic barriers were to be removed. When two populations can no longer interbreed and produce fertile offspring, they have become two distinct species.

Time Frame for Evolution

Although scientists generally agree on the basic factors involved in evolutionary change, there is some disagreement about the time frame in which such change occurs.

According to Darwin's original theory, evolutionary change occurs very gradually and continuously over the course of **geologic time** (millions of years). This theory, called *gradualism*, proposes that new species develop as a result of the gradual accumulation of small genetic variations that eventually, together, cause reproductive isolation and lead to speciation.

The more recent theory of *punctuated equilibrium* proposes that most species have long periods (several million years) of relative stability, or stasis, interrupted by geologically brief periods during which major changes occur, possibly leading to the evolution of new species (Figure 8-5, page 124). In this way, drastic environmental changes, for example, a global cooling event, could cause species to evolve—or become extinct.

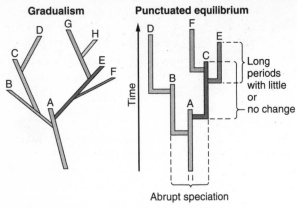

Gradualism

D G H
C   E
B   F
   A

Punctuated equilibrium

D   F
         E
      C
   B
   A
Time →

Long
periods
with little
or
no change

Abrupt speciation

Figure 8-5. Two evolutionary processes: gradualism and punctuated equilibrium.

In the fossil records of some evolutionary lineages, there are transitional forms that support the theory of gradualism. However, in many evolutionary lineages, there is an apparent lack of transitional forms, which better supports the theory of punctuated equilibrium.

Impact of Humans on Natural Selection

It has been found that some insects have a genetic mutation that makes them resistant to the effects of insecticides (the group of **pesticides** developed to kill insect pests). Before the widespread use of insecticides, this trait was of no particular survival value. With the increased use of insecticides, however, this trait developed a very high survival value. Because the insects that are resistant to insecticides have survived and reproduced, the frequency of insecticide resistance has increased greatly in insect populations.

Resistance to **antibiotics** (drugs that fight bacterial infections) in populations of bacteria has followed the same pattern. The frequency of resistant individuals in bacterial populations has increased with the increasing use of certain antibiotics.

It is important to note that resistance to insecticides and antibiotics did not arise as a result of exposure to these substances. The traits were already present in some members of the organisms' populations, and the insecticides and antibiotics simply acted as the selecting agents.

Humans and Artificial Selection

In nature, there is no particular direction in which each species must evolve. Over time, many variations appear within the populations of organisms, similar to the branching of twigs on a tree. Natural selection continuously "prunes" these branches, or lineages, eliminating those with unfavorable adaptations while letting those with favorable adaptations survive.

In contrast, humans can and do have an effect on the inheritance of traits in some populations of organisms. For example, human actions, such as pesticide use, have led to unexpected changes in the genetic makeup of some insect populations. Modern humans have intentionally altered the traits of many plant and animal species, as well. In the process of domesticating organisms, people have selectively bred plants and animals for desired traits. In such cases, it is a person, not the environment, that is the selecting agent. Advances in **biotechnology**, or bioengineering techniques, have also had an impact on the genetic traits of some plant and animal populations.

QUESTIONS

PART A

34. A population of mosquitoes is sprayed with a new insecticide. Most of the mosquitoes are killed, but a few survive. In the next generation, the spraying continues, but still more mosquitoes hatch that are immune to the insecticide. How could these results be explained according to the present concept of evolution? (1) The insecticide caused a mutation in the mosquitoes. (2) The mosquitoes learned how to fight the insecticide. (3) A few mosquitoes in the first population were resistant and transmitted this resistance to their offspring. (4) The insecticide caused the mosquitoes to develop an immune response, which was inherited.

35. What would be the most likely effect of geographic isolation on a population? (1) It has no effect on variations in the species. (2) It favors the production of new species. (3) It prevents the occurrence of mutations. (4) It encourages the mixing of gene pools.

36. Two organisms can be considered to be of different species if they (1) cannot mate with each other and produce fertile offspring (2) live in two different geographical areas (3) mutate at different rates depending on their environment (4) have genes drawn from the same gene pool

37. Certain strains of bacteria that were susceptible to penicillin have now become resistant. The probable explanation for this is that (1) the gene mutation rate must have increased naturally (2) the strains have become resistant because they needed to do so for survival (3) a mutation that gave some of them resistance was passed on to succeeding generations because it had high survival value (4) the penicillin influenced the bacterial pattern of mating

38. The continents of Africa and South America were once a single landmass but have drifted

apart over millions of years. The monkeys of both continents, although similar, show several genetic differences from each other. Which factor is probably the most important for causing and maintaining these differences? (1) fossil records (2) comparative anatomy (3) use and disuse (4) geographic isolation

39. A change in the frequency of any mutant allele in a population most likely depends on the (1) size of the organisms possessing the mutant allele (2) adaptive value of the trait associated with the mutant allele (3) degree of dominance of the mutant allele (4) degree of recessiveness of the mutant allele

40. Modern evolutionary biologists have accepted the main ideas of Darwin's theory of evolution but have added genetic information that gives a scientific explanation for (1) overproduction (2) the struggle for existence (3) the survival of the fittest (4) variations

41. As a result of sexual reproduction, the potential for evolutionary change in plants and animals is greatly increased because (1) the offspring show more variability than those from asexual reproduction (2) characteristics change less frequently than in asexual reproduction (3) environmental changes never affect organisms produced by asexual reproduction (4) two parents have fewer offspring than one parent

42. Populations of a species may develop traits that are different from each other if they are geographically isolated for sufficient lengths of time. The most likely explanation for these differences is that (1) acquired traits cannot be inherited by the offspring (2) the environmental conditions in the two areas are identical (3) mutations and selective forces will be different in the two populations (4) mutations will be the same in both populations

PART B-2

43. State how genetic variations and natural selection in a population can lead to the evolution of a new species.

44. How did Darwin explain the evolution of 14 finch species from one ancestral finch species? Use the terms *geographic isolation* and *reproductive isolation* in your answer.

PART C

Base your answers to questions 45 through 48 on the following information and data table.

A population of snails was living on a sandy beach. The snails' shells appeared in two colors: tan or black. The sand on the beach was a tan color. One day, a volcano in a nearby mountain range erupted, spewing out tons of ash and debris. The ash and debris coated the sand on the beach, blackening it. Biologists had kept careful records of the snail population before and after the volcanic eruption. Their data are presented in the table below.

Time	Number of Tan Snails	Number of Black Snails
Before volcano erupted	6000	50
After volcano erupted (one year later)	400	3000

45. Explain why the numbers of tan snails and black snails changed.

46. How does this event support the idea of evolution by natural selection?

47. Give one reason why the tan snails might disappear within a few years.

48. Using the data in the preceding table, prepare a bar graph that shows the information on snail populations before and after the volcanic eruption.

49. A species of wildflower grows in a meadow. The flowers are of two color varieties: yellow and purple. There are about equal numbers of yellow flowers and purple flowers. A biologist observes that bees frequently visit the yellow flowers but seldom go to the purple ones.

Use the above data and your knowledge of biology to write a brief experimental procedure that addresses the following:

• a question prompted by the information given;

• a hypothesis that addresses your question;

• a brief experimental procedure that could be used to test your hypothesis;

• a description of the main selecting force on the flowers in this meadow;

• a prediction of what may happen to this population of wildflowers in 50 years.

50. Describe how the continued widespread use of antibiotics may result in the evolution of more resistant strains of bacteria. How does the antibiotic act as a selecting agent? How does this illustrate the concept of natural selection?

Base your answers to questions 51 through 54 on the information below and on your knowledge of biology. Source: Science News *(May 21, 2005): vol. 167, no. 21, p. 324.*

New Mammals: Coincidence [and] Shopping Yield Two Species

After 21 years without a new kind of monkey being reported in Africa, two research teams working independently in different mountain ranges have described the same novel species. And other researchers, after poking through meat for sale in Southeast Asia, report a rodent that they say justifies a new family among mammals, the first in 31 years.

The monkey species, now called the highland mangabey or *Lophocebus kipunji*, has turned up at locations in southern Tanzania 370 kilometers apart. One discovery came from scientists' curiosity about stories around Mount Rungwe of an elusive monkey. In December 2003, Tim Davenport of Mbeya, Tanzania, who works for the New York–based Wildlife Conservation Society, and his team got a good-enough look to recognize it as a new species.

Meanwhile, ornithologists had told Trevor Jones at Udzungwa Mountains National Park in Tanzania that they had spotted sanje mangabeys, an endangered monkey, in a remote forest. But as soon as Jones saw one of the mangabeys there, he says that he knew the brownish color and high crest of hair were all wrong for a sanje mangabey. "I was immediately gobsmacked," he says.

Last October, in Dar es Salaam, one of Jones' colleagues happened to be in the same hotel as Davenport. Jones says that a conversation in the bar, with veiled hints of working on "something a bit special," escalated to revelations that both teams had found the same mangabey species. The Jones team withdrew a paper that it had previously submitted, and the two groups united to describe the new species in the May 20 *Science*.

There's "no question" about it being a new species, comments primate systematist Colin Groves of the Australian National University in Canberra. During the past 2 decades, monkey species not previously described by scientists have turned up only rarely, he says, and the finding of one in Africa surprised him. "The startling new discoveries have mainly been in Asia," he notes.

Loggers are cutting down the forests where the highland mangabey was found near Mount Rungwe, and Davenport calls for immediate protection of the monkey's already fragmented habitat.

The other new species announced this month, an unusual rodent, was reported by a team surveying biodiversity in the forests of Khammouan province in the Lao People's Democratic Republic. The Wildlife Conservation Society–sponsored survey included routine trips through village food markets. Starting in 1996, Robert J. Timmins and others occasionally bought what the local people call the *kha-nyou*. Dark fur covers a somewhat ratlike body about 25 centimeters long with a furry tail.

The biodiversity surveyors sent market specimens to Paulina Jenkins of the Natural History Museum in London. Details of the bones and teeth, plus DNA analysis, place the animal in the new family Laonastidae, the researchers argue in the latest quarterly issue of *Systematics and Biodiversity*, dated December 2004. They've christened the rodent *Laonastes aenigmamus*.

Dorothée Huchon, a rodent specialist at Tel Aviv University in Israel, says that the DNA evidence hasn't yet convinced her that this should be a new family. However, the rodent group "is full of surprises," she says.

51. What two features about the supposed sanje monkey (seen in the Udzungwa Mountains National Park) aroused Trevor Jones' suspicions that this could be a new species of mangabey?

52. How can the scientists be sure that this monkey is indeed a new species of mangabey?

53. Why is it critical that the highland mangabey's habitat be protected? How is the area now threatened?

54. What three pieces of evidence indicate that the new rodent species *Laonastes aenigmamus* probably should be placed in a new taxonomic family?

CHAPTER 9

Humans and the Environment

Humans, more than any other organisms, have the capacity to change the environment. Some human activities have a negative effect on the environment, while other activities have a positive effect. Many other living things can make changes to their environment in order to ensure their survival, such as beavers building dams. Yet no creature can make the dramatic, large-scale changes that humans can; and these changes usually have an effect on other organisms as well.

PEOPLE AFFECT THE ENVIRONMENT

Some human activities have upset the natural balance of ecosystems. These activities have brought about undesirable and lasting changes in one or more of the biotic or abiotic factors in some ecosystems, harming humans and other living things.

Human Population Growth. The human population of Earth is increasing at a rapid rate (Figure 9-1). A major factor in this increase involves medical advances that have increased human survival rates and the average life span. In most parts of the world, population growth is no longer limited by disease to the extent that it was in the past. However, in many places, the population has grown faster than the food-producing capacity, resulting in hunger and starvation. In addition, as the human population grows, more people move into and alter wilderness areas, resulting in a loss of natural habitat for wildlife.

Human Activities. Some human activities have led to the endangerment or extinction of numerous species of plants and animals, and also

have produced less favorable living conditions for many species. Such activities include over-hunting, importation of organisms, exploitation of wild organisms, poor land-use practices, and technological oversights.

Uncontrolled hunting, fishing, and trapping, which still occur in many parts of the world, have resulted in the extinction of some species and the endangerment of others. Several vertebrate species, such as the passenger pigeon and Steller's sea cow, have been hunted to extinction already; hundreds of other animals are currently listed as threatened or endangered as a result of such human activities.

Humans have both accidentally and intentionally imported species into areas where they have no natural enemies or competition. These imported organisms have increased in numbers, leading to the disruption of existing ecosystems. Imported organisms that have caused serious damage include the European starling, Japanese beetle, gypsy moth, zebra mussel, and various plant species, such as the purple loosestrife. This plant thrives in the wild because few animals eat it; and it out-competes native plant species, resulting in a loss of biodiversity.

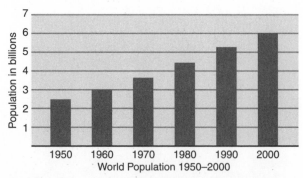

Figure 9-1. Human population growth over the past 50 years.

People have exploited plants and animals for their own use for centuries, often with negative impacts on wild populations. For example, the extensive cutting down, or **deforestation**, of tropical rain forests has led to habitat loss for wildlife and erosion of topsoil; elephants and walrus have been over-hunted for their ivory tusks; tropical parrots and monkeys have been captured and sold as pets.

The increased building of cities and suburbs has reduced the amount of farmland and natural habitats, threatening the existence of various native plant and animal species. Overgrazing and poor agricultural practices have caused valuable soil nutrients and topsoil to be lost, or **depleted**.

Some technological developments have contributed to the pollution of air, land, and water. In many areas, chemical wastes from homes, factories, and farmlands have polluted the water. Major chemical wastes include phosphates, heavy metals, and PCBs (an industrial by-product). Radioactive materials have been dumped or have leaked into the water supply from factories and waste-storage areas. Such chemical and radioactive wastes are **toxins** that can cause harm to people and wildlife alike.

The increased temperature, or *thermal pollution*, of river water occurs when water is taken from a river and used for cooling in factories and then returned to the waterway. Untreated sewage, which contains harmful bacteria, has been dumped into rivers and oceans. Water pollutants have killed fish and other animals, as well as plant life.

Exhaust gases from the burning of fossil fuels in factories, automobiles, and other places have polluted the air. The major air pollutants include carbon dioxide, carbon monoxide, hydrocarbons, and particulate matter. The increasing levels of carbon dioxide and other greenhouse gases in the atmosphere have been linked to **global warming** trends. (The destruction of rain forests also adds to an increase in atmospheric carbon dioxide, because there are fewer trees taking in the gas.) Compounds used in aerosol sprays have weakened Earth's **ozone shield**, allowing more ultraviolet radiation to penetrate the atmosphere, causing harm to both plants and animals. Nitrogen oxides and sulfur dioxides are gaseous pollutants that combine with water vapor in the atmosphere, forming acids (Figure 9-2). Precipitation of these acids, called *acid rain*, kills plants and lowers the pH of lakes and ponds, thereby harming and killing aquatic wildlife.

Several *biocides* (pesticides and herbicides) that are used to kill insects and to prevent the growth of weeds have had negative effects on the environment. Biocides have contaminated the soil, air, and water supplies. The **residue**, or chemical remains, of their use has also entered food chains and caused harm to some organisms, thereby disrupting whole food webs. For example, the pesticide DDT was linked to reproductive failure and population decrease among bald eagles and peregrine falcons. Its use has been banned in the United States, and eagle and falcon populations are now on the increase.

Technological developments have resulted in the increased production of solid, chemical, and nuclear wastes. Disposal of these wastes, many of them highly toxic, is a major problem. In addition, disposal of household garbage is becoming a problem, as more landfills are filled to capacity and shut down.

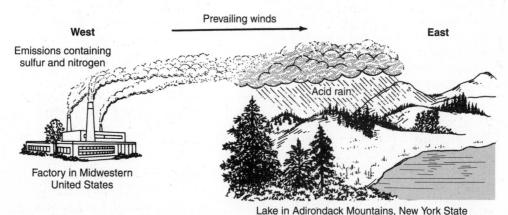

Figure 9-2. Nitrogen oxides and sulfur dioxides combine with water vapor in the atmosphere, thus forming acid rain.

QUESTIONS

1. An increased burning of coal would cause additional tons of sulfur dioxide to be released into the atmosphere, which could increase environmental problems associated with (1) acid rain (2) PCBs (3) DDT (4) dioxin

2. The number of African elephants has been greatly reduced by poachers who kill the animals for their ivory tusks. This negative aspect of human involvement in the ecosystem could best be described as (1) poor land-use management (2) importation of organisms (3) poor agricultural practices (4) exploitation of wildlife

3. Recent evidence indicates that lakes in large areas of upstate New York are being affected by acid rain. The major effect of acid rain in these lakes is the (1) increase in game fish population levels (2) stimulation of a rapid rate of evolution (3) elimination of various species of aquatic wildlife (4) increase in local agricultural productivity

4. Compared to other organisms, humans have had the greatest ecological effect on the biosphere due to their (1) internal bony skeleton (2) homeostatic regulation of metabolism (3) adaptations for respiration (4) ability to modify the environment

5. The rapid increase in the human population over the past few hundred years has been due mainly to (1) increasing levels of air and water pollution (2) depletion of topsoil from farmable lands (3) medical advances that increase survival rates (4) increasing resistance levels of insect species

6. When plant and animal species are introduced into a new area, they often become pests in the new habitat, even though they were not pests in their native habitats. The most probable reason for this is that, in their new habitat, they (1) have fewer natural enemies (2) have a much lower mutation rate (3) develop better resistance to the new climate (4) learn to use different foods

7. Recent studies have found traces of the insecticide DDT accumulated in the fat tissue of many wild animals. A correct explanation for this accumulation is that (1) fat tissue absorbs DDT directly from the air (2) fat tissue cells secrete DDT (3) DDT is needed for proper metabolic functioning (4) DDT is passed along in many food chains

8. Which factor is a major cause of global warming? (1) increased burning of fuels (2) increased number of green plants (3) decreased mineral availability (4) decreased carbon dioxide in the atmosphere

9. If humans remove carnivores such as wolves and coyotes from an ecosystem, what will probably be the first noticeable result? (1) The natural prey will die off. (2) Certain plant populations will increase. (3) Certain herbivores will exceed carrying capacity. (4) The decomposers will fill the predator niche.

10. Which situation has had the most harmful effect on the ecosystems of Earth? (1) use of air pollution controls (2) use of natural predators to control insect pests (3) recycling glass, plastic, and metals (4) the increasing human population

11. Which factor is primarily responsible for the destruction of the greatest number of habitats? (1) human population growth (2) decreased use of renewable resources (3) spread of predatory insects (4) epidemic diseases

12. A new automobile manufacturing plant is opening in a certain town. It will have some negative effects on the environment. This is a trade-off that the town officials had to consider carefully before giving final approval. They most likely gave their approval because, along with the negative effects, there would be the (1) release of pollutants into the environment (2) creation of new employment opportunities (3) decrease of property values in the area around the plant (4) increase of automobile traffic in the area around the plant

13. The importation of organisms such as the Japanese beetle and gypsy moth to areas where they have no natural enemies best illustrates (1) the use of abiotic factors to reduce pest species (2) the selection of species to mate with each other to produce a new variety (3) attempts by humans to protect extinct species (4) a human activity that disrupts existing, native ecosystems

14. A farmer has been growing only corn in his fields for several years. Each year the corn stalks were cut off near the ground and processed as food for cattle. The farmer observed that with each passing year, corn production in his fields decreased. Explain why removing the dead corn stalks reduced corn production in these fields.

Base your answers to questions 15 through 17 on the following information and on your knowledge of biology.

A group of western ranchers was very concerned about a local wolf population that they thought was preying on their livestock. After several attempts, they persuaded a court to authorize a hunt in order to eliminate the wolf population in their area. The court set a quota of 200 wolves, to be eliminated over the course of six months.

15. State a negative effect that the removal of these wolves might have on the habitat.

16. How would this hunt affect the biodiversity of the area's natural ecosystem?

17. Suppose that you are a local environmental official in the area. Propose a plan that would preserve the wolves yet keep them away from the ranchers' livestock.

PART C

Base your answers to questions 18 and 19 on the following information and graph.

Reducing toxic chemicals released into the environment often requires laws. When making decisions about whether or not to support the passing of such laws, individuals must weigh the benefits against the potential risks if the law is not passed.

The amounts of toxic chemicals released into the environment of New York State over a ten-year period are shown in the graph below.

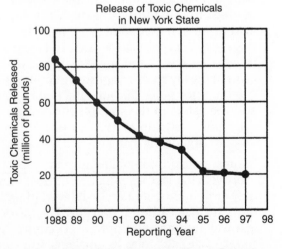

Release of Toxic Chemicals in New York State

18. State one possible negative effect of passing a law to reduce the release of toxic chemicals.

19. State one possible explanation for why the amount of toxic chemicals released remained relatively constant between 1995 and 1997.

PEOPLE PROTECT THE ENVIRONMENT

People are becoming increasingly aware of the negative effects of some of their activities on the environment. As a result, they are making many efforts to correct past damage and avoid future harmful effects.

Population Control. Methods for limiting the high rate of human population growth have been, and continue to be, developed. For example, in some countries, policies have been adopted to limit family size. Also, many people make their own choice to limit the number of offspring they have by using birth-control methods to prevent conception.

Conservation of Resources. Measures have been taken to conserve water, fossil fuels (oil, coal, and natural gas), and other natural resources (such as trees and wildlife). Reforestation projects help prevent further loss of trees. This is good for the environment because trees hold topsoil in place and take in excess carbon dioxide from the atmosphere. Planting of cover crops helps conserve topsoil, too. People are now realizing the economic and environmental benefits of recycling various materials, such as paper, metals, glass, and plastic. In many cities, laws have been enacted to make people recycle household items that used to be thrown away and dumped into landfills. Failure to comply with such laws can result in stiff fines and penalties.

Pollution Control. Laws have been passed to control the pollution of air and water. New techniques for limiting pollution from cars and factories, and for better sanitation and disposal of hazardous wastes, have been developed. In recent years, new types of automobiles called *hybrids* have been developed that use a combination of gasoline and electricity for power. The widespread use of these vehicles could eventually decrease pollution of the air and our dependence on fossil fuels.

Species Preservation. Endangered species are being protected, and efforts are being undertaken to increase their populations in the wild and in captivity. This is accomplished through captive breeding programs, protection of wild habitats, and the establishment of wildlife refuges and national parks. Management of various forms of wildlife also includes laws that regulate hunting and fishing.

Examples of animals that are, or were, endangered but are now increasing in numbers include

Figure 9-3. Bald eagle populations are now on the increase due to protective measures.

the American bison, alligator, whooping crane, fur seal, and bald eagle (Figure 9-3). However, the future of many species is still very much in doubt.

Biological Control. Biological control of insect pests reduces the use of chemical pesticides. One method of biological control involves the use of sex hormones to attract and trap insect pests. Another method of biological control involves the use of natural parasites that kill harmful insects. Biological control methods are less likely than chemical methods to affect species that are beneficial to humans, disrupt food webs, or contaminate the land.

The Future

A greater awareness of ecological principles and careful use of energy and other natural resources will help to ensure a suitable environment for future generations. There are many environmental organizations that are involved in the effort to conserve wilderness areas, protect threatened and endangered species, and reduce environmental pollution. It is the responsibility of every person who can to try to help in the preservation of the planet—for now and for the future.

QUESTIONS

PART A

20. Which accomplishment by people has had the most positive ecological effect on the environment? (1) the importation of organisms such as the starling and Japanese beetle into the United States (2) reforestation efforts and planting of cover crops to prevent soil erosion (3) the extinction or near extinction of many predators to protect prey animals (4) the use of pesticides and other chemical compounds to reduce the insect population

21. When a garden became infested with a large population of aphids, some ladybird beetles were introduced into the community as predators on the aphids. The resultant decrease in the aphid population was due to (1) biological control (2) parthenogenesis (3) vegetative propagation (4) chemical control

22. Gypsy moth infestations in rural areas of New York State may pose a potentially serious threat to many forested areas. Which would probably be the most ecologically sound method of gypsy moth control? (1) widespread application of DDT (2) introduction of a biological control (3) removal of its forest habitat (4) contamination of its food sources

23. Which illustrates the human population's increased understanding of, and concern for, ecological interrelationships? (1) importing organisms in order to disrupt existing ecosystems (2) allowing the air to be polluted only by those industries that promote technology (3) removing natural resources from Earth at a rate equal to or greater than the needs of an increasing population (4) developing wildlife game laws that limit the number of organisms that may be hunted each year

24. Some homeowners mow their lawns during the summer, collect the cut grass, then dispose of it in a landfill. Instead of taking the cuttings to a landfill, it might better for the environment if they were to (1) leave the cuttings to decompose in the lawn and form materials that enrich the soil (2) spray the cuttings in the lawn with imported microbes that use them for food (3) burn the cuttings and add the ashes to the soil (4) throw the cuttings into a stream or river to provide extra food for organisms living there

25. In most states, automobiles must be inspected every year to make sure that the exhaust fumes they emit do not contain high levels of pollutants such as carbon monoxide. This process is a way humans attempt to (1) control the water cycle (2) recycle nutrients from one ecosystem to another (3) control energy flow in natural ecosystems (4) maintain the quality of the atmosphere

26. A greater stability of the biosphere would most likely result from (1) decreased finite

resources (2) increased deforestation (3) increased biodiversity (4) decreased consumer populations

27. State two positive and two negative effects that humans have had on the natural environment.

28. Describe two benefits of recycling materials rather than throwing them away. In your response, be sure to discuss the

- effect on the environment;
- costs involved;
- effect on people.

29. Carbon dioxide is known to be a greenhouse gas that contributes to global warming. Due largely to human activity, such as the burning of fossil fuels, the amount of carbon dioxide in the atmosphere continues to increase significantly. Discuss how planting more trees could help reduce the amount of carbon dioxide in the atmosphere. How might this slow down the rate of global warming?

30. Tropical rain forests around the world are being cleared at an alarming rate to make room for the increasing human population and its needs. Describe two reasons why the remaining tropical forests must be preserved. Suggest a plan by which people might be able to both protect and utilize tropical rain forests in a way that is sustainable.

31. The graph below shows the percentage of solid wastes recycled in New York State between 1987 and 1997. Discuss the effects of recycling. In your answer be sure to state:

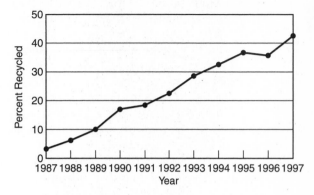

- what recycling is and give *one* example of a material that is often recycled;
- *one* specific positive effect that recycling has on the environment;
- *one* specific reason the percentage of solid wastes recycled increased between 1987 and 1997.

32. Human activities continue to place strains on the environment. One of these strains is the loss of biodiversity. Explain what this problem is and describe some ways that humans are involved in both the problem and the possible solutions. In your answer be sure to state:

- the meaning of the term *biodiversity*;
- *one* negative effect on humans if biodiversity continues to be lost;
- *one* practice that could be used to preserve biodiversity in New York State.

Base your answers to questions 33 through 36 on the passage below and on your knowledge of biology. Source: Living Environment Regents Exam—August 2004, p. 12.

Fighting Pollution with Bacteria

You may think that all bacteria are harmful. Think again! Some bacteria are working to clean up the damage humans have caused to the environment.

In 1989, the oil tanker *Exxon Valdez* hit ground and a hole was ripped in its hull. Millions of gallons of crude oil spread along the coast of Alaska. In some places, the oil soaked 2 feet deep into the beaches. There seemed to be no way to clean up the spill. Then scientists decided to enlist the help of bacteria that are found naturally on Alaskan beaches. Some of these bacteria break down hydrocarbons (molecules found in oil) into simpler, less harmful substances such as carbon dioxide and water.

The problem was that there were not enough of these bacteria to handle the huge amount of oil. To make the bacteria multiply faster, the scientists sprayed a chemical that acted as a fertilizer along 70 miles of coastline. Within 15 days, the number of bacteria had tripled. The beaches that had been treated with the chemical were much cleaner than those that had not. Without this bacterial activity, Alaska's beaches might still be covered with oil.

This process of using organisms to eliminate toxic materials is called *bioremediation*. Bioremediation is being used to clean up gasoline that leaks into the soil under gas stations. At factories that process wood pulp, scientists are using microorganisms to break down phenols (a poisonous by-product of the process) into harmless salts. Bacteria also can break down acid drainage that seeps out of abandoned coal mines, and explosives, such as TNT. Bacteria are used in sewage treatment plants to clean water. Bacteria also reduce acid rain by removing sulfur from coal before it is burned.

Because Americans produce more than 600 million tons of toxic waste a year, bioremediation may soon become a big business. If scientists can identify microorganisms that attack all the kinds of waste we produce, expensive treatment plants and dangerous toxic dumps might be put out of business.

33. A chemical was sprayed along the Alaskan coastline in order to (1) introduce new bacteria to the beaches (2) break the oil down into harmless salts (3) increase the population of helpful bacteria (4) wash away oil that had been spilled

34. Which statement does *not* represent an example of bioremediation? (1) Duckweed removes heavy metals from ponds and lakes. (2) Ladybugs eliminate insect pests from plants. (3) Bacteria break down hydrocarbons in oil. (4) Ragweed plants remove lead from the ground around factory sites.

35. State one economic advantage of using bioremediation on toxic wastes.

36. Describe one biological problem that may result from using microorganisms to fight pollution.

CHAPTER 10

Laboratory Skills and Part D Labs

As part of the Regents Living Environment course, students are expected to master a number of specific science-related skills. Some of these skills involve application of the scientific method, while others are actual laboratory techniques and procedures.

Skills Using the Scientific Method

- Formulate a question or define a problem for investigation and develop a hypothesis to be tested in an investigation.
- Distinguish between controls and variables in an experiment.
- Collect, organize, and graph data.
- Make predictions based on experimental data.
- Formulate generalizations or conclusions based on the investigation.
- Apply the conclusion to other experimental situations.

Skills Involving Laboratory Procedures

- Given a laboratory problem, select suitable lab materials, safety equipment, and appropriate observation methods.
- Demonstrate safety skills in heating materials in test tubes or beakers, use of chemicals, and handling dissection equipment.
- Identify the parts of a compound light microscope and their functions. Use the microscope correctly under low power and high power.
- Determine the size of microscopic specimens in micrometers.

- Prepare wet mounts of plant and animal cells and apply stains, including iodine and methylene blue.
- With the use of a compound light microscope, identify cell parts, including the nucleus, cytoplasm, chloroplasts, and cell walls.
- Use indicators, such as pH paper, Benedict solution (or Fehling solution), iodine solution (or Lugol solution), and bromthymol blue. Interpret changes shown by the indicators.
- Use measurement instruments, such as metric rulers, Celsius thermometers, and graduated cylinders.
- Dissect plant or animal specimens, exposing major structures for examination.

The Scientific Method

Defining a Problem and Developing a Hypothesis. Scientists do research to answer a question or to solve a problem. The first step in planning a research project is to define the problem to be solved; this is usually stated in the form of a question. The next step is to develop a possible solution to the problem. This proposed explanation, or *hypothesis*, is the statement that identifies the factor to be tested in the experiment.

For example, a scientist interested in studying the enzyme amylase might want to measure the rate of enzyme action at various temperatures. The basic hypothesis for such an experiment would be that the rate at which amylase hydrolyzes, or breaks down, starch is affected by temperature.

Designing and Conducting an Experiment.

Biologists use controlled experiments when doing research. In a controlled experiment, there are actually two setups: an experimental setup and a control setup. The experimental and control setups are identical except for the single factor, or *variable*, that is being tested. Any changes observed during the experiment can then be explained in terms of the variable factor. In an experiment to determine the effect of temperature on the rate of action of the enzyme amylase, temperature is the variable.

A basic controlled experiment would use two setups—one containing a starch solution only, the other containing exactly the same amount of the same starch solution plus the enzyme amylase. Both setups would then be tested at various temperatures to determine how much starch had undergone hydrolysis. The setup with no enzyme is the control; the setup with the enzyme is the experimental one. The control can show that no hydrolysis occurs without the enzyme. In the experimental setup, all conditions are kept constant except temperature. Thus, the scientist knows that changes in the rate of hydrolysis are caused by the effects of temperature on the enzyme amylase.

Collecting, Organizing, and Graphing Data.

During an experiment, the scientist collects data. These data are the results of the experiment. The data may be recorded in a log in the form of a chart or data table. Sometimes the results are plotted on a graph. Scientists also use computers to record and organize experimental results.

In an experiment to determine the rate of action of amylase at various temperatures, the data collected might be written in a table, as shown in Figure 10-1.

The relationship between two varying factors can also be shown clearly on a line graph. The graph in Figure 10-2 shows the same information as the data table in Figure 10-1.

Temperature (°C)	Grams of starch hydrolyzed per minute
0	0.0
10	0.2
20	0.4
30	0.8
40	1.0
50	0.3
60	0.2

Figure 10-1. A data table.

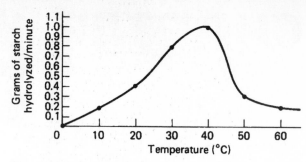

Figure 10-2. A line graph.

Making Predictions Based on Experimental Data.

Scientists may make predictions based on experimental data. The validity of these predictions can then be tested by further experimentation.

For example, on the basis of the data shown in Figure 10-1, a scientist might predict that the number of grams of starch hydrolyzed at normal body temperature (37°C) would be between 0.8 and 1.0 gram/minute. Further measurements might show that the prediction was correct, or they might show that at 37°C the rate was higher than 1.0 gram/minute. Scientists must be extremely careful not to make any assumptions that are not supported by the data.

Making Generalizations and Drawing Conclusions.

The results of an experiment are collected and analyzed. For a conclusion to be valid, the experiment must be repeated many times, obtain similar results, and all the results must be included in the analysis. The conclusion is based solely on the experimental data.

In the experiment on the effect of temperature on the rate of action of amylase, the data in the table show that the enzyme functions most efficiently at 40°C. However, if measurements were made only at 10° intervals, you could not say definitely that 40°C is the optimum temperature for amylase without making measurements at other intermediate temperatures. Still, it is probably safe to conclude that the optimum temperature is close to 40°C.

QUESTIONS
MULTIPLE CHOICE

1. The diagram on page 137 represents a setup at the beginning of a laboratory investigation.

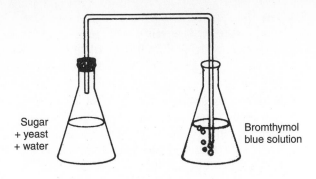

Sugar + yeast + water

Bromthymol blue solution

Which hypothesis would most likely be supported by observing and collecting data from this investigation? (1) The fermentation of a yeast-sugar solution results in the production of carbon dioxide. (2) Yeast cells contain simple sugars. (3) Oxygen is released when a yeast-sugar solution is illuminated with green light. (4) Yeast cells contain starches.

Base your answers to questions 2 and 3 on the information and data table below.

A green plant was placed in a test tube, and a light was placed at varying distances from the plant. The bubbles of oxygen given off by the plant were counted. The table shows the data collected during this experiment.

Distance of light from plant (cm)	Number of bubbles per minute
10	60
20	25
30	10
40	5

2. A variable in this investigation is the (1) color of the light used (2) distance between the light and the plant (3) size of the test tube (4) type of plant used

3. Which conclusion can be drawn from this investigation? (1) As the distance from the light increases, the number of bubbles produced decreases. (2) As the distance from the light increases, the number of bubbles produced increases. (3) As the distance from the light decreases, the number of bubbles produced decreases. (4) There is no relationship between the number of bubbles produced and the distance of the plant from the light.

Base your answers to questions 4 through 6 on the following information, diagram, and data table, and on your knowledge of biology.

A student is studying the effect of temperature on the hydrolytic action of the enzyme gastric protease, which is contained in gastric fluid. An investigation is set up using five identical test tubes, each containing 40 milliliters of gastric fluid and 20 millimeters of glass tubing filled with cooked egg white, as shown in the diagram below. After 48 hours, the amount of egg white hydrolyzed in each tube was measured. The data collected are shown in the data table below.

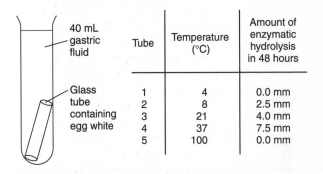

Tube	Temperature (°C)	Amount of enzymatic hydrolysis in 48 hours
1	4	0.0 mm
2	8	2.5 mm
3	21	4.0 mm
4	37	7.5 mm
5	100	0.0 mm

4. Which is the variable in this investigation? (1) gastric fluid (2) length of glass tubing (3) temperature (4) time

5. If an additional test tube were set up identical to the other test tubes and placed at a temperature of 15°C for 48 hours, what amount of hydrolysis might be expected? (1) less than 2.5 mm (2) between 2.5 mm and 4.0 mm (3) between 4.0 mm and 7.5 mm (4) more than 7.5 mm

6. Which set of axes would produce the best graph for plotting the data from the results of this investigation? (1) 1 (2) 2 (3) 3 (4) 4

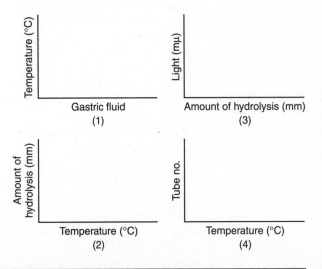

Base your answers to questions 7 through 11 on the information and two charts below and on your knowledge of biology.

Chart I shows the percentages of certain materials in the blood entering the kidney and the percentages of the same materials in the urine leaving the body. Chart II shows the number of molecules in the beginning and at the end of the kidney tubule for every 100 molecules of each substance entering the glomerulus.

Chart I

Substance	% of blood	% of urine
Protein	7.0	7.0
Water	91.5	96.0
Glucose	0.1	0.0
Sodium	0.33	0.29
Potassium	0.02	0.24
Urea	0.03	2.7

Chart II

	Number of Molecules		
Substance	In blood entering glomerulus	Beginning of tubule	End of tubule
Protein	100	0	0
Water	100	30	1
Glucose	100	20	0
Sodium	100	30	1
Potassium	100	23	12
Urea	100	50	90

7. According to Chart I, which substance is more highly concentrated in the urine than in the blood? (1) water (2) sodium (3) protein (4) glucose

8. According to Charts I and II, which substance enters the tubules but does *not* appear in the urine leaving the body? (1) protein (2) water (3) glucose (4) potassium

9. According to the data, which substance did *not* pass out of the blood into the tubule? (1) water (2) urea (3) glucose (4) protein

10. The data in the two charts would best aid a biologist in understanding the function of the (1) heart of a frog (2) nephron of a human (3) nerve cell of a fish (4) contractile vacuole of a paramecium

11. Which substances enter the tubule and then are reabsorbed back into the blood as they pass through the tubule? (1) urea and potassium (2) water and sodium (3) urea and protein (4) protein and glucose

Base your answers to questions 12 through 14 on the information provided by the graph below. The graph shows the average growth rate for 38 pairs of newborn rats. One member of each pair was injected with anterior pituitary extract. The other member of each pair served as a control.

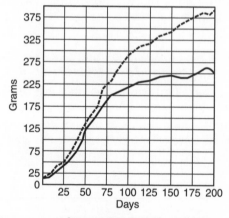

—— Average growth of 38 untreated littermates (control)

---- Average growth of 38 rats injected with anterior pituitary extract (experimental)

12. At 75 days, what was the average weight of the rats injected with pituitary extract? (1) 65 grams (2) 125 grams (3) 200 grams (4) 225 grams

13. Based on the graph, it can be correctly concluded that the pituitary extract (1) is essential for life (2) determines when a rat will be born (3) affects the growth of rats (4) affects the growth of all animals

14. The graph shows the relationship between the weight of treated and untreated rats and the (1) age of the rats (2) sex of the rats (3) size of the rats' pituitary glands (4) type of food fed to the rats

Base your answers to questions 15 and 16 on the following information, diagrams, and data

table, and on your knowledge of laboratory procedures used in biology.

Diagrams *A* through *E* show the general appearance of five tree fruits that were used by a science class in an experiment to determine the length of time necessary for each type of fruit to fall from a second-floor balcony to the lobby floor of their school. One hundred fruits of each type were selected by the students, and the average time of fall for each type of fruit is shown in the table below.

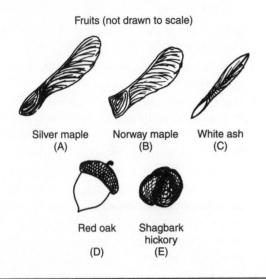

Fruits (not drawn to scale)

Silver maple (A)   Norway maple (B)   White ash (C)

Red oak (D)   Shagbark hickory (E)

Tree type	Average fall time of 100 fruits
Silver maple	3.2 sec
Norway maple	4.9 sec
White ash	1.5 sec
Red oak	0.8 sec
Shagbark hickory	0.8 sec

15. Based on this experimental evidence, what inference seems most likely to be true concerning the distribution of these fruits during windstorms in nature? (1) Silver maple fruits would land closer to the base of their parent tree than would shagbark hickory fruits. (2) White ash fruits would land farther from the base of their parent tree than would silver maple fruits. (3) White ash fruits would land closer to the base of their parent tree than would shagbark hickory fruits. (4) Norway maple fruits would land farther from the base of their parent tree than would silver maple fruits.

16. Which graph best shows the average fall time for each fruit type tested during this experiment? (1) 1 (2) 2 (3) 3 (4) 4

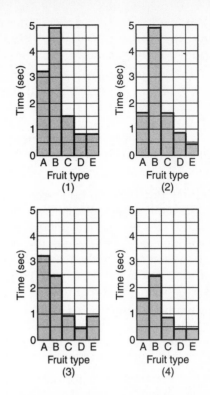

17. The graph below was developed as a result of an investigation of bacterial counts of three identical cultures grown at different temperatures. Which conclusion might be correctly drawn from this graph?

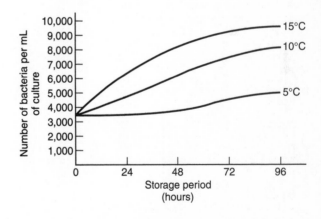

(1) The culture contains no bacteria. (2) Refrigeration retards bacterial reproduction. (3) Temperature is unrelated to the bacteria reproduction rate. (4) Bacteria cannot grow at a temperature of 5°C.

Base your answers to questions 18 through 20 on the graphs on page 140, which show data on some environmental factors affecting a large New York lake.

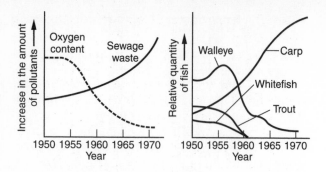

18. Which relationship can be correctly inferred from the data presented? (1) As sewage waste increases, oxygen content decreases. (2) As sewage increases, oxygen content increases. (3) As oxygen content decreases, carp population decreases. (4) As oxygen content decreases, trout population increases.

19. The greatest change in the lake's whitefish population occurred between which years? (1) 1950 and 1955 (2) 1955 and 1960 (3) 1960 and 1965 (4) 1965 and 1970

20. Which of the fish species appears able to withstand the greatest degree of oxygen depletion? (1) trout (2) carp (3) walleye (4) whitefish

Laboratory Procedures

Selecting Suitable Lab Equipment. Knowledge of the correct lab equipment is essential for planning and carrying out an experiment. Figure 10-3 illustrates some basic laboratory equipment that you should know.

Safety in the Laboratory. Following are some safety precautions that you should practice in the laboratory.

- Do not handle chemicals or equipment unless you are told by your teacher to do so.

- If any of your lab equipment appears to be broken or unusual, do not use it. Report it to your teacher.

- Report any personal injury or damage to clothing to your teacher immediately.

- Report any unsafe activities to your teacher immediately.

- Wear appropriate safety equipment, such as goggles and apron. Tie back long hair; secure dangling jewelry and loose sleeves.

- Never taste or directly inhale unknown chemicals. Never eat or drink in the lab.

- Never pour chemicals back into stock bottles; never exchange bottle stoppers.

- When heating a liquid in a test tube, always wear safety goggles and point the opening away from yourself and all others.

- Handle all sharp instruments with care, moving slowly and deliberately.

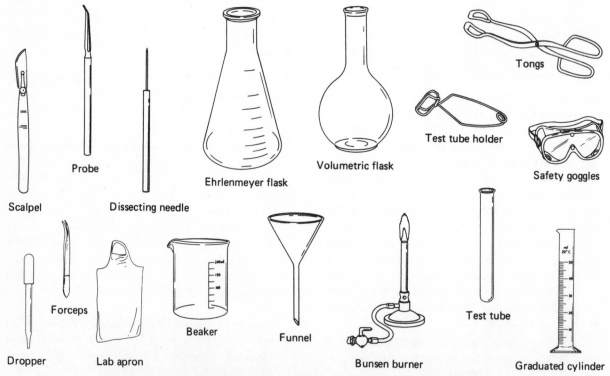

Figure 10-3. Examples of basic laboratory equipment (not drawn to scale).

Using a Compound Light Microscope.

Review the parts of the compound light microscope and their functions by studying Figure 2-2 and Table 2-1 (in Chapter 2).

In using the compound microscope, the observer should begin by viewing the specimen with the low-power objective, focusing first with the coarse adjustment, then with the fine adjustment. The objectives can then be switched from low power to high power. All focusing under high power should be done with the fine adjustment. The field appears dimmer under high power than under low power. Opening the diaphragm allows more light to reach the specimen.

The image of an object seen under the microscope is enlarged, reversed (backward), and inverted (upside down). When viewed through the microscope, an organism that appears to be moving to the right is actually moving to the left. An organism that appears to be moving toward the observer, or up, is actually moving away from the observer, or down.

Determining the Size of Microscopic Specimens.

To determine the size of a specimen being examined under a microscope, you must know the diameter of the microscope field. You can actually measure the field diameter with a clear plastic centimeter ruler. Place the ruler over the opening in the stage of the microscope, as shown in Figure 10-4. Focus on the ruler markings and adjust the position of the ruler so that a millimeter marking is at the left.

Once you have estimated the field diameter under low power, you can estimate the size of specimens observed under low power by how much of the field they cover. For example, if the diameter of the field is 1.5 mm and a specimen is about one-third the diameter of the field, the specimen is about 0.5 mm in length.

The unit most commonly used in measuring microscopic specimens is the *micrometer*, symbol μm, which is one-thousandth of a millimeter.

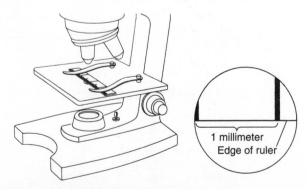

Figure 10-4. Measuring with a microscope.

Note: The micrometer is also referred to as a *micron*.

$$1 \text{ mm} = 1000 \text{ } \mu m \qquad 1 \text{ } \mu m = 0.001 \text{ mm}$$

In the example above, the field diameter is 1.5 mm, which is equal to 1500 μm. The specimen is 0.5 mm long, which equals 500 μm.

When you switch from low power to high power, the field diameter decreases. For example, if the magnification under low power is 100× and under high power is 400× then the field diameter under high power will be one-fourth that under low power. If the low-power magnification is 100× and the high-power magnification is 500× then the diameter of the high-power field will be one-fifth that of the low-power field.

Preparing a Wet Mount and Applying Stains.

A wet mount is a temporary slide preparation used for viewing specimens with a compound light microscope. Any specimen to be examined must be thin enough for light to pass through it.

The preparation of a wet mount involves the following steps:

1. Use a medicine dropper to put a drop of water in the center of the slide.
2. Place the tissue or organism to be examined in the drop of water on the slide.
3. Cover the specimen with a coverslip, as shown in Figure 10-5 on page 142.
4. To stain the section, add a drop of iodine solution or methylene blue at one edge of the coverslip. Touch a small piece of paper towel to the opposite side of the coverslip to draw the stain across the slide and through the specimen.

Identifying Cell Parts with a Compound Light Microscope.

Review the structure of plant cells and animal cells (Figure 2-1) and the functions of cell organelles (Chapter 2).

Unstained cells viewed with a compound light microscope show relatively little detail. The use of stains, such as iodine or methylene blue, enhances contrast. With such stains, the nucleus becomes clearly visible, and in plant and algal cells the cell wall becomes visible, too. Chloroplasts are visible as small oval green structures. Most other cell organelles, including mitochondria and the endoplasmic reticulum, are not visible with the compound light microscope.

Using Indicators and Interpreting Changes.

Indicators are used to test for the presence of specific substances or chemical characteristics.

Litmus paper is an indicator used to determine whether a solution is an acid or a base. An

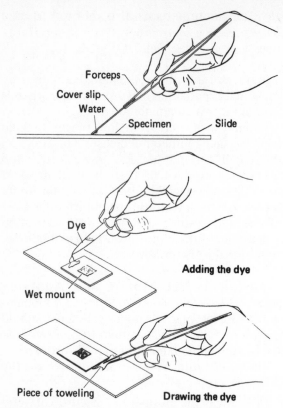

Figure 10-5. Making a wet mount (above) and staining a specimen.

acid turns blue litmus paper red. A base turns red litmus paper blue.

pH paper is an indicator that is used to determine the actual pH of a solution. When a piece of pH paper is dipped into a test solution, it changes color. The color of the pH paper is then matched against a color chart, which shows the pH.

Bromthymol blue is an indicator used to detect carbon dioxide. In the presence of carbon dioxide, bromthymol blue turns to bromthymol yellow. If the carbon dioxide is removed, the indicator changes back to bromthymol blue.

Benedict solution is an indicator used to test for simple sugars. When heated in the presence of simple sugars, Benedict solution turns from blue to yellow, green, or brick red, depending on the sugar concentration. *Fehling solution* also may be used to test for simple sugars.

Lugol, or *iodine, solution* is an indicator used to test for starch. In the presence of starch, Lugol solution turns from red to blue-black.

Biuret solution is an indicator used to test for protein. In the presence of protein, Biuret solution turns from light blue to purple.

Using Measurement Instruments. The following tools are used for making scientific measurements.

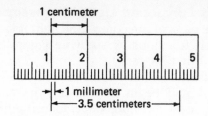

Figure 10-6. A centimeter ruler.

- *Metric ruler.* The basic unit of length in the metric system is the meter, abbreviated m. One meter contains 100 centimeters (cm). As shown in Figure 10-6, metric rulers are generally calibrated in centimeters and millimeters (mm). Each centimeter contains 10 millimeters, thus each meter is equal to 1000 mm.

- *Celsius thermometer.* In the metric system, temperature is commonly measured in degrees Celsius. On the Celsius scale, 0°C is the freezing point of water, 21°C is room temperature, and 100°C is the boiling point of water. Figure 10-7 shows a thermometer calibrated in degrees Celsius. Note that each degree is marked by a short line (such as 37°C), and every tenth degree is labeled with the number (such as 30°C and 40°C).

- *Graduated cylinder.* The basic unit used for measuring the volume of a liquid in the metric system is the liter, abbreviated L. A liter contains 1000 milliliters (mL). Most laboratory measurements involve milliliters rather than liters.

The volume of a liquid is frequently measured in graduated cylinders, which come in many sizes. When you need an accurately measured amount of liquid, use a graduated cylinder of appropriate size—that is, to measure 5 mL of liquid, use a 10-mL graduated cylinder, not a 1000-mL graduated cylinder.

The surface of water and similar liquids curves upward along the sides of a cylinder. This curved surface, or *meniscus*, is caused by the strong attraction of liquid molecules to the glass surface (Figure 10-8). For an accurate measurement, the reading should be done at eye level, and the

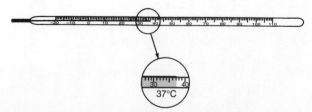

Figure 10-7. A Celsius thermometer.

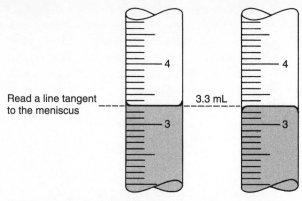

Figure 10-8. Measuring with a graduated cylinder.

Read a line tangent to the meniscus

3.3 mL

measurement should be read at the bottom of the meniscus. With other types of liquids, the meniscus curves the other way (that is, downward); in such cases, the measurement should be read across the top of the meniscus.

Dissecting Plant and Animal Specimens.

Dissections are done to expose major structures for examination. The specimen is generally placed in a dissection pan and fastened down with pins. While doing a dissection, you should be very careful with the dissection instruments, which are sharp. Scalpels, forceps, scissors, and stereomicroscopes are used. You should also be careful in cutting into and handling the specimen so that you do not damage important structures. Follow all instructions and record your observations by making labeled diagrams as you proceed with the dissection.

QUESTIONS
MULTIPLE CHOICE

Base your answers to questions 21 through 23 on the four sets of laboratory materials listed below and on your knowledge of biology.

Set A	Set B	Set C	Set D
Light source	Droppers	Scalpel	Compound microscope
Colored filters	Benedict solution	Forceps	Glass slides
Test tubes	Iodine	Scissors	Water
Test-tube stand	Test tubes	Pan with wax bottom	Forceps
	Test-tube rack	Stereo-microscope	
	Heat source	Pins	
	Goggles	Goggles	

21. Which set should a student select to test for the presence of a carbohydrate in food? (1) Set *A* (2) Set *B* (3) Set *C* (4) Set *D*

22. Which set should a student select to determine the location of the ovules in the ovary of a flower? (1) Set *A* (2) Set *B* (3) Set *C* (4) Set *D*

23. Which set should a student use to observe chloroplasts in elodea (a water plant)? (1) Set *A* (2) Set *B* (3) Set *C* (4) Set *D*

24. To view cells under the high power of a compound light microscope, a student places a slide of the cells on the stage and moves the stage clips over to secure the slide. She then moves the high-power objective into place and focuses on the slide with the coarse adjustment. Two steps in this procedure are incorrect. For this procedure to be correct, she should have focused under (1) low power using coarse and fine adjustments and then under high power using only the fine adjustment (2) high power first, then low power using only the fine adjustment (3) low power using the coarse and fine adjustments and then under high power using coarse and fine adjustments (4) low power using the fine adjustment and then under high power using only the fine adjustment

Base your answers to questions 25 and 26 on the following diagram of a compound light microscope.

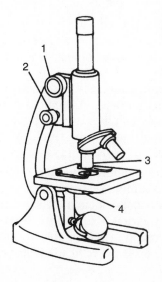

25. The part labeled 1 is used to (1) increase the amount of light reaching the specimen (2) focus with the high-power objective (3) hold the lenses in place (4) focus with the low-power objective

26. To adjust the amount of light reaching the specimen, you would use the part labeled (1) 1 (2) 2 (3) 3 (4) 4

Base your answers to questions 27 through 29 on the information below and on your knowledge of biology.

A student prepares a wet mount of onion epidermis and observes it under three powers of magnification with a compound light microscope (40×, 100×, and 400×).

27. An adjustment should be made to allow more light to pass through the specimen when the student changes the magnification from (1) 100× to 400× (2) 400× to 100× (3) 400× to 40× (4) 100× to 40×

28. Iodine stain is added to the slide. Under 400× magnification, the student should be able to observe a (1) mitochondrion (2) nucleus (3) ribosome (4) centriole

29. A specimen that is suitable for observation under this microscope should be (1) stained with Benedict solution (2) moving and respiring (3) alive and reproducing (4) thin and transparent

30. A microscope is supplied with 10× and 15× eyepieces, and with 10× and 44× objectives. What is the maximum magnification that can be obtained from this microscope? (1) 59× (2) 150× (3) 440× (4) 660×

31. Under low power (100×), a row of eight cells can fit across the field of a certain microscope. How many of these cells could be viewed in the high power (400×) visual field of this microscope? (1) 1 (2) 2 (3) 8 (4) 32

32. A compound light microscope has a 10× ocular, a 10× low-power objective, and a 40× high-power objective. A student noted that under high power, four cells end to end extended across the diameter of the field. If the microscope were switched to low power, approximately how many cells would fit across the field? (1) 1 (2) 8 (3) 16 (4) 4

33. The diagram below shows a section of a metric ruler scale as seen through a compound light microscope. If each division represents 1 millimeter, what is the approximate width of the microscope's field of view in micrometers (µm)? (1) 3700 (2) 4200 (3) 4500 (4) 5000

Base your answers to questions 34 through 37 on your knowledge of biology and on the diagrams below, which represent fields of view under the low power of the same compound microscope (100×). Diagram A shows the millimeter divisions of a plastic ruler, and diagram B shows a sample of stained onion epidermal cells.

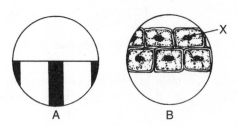

34. Structure *X* in diagram *B* was most likely stained by adding (1) water (2) iodine solution (3) Benedict solution (4) bromthymol blue

35. Structure *X* in diagram *B* indicates (1) a nucleus (2) a mitochondrion (3) the cell wall (4) the cytoplasm

36. The diameter of the field of vision in diagram *A* is approximately (1) 500 µm (2) 1000 µm (3) 1500 µm (4) 2000 µm

37. What is the approximate length of each onion epidermal cell in field *B*? (1) 200 µm (2) 660 µm (3) 1000 µm (4) 2500 µm

38. Iodine solution is used to test for the presence of (1) proteins (2) simple sugars (3) oxygen (4) starch

39. In the presence of carbon dioxide, bromthymol blue (1) shows no color change (2) turns yellow (3) turns blue-black (4) turns red-orange

40. Benedict solution is used to test for (1) disaccharides (2) oxygen (3) starch (4) simple sugars

41. Which piece of equipment should be used to transfer a protist onto a microscope slide? (1) scissors (2) dissecting needles (3) medicine dropper (4) forceps

42. While a student is heating a liquid in a test tube, the mouth of the tube should always be (1) corked with a rubber stopper (2) pointed toward the student (3) allowed to cool off (4) aimed away from everybody

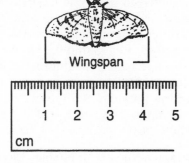

43. Which row in the chart below best represents the ratio of body length to wingspan of the peppered moth? (1) 1 (2) 2 (3) 3 (4) 4

Row	Body Length:Wingspan
(1)	1:1
(2)	2:1
(3)	1:2
(4)	2:2

Part D Mandated Laboratories

On Part D of the Living Environment (Biology) Regents, you will be assessed on your knowledge and understanding of four mandatory lab exercises. These investigations are based on the learning standards contained in the Living Environment Core Curriculum and test a variety of skills. Below is a description of each of the mandated labs scheduled for the 2006 and 2007 administration of the Living Environment Regents exam.

Lab #1: Relationships and Biodiversity

In this lab, you will be performing a number of tests to determine evolutionary relationships between an environmentally important but endangered plant, *Botana curus*, and three related plants. The goal is to identify the plant most closely related to *B. curus*. The tests you will perform include structural and molecular evidence for relationship. They are:

- comparison of the four plants' structures, including types of leaves and stems;
- comparison of the four plants' seeds in terms of size and appearance;
- microscopic examination of the plants' stem cross sections;
- paper chromatography of pigment extracts from the four plants;

- chemical test for the presence of an important enzyme in each plant;
- simulation of a gel electrophoresis of DNA from the four plants;
- translating the DNA from each plant into the corresponding protein product, using a Universal Genetic Code Chart.

The plant most closely related to *B. curus* will show many similarities to it in the tests you will perform. Therefore, the more "matches" between the unknown relative and *B. curus*, the more closely related the two plants are. The most significant tests are those for genetic relationships; thus the plant most closely related to *B. curus* will have DNA very similar to it.

Lab #2: Making Connections

In this activity, you will learn how to measure and record your normal pulse rate and how it may change under certain circumstances, such as exercising. These data will then be graphed and analyzed. You will also explore the effect of repeated exercise on muscle fatigue by squeezing a clothespin as many times as you can in one minute and then repeating the procedure a second time. The purpose of this part of the lab is to determine the effect of exercise on muscles. The second part of the lab exercise involves analyzing two different but opposing views about the effect of exercise on muscle fatigue. Two students make the following claims: one states that exercise causes a reduction in muscle activity and that rest is necessary before muscles can work effectively again; the second claims that exercise actually enhances the function of muscles. For this part, you will take one of the positions and design a scientific investigation to support or refute the claim. In addition, you may be asked by your teacher to present an oral report of your findings to the class.

Lab #3: The Beaks of Finches

This lab explores adaptations for survival (natural selection). In the activity, you will use some type of common household or lab equipment that has occluding parts (that is, parts that come together) to simulate the beak of a finch. Some examples of these tools are pliers, forceps, test-tube holders, hair clips, or salad tongs. These devices represent the different types of finch beaks. They will be used to pick up as many seeds as possible, just one at a time, in a 30-second time period and then transfer them to a storage dish (representing the bird's stomach). You will run trials both with and without competition. During the lab, it also may be necessary for your finch to switch to a different type of food if it was not successful in eating enough of the first type of seed. Survival of your finch depends on picking up and storing at least 13 seeds in the 30-second interval. As the lab is run, you will become aware that some beaks are more adaptive than others, thus allowing their "owners" to survive (i.e., eat enough seeds). This activity is meant to simulate what occurs in natural selection; that is, those organisms having the most adaptive traits for their environment survive, while those with less adaptive traits tend to die out. In some cases, in order to survive, organisms may migrate to a different environment in which they are better adapted for survival.

Lab #4: Diffusion Through a Membrane

In this lab activity, you will explore the permeability of a membrane and the process of diffusion. You will create an artificial cell using dialysis tubing or a thin, plastic bag. This "cell" will contain a mixture of a glucose and starch solution. The "cell" will be placed in a water bath that contains starch indicator (iodine solution). After the "cell" has been in the water for at least 20 minutes, you will use chemical indicators to check for diffusion of molecules into and out of the "cell." The chemical indicators include one that tests for glucose (blue Benedict solution) and one that tests for starch (iodine solution). Recall that Benedict solution, when heated in the presence of simple sugars such as glucose, will change from blue to green to yellow to orange to red as the concentration of sugar increases. Iodine solution changes from yellow (amber) to blue-black in the presence of starch. Iodine does not have to be heated to show that starch is present.

In the second part of the lab, you will examine the effects of osmosis on a red onion cell. By preparing a wet mount of onion tissue and surrounding it with 10% salt solution, you should be able to observe and explain the effect this solution has on the cells. Next, replacing the salt solution on the slide with distilled water should produce another effect on the cells. You will observe these effects on the onion cells through a microscope and draw them.

QUESTIONS

MULTIPLE CHOICE AND CONSTRUCTED RESPONSE

Base your answer to question 44 on the following information and data table.

To determine which colors of light are best used by plants for photosynthesis, three types of underwater green plants of similar mass were subjected to the same intensity of light of different colors for the same amount of time. All other environmental conditions were kept the same. After 15 minutes, a video camera was used to record the number of bubbles of gas (oxygen) each plant gave off in a 30-second time period. Each type of plant was tested six times. The average of the data for each plant type is shown in the table below.

Average Number of Bubbles Given Off in 30 Seconds

Plant Type	Red Light	Yellow Light	Green Light	Blue Light
Plant #1	35	11	5	47
Plant #2	48	8	2	63
Plant #3	28	9	6	39

44. Which statement is a valid inference based on the data? (1) Each plant carried on photosynthesis best in a different color of light. (2) Red light is better for photosynthesis than blue light. (3) These types of plants make food at the fastest rates with red and blue light. (4) Water must filter out red and green light.

Base your answers to questions 45 through 48 on the information below and on your knowledge of biology.

Three students each added equal volumes of pond water to four beakers and placed each beaker in a different water bath. Each student maintained the water baths at temperatures shown in the data table. The students then added an equal number of water fleas to each of their four beakers. After one hour, the students used microscopes to determine the average heart rate of the water fleas. The procedure was repeated for a total of three trials at each temperature. The results of the investigation are summarized in the data table below.

Water Flea Heart Rate

Water Temperature (°C)	Average Water Flea Heart Rate (beats/minute)
5	40
15	119
25	205
35	280

45. Use the information in the data table to construct a line graph on the grid provided below. Mark an appropriate scale on each labeled axis.

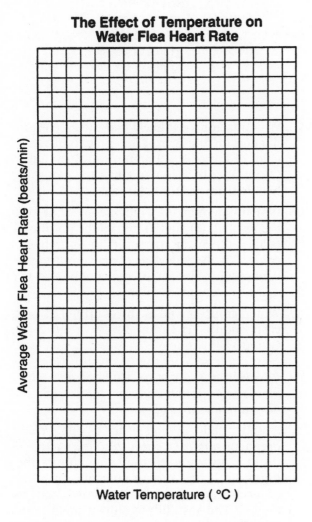

The Effect of Temperature on Water Flea Heart Rate

Average Water Flea Heart Rate (beats/min)

Water Temperature (°C)

46. Plot the data for the average heart rates on the grid; surround each point with a small circle and connect the points.

47. The independent variable in this investigation is the (1) number of trials (2) number of water fleas (3) temperature of the water (4) average heart rate

48. State the relationship between temperature and heart rate in water fleas.

Base your answers to questions 49 through 52 on the following information and diagram and on your knowledge of biology.

The diagram shows the results of a test that was done using DNA samples from three bears of different species. Each DNA sample was cut into fragments using a specific enzyme and placed in the wells as indicated

below. The DNA fragments were then separated using gel electrophoresis.

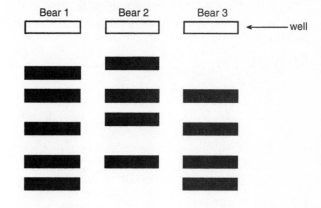

49. Which *two* bears are most closely related? Support your answer with data from the test results.

50. Identify one additional way to determine the evolutionary relationship of these bears.

51. Gel electrophoresis was used to separate the bears' DNA fragments on the basis of their (1) size (2) color (3) functions (4) chromosomes

52. Identify one procedure, other than electrophoresis, that is used in the laboratory to separate the different types of molecules in a liquid mixture.

53. On a television talk show, a guest claims that people who exercise vigorously for 15 minutes or more every day are able to solve math problems more rapidly than people who have no vigorous exercise in their daily routine. Describe a controlled experiment that could be conducted to test this claim. In your description be sure to state:

- the purpose of the experiment;
- why the sample to be used should be large;
- how the experimental group and the control group will be treated;
- the specific data to be collected during the experiment;
- *one* way to determine if the results support the claim.

Base your answer to question 54 on the following information.

A student measures his pulse rate while he is watching television and records it. Next, he walks to a friend's house nearby and, when he arrives, measures and records his pulse rate again. He and his friend then decide to run to the mall a few blocks away. On arriving at the mall, the student measures and records his pulse rate once again. Finally, after sitting and talking for a half hour, the student measures and records his pulse rate one last time.

54. Which graph below best illustrates the expected changes in the student's pulse rate according to the activities described above? (1) 1 (2) 2 (3) 3 (4) 4

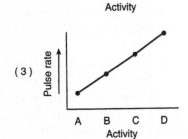

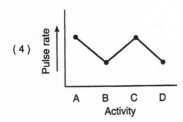

Base your answers to questions 55 through 57 on the following information and chart and on your knowledge of biology.

It has been hypothesized that a chemical known as BW prevents colds. To test this hypothesis, 20,000 volunteers were divided into four groups (of 5000 each). Each volunteer took a small pill every morning for one year. The contents of the pill taken by the members of each group are shown in the chart.

Group	Number of Volunteers	Contents of Pill	% Developing Colds
1	5000	5 grams of sugar	20
2	5000	5 grams of sugar 1 gram of BW	19
3	5000	5 grams of sugar 3 grams of BW	21
4	5000	5 grams of sugar 9 grams of BW	15

55. Which factor most likely had the greatest influence on these experimental results? (1) color of the pills (2) amount of sugar added (3) number of volunteers in each group (4) health history of the volunteers

56. Which statement is a valid inference based on the results? (1) Sugar reduced the number of colds. (2) Sugar increased the number of colds. (3) BW is always effective in the prevention of colds. (4) BW may not be effective in the prevention of colds.

57. Which group served as the control in this investigation? (1) Group 1 (2) Group 2 (3) Group 3 (4) Group 4

Base your answers to questions 58 through 61 on the information and data table below and on your knowledge of biology.

A student added two species of single-celled organisms, *Paramecium caudatum* and *Didinium nasutum*, to the same culture medium. Each day, the number of individuals of each species was determined and recorded. The results are shown in the data table below.

Culture Population

Day	Number of Paramecium	Number of Didinium
0	25	2
1	60	5
2	150	10
3	50	30
4	25	20
5	0	2
6	0	0

58. Use the information in the table to construct a line graph on a copy of the grid on page 149. Mark a scale on the axis labeled "Number of Individuals" that is appropriate for the plotted

Didinium population and for plotting the *Paramecium* population.

Culture Population

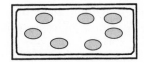

Key

Didinium ⊙

Paramecium △

59. Plot the data for *Paramecium* on the grid. Surround each data point with a small triangle and connect the points.

60. What evidence in the data indicates that *Didinium* could be a predator of the *Paramecium*?

61. State *two* possible reasons why the two populations died off between days 4 and 6.

62. Molecules *A* and *B* are both organic molecules found in many cells. When tested, it is found that molecule *A* cannot pass through a cell membrane, but molecule *B* easily passes through. State one way the two molecules could differ that would account for the differences in their ability to pass through the cell membrane.

63. If vegetables become wilted, they can often be made crisp again by soaking them in water. However, they may lose a few nutrients during this process. Using the concept of diffusion and concentration, state why some nutrients would leave the plant cell.

Base your answer to the following question on the information and diagram below.

Elodea is a plant that lives in freshwater. The diagram represents one Elodea leaf cell in its normal freshwater environment.

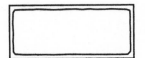

Elodea cell in freshwater

64. Predict how the contents of the Elodea cell would change if the cell were placed in salt-water for several minutes by completing, in your notebook, a copy of the diagram "Elodea cell in saltwater," shown below. Label the location of the cell membrane.

Elodea cell in saltwater

65. A scientist conducted an experiment in which he fed mice large amounts of the amino acid cysteine. He observed that this amino acid protected mouse chromosomes from damage by toxic chemicals. The scientist then claimed that cysteine, added to the diet of all animals, would protect their chromosomes from damage. State whether or not this is a valid claim. Support your answer.

Base your answers to questions 66 through 68 on the information and diagram below.

An investigation was carried out using the two setups shown below. Other than the difference shown in the diagram, all other conditions were identical.

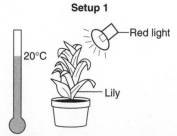

Setup 1

20°C — Red light — Lily

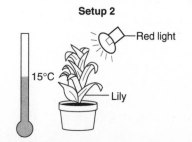

Setup 2

15°C — Red light — Lily

66. State one possible hypothesis that could be tested using these two setups.

67. What data should be collected in order to test the hypothesis you stated for question 66?

68. Describe one change that could be made in the investigation to improve it.

Base your answer to question 69 on the following information.

An experiment was designed to see what effects ibuprofen would have on laboratory mice. Large numbers of male mice and an equal number of female mice were used in this investigation. The male mice were placed in an area with food and water. The female mice were placed in a separate area of the same size. The female mice were given additional food and water. The males were each given 100 milligrams of ibuprofen each day, mixed with their food, and the females were each given 50 milligrams of ibuprofen each day, mixed with their food.

69. Identify *two* errors that were made in the design of this investigation.

Appendix

Living Environment Part D— Sample Lab Questions

Living Environment Part D: Sample Questions

Beginning with the June 2004 administration, the Regents Examination in Living Environment will include a new section, Part D. The questions on Part D will consist of a combination of multiple-choice and open-ended questions related to at least three of the four required living environment laboratory activities and will comprise approximately 15% of the examination.

These sample questions are provided to help teachers and students become familiar with the format of questions for this part of the examination. They provide examples of ways the required laboratory experiences may be assessed. A rating guide is also included.

Sample Items Related to Lab Activity #1: *Relationships and Biodiversity*

1   In the *Relationships and Biodiversity* laboratory activity, students were instructed to use a clean dropper to place each of four different samples of plant extracts on the chromatography paper. A student used the same dropper for each sample without cleaning it between each use. State one way this student's final chromatogram would be different from a chromatogram that resulted from using the correct procedure.   [1]

2   State one reason that safety goggles were required during the indicator test for enzyme *M*.   [1]

Base your answers to questions 3 through 6 on the information and data table below and on your knowledge of biology.

A student was told that three different plant species are very closely related. She was provided with a short segment of the same portion of the DNA molecule that coded for enzyme X from each of the three species.

Information Regarding Enzyme X

DNA sequence from plant species A	CAC	GTG	GAC
Amino acid sequence for enzyme X coded for by that DNA	Val	His	Leu
DNA sequence from plant species B	CAT	GTG	CAA
Sequence of bases in mRNA produced by that DNA	_____	_____	_____
Amino acid sequence for enzyme X coded for by the DNA	Val	His	Val
DNA sequence from plant species C	CAG	GTA	CAG
Sequence of bases in mRNA produced by that DNA	GUC	CAU	GUC
Amino acid sequence for enzyme X coded for by the DNA	_____	_____	_____

3   The correct sequence of mRNA bases for plant species B is

(1) GUA CAC GUU
(2) GTA CAC GTT
(3) CAU GUG CAA
(4) TCG TGT ACC

4   Use the mRNA Codon Chart on the next page to determine the amino acid sequence for enzyme X in plant species C and record the sequence in the appropriate place in the data table.   [1]

5   Is it possible to determine whether species B or species C is more closely related to species A by comparing the amino acid sequences that would result from the three given DNA sequences? Support your answer.   [1]

6   Determine whether species B or species C appears more closely related to species A. Support your answer with data from the data table. (*Base your answer only on the DNA sequences provided* for enzyme X in these three plant species.)   [1]

Universal Genetic Code Chart

Messenger RNA codons and the amino acids they code for.

		SECOND BASE			
	U	**C**	**A**	**G**	
U	UUU } PHE UUC UUA } LEU UUG	UCU UCC } SER UCA UCG	UAU } TYR UAC UAA } STOP UAG	UGU } CYS UGC UGA } STOP UGG } TRP	U C A G
C	CUU CUC } LEU CUA CUG	CCU CCC } PRO CCA CCG	CAU } HIS CAC CAA } GLN CAG	CGU CGC } ARG CHA CGG	U C A G
A	AUU AUC } ILE AUA AUG } MET or START	ACU ACC } THR ACA ACG	AAU } ASN AAC AAA } LYS AAG	AGU } SER AGC AGA } ARG AGG	U C A G
G	GUU GUC } VAL GUA GUG	GCU GCC } ALA GCA GCG	GAU } ASP GAC GAA } GLU GAG	GGU GGC } GLY GGA GGG	U C A G

FIRST BASE (left side) · **THIRD BASE** (right side)

SECOND BASE

Base your answers to questions 7 through 9 on the information and data table below and on your knowledge of biology.

In the Making Connections laboratory activity, a group of students obtained the following data:

Student Tested	Pulse Rate at Rest	Pulse Rate After Exercising
1	70	97
2	75	106
3	84	120
4	60	91
5	78	122

7 Explain how this change in pulse rate is associated with homeostasis in muscle cells. [1]

8 Identify the system of the human body whose functioning is represented by this data. [1]

9 Identify *one* other system of the human body whose functioning would be expected to be altered as a direct result of the exercise. Describe how this system would most likely be altered. [1]

Base your answers to question 10 and 11 on the information below and on your knowledge of biology.

A biology class performed an investigation to determine the influence of exercise on pulse rate. During the investigation, one group of twelve students, Group *A*, counted how many times they could squeeze a clothespin in a 1-minute period, then exercised for 4 minutes, and repeated the clothespin squeeze for an additional 1 minute. Another group of twelve students, Group *B*, also counted how many times they could squeeze a clothespin in a 1-minute period, but then they rested for 4 minutes, and repeated the clothespin squeeze for an additional 1 minute. The data table below shows the average results obtained by the students.

Effect of Exercise on Number of Clothespin Squeezes

Groups of Student	Average Number of Clothespin Squeezes During First Minute	Average Number of Clothespin Squeezes During Second Minute
Group A (exercise)	75	79
Group B (rest)	74	68

10   State *two* specific examples from the description of the investigation and the data table that support this investigation being a well-designed experiment.   [2]

11   The chart below shows relative blood flow through various organs during exercise and at rest.

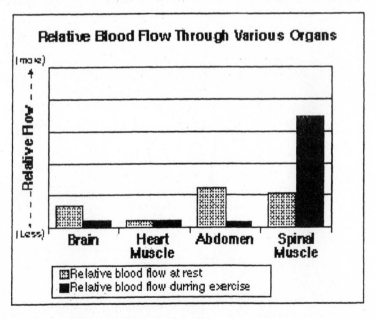

Using information from both the data table and the chart, explain how muscle fatigue and blood circulation could account for the results the students obtained.   [2]

Base your answers to questions 12 and 13 on the diagrams below and on your knowledge of biology.

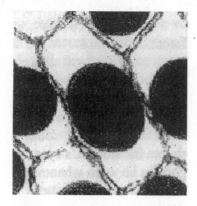 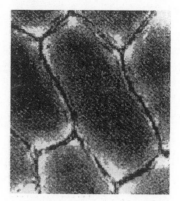

Diagram 1: red onion cells   Diagram 2: red onion cells

12 Describe how to prepare a wet-mount slide of red onion cells with the cell membrane shrinking away from the cell wall, as shown in diagram 1. The following materials are available: microscope slide, pipettes, cover slips, paper towels, water, salt solution, and red onion sections. [3]

13 List the laboratory procedures to follow that would cause the cells in diagram 1 to resemble the cells in diagram 2.   [2]

14 A student places an artificial cell, similar to the one used in the laboratory activity *Diffusion Through a Membrane*, in a beaker containing water. The artificial cell contains starch and sugar. A starch indicator is added to the water in the beaker. Explain how the student will know if the starch is able to diffuse out of the artificial cell.   [1]

Scoring Guide for Sample Part D Questions

1   Allow 1 credit for stating one way the student's final chromatogram would be different from a chromatogram that resulted from using the correct procedure. Acceptable responses include, but are not limited to:

— The number of bands could be different.
— The size of the spots would differ.

2   Allow 1 credit for stating one reason that safety goggles were required during the indicator test for enzyme *M*. Acceptable responses include, but are not limited to:

— Some of the chemicals might splash into the student's eyes.
— It is a laboratory requirement that goggles be worn whenever chemicals are used in the lab.

3   **1**

4   Allow 1 credit for **Val   His   Val**

5   Allow 1 credit for indicating that it is not possible to determine whether species *B* or species *C* is more closely related to species *A* by comparing the amino acid sequences, and providing an explanation that supports this response. Acceptable responses include, but are not limited to:

— The amino acid sequences do not make it possible to determine whether species *B* or species *C* is more closely related to species *A*, because both *A* and *B*, and *A* and *C* only differ by one amino acid.

Allow credit for an answer that is consistent with the student's response to question 4.

5   Allow 1 credit for using data from the data table to indicate whether species *B* or species *C* appears to be more closely related to species *A*. Acceptable responses include but are not limited to:

— Species *A* and *B* are the closest because they have only three differences in their DNA code, while species *A* and *C* have four differences.
— Species *A* and *B* are more closely related because there are not as many differences between their DNA sequences as there are between species *A* and *C*.

7 Allow 1 credit for explaining how this change in pulse rate is associated with homeostasis in muscle cells. Acceptable responses include but are not limited to:

— Muscle cells produce more carbon dioxide when they are active than when they are not very active. The increased blood flow carries away the extra carbon dioxide to the lungs, where it can be excreted.
— Muscle cells use more oxygen when they are active than when they are not very active. An increased circulation rate brings more oxygen to the muscle cells.
— Muscle cells produce more heat when they are active than when they are not very active. The increased blood flow helps carry away the excess heat to the skin, where it is lost to the surroundings.

8 Allow 1 credit for indicating that the functioning of the circulatory system is represented by the data.

9 Allow 1 credit for identifying one other system of the human body whose functioning would be expected to be altered as a direct result of the exercise *and* how it would be most likely be altered. Acceptable responses include, but are not limited to:

— respiratory system – increase the intake of oxygen
— respiratory system – increase the exhalation of carbon dioxide
— excretory system – increased sweating

10 Allow a maximum of 2 credits, 1 credit for each of two specific examples that support this investigation being a well-designed experiment. Acceptable responses include but are not limited to:

— It has two groups for comparison (experimental and control).
— It is a controlled experiment.
— Each group did exactly the same thing except for the exercise/rest part.
— Each group contained the same number of students.

11 Allow a maximum of 2 credits, 1 credit for indicating that muscle fatigue and/or waste buildup was the reason for group *A*'s results, and 1 credit for indicating that increased circulation from exercise was the reason for group *B's* results.

Example of a 2-credit response:

The squeezing caused the muscles to become tired as waste products of muscle activity built up. The increased circulation shown in the second table (blood flow to skeletal muscle during exercise) helped the students in group *B* carry the wastes away from their muscle cells so they could continue to function efficiently. Students in group *A* did not increase their circulation rate, so the wastes interfered with the ability to keep squeezing the clothespin.

2 Allow a maximum of 3 credits for correctly describing how to prepare a slide showing red onion cells with the cell membrane shrinking away from the cell wall. The responses must include:

•   a simple description of how the onion cells will be obtained

- a description of how a wet-mount slide is made
- an indication that salt solution must be used

Acceptable 3-credit responses include, but are not limited to:

— Break the onion piece and peel off the skin. Put it on a slide and add water. Add a coverslip. Place a piece of paper towel on the edge of the coverslip and add salt solution to the other side. After adding a few drops of salt solution, use a microscope to observe the slide.

— From the onion section obtain a piece of onion skin and put it on a slide. Use a pipette to add some salt solution, then add a coverslip on the onion skin.

13   Allow a maximum of 2 credits for listing the procedures that should be followed to cause the cells in diagram 1 to resemble the cells in diagram 2. The responses must include:

- adding water to the slide[1]
- using a technique that will cause the salt to be rinsed away and replaced with water [1].

Acceptable 2-credit responses include, but are not limited to:

— Remove the slide from the microscope then place a piece of paper towel on the edge of the coverslip and add distilled water (or just water) to the other side. After adding many drops of water put the slide back on the microscope to observe the change. If necessary, repeat the process adding even more water.

— Remove the slide from the microscope, remove the coverslip and carefully rinse the onion skin with water to remove the salt, put the coverslip back on the slide.

14   Allow 1 credit for explaining how the student will know if the starch is able to diffuse out of the artificial cell. Acceptable responses include but are not limited to:

— If the starch diffuses out, the indicator solution will turn black or blueblack.
— If the starch is able to diffuse out, the starch indicator will change color.

Science, Technology, and Society

The Eastern Timber Wolf: Is It Welcome Back East?

Howls of the eastern timber wolf once filled the air in New England forests. The eastern timber wolf once roamed from New England to the Great Lakes, and from southeastern Canada to the Hudson Bay. A subspecies of the gray wolf, the eastern timber wolf varies from black to gray to white in coat color.

In the 1600s, European settlers in New England had a deep fear of wolves, which were blamed for wiping out entire populations of livestock. Beginning in 1630, the Massachusetts Bay Colony and others paid hunters to trap, shoot, and poison wolves. Records show that the last wolf in the region was killed in New York in 1897.

Today, the eastern timber wolf survives in just a fraction of its original range. All that remain are a population of about 2000 wolves near Lake Superior in Minnesota and a few hundred wolves spread through Michigan and Wisconsin.

The wolf became protected under the Endangered Species Act in 1974. A species is considered endangered if it is at risk of becoming extinct in all or a large part of its range. Being classified as endangered offers a species special protection by the federal government. That means it is a crime to kill a plant or animal that is endangered.

In April of 2004, the U.S. Fish and Wildlife Service reclassified the wolf from endangered to threatened. A species is considered threatened if it is likely to become endangered in the near future. The level of federal protection offered to a threatened species is much less than that for an endangered species.

As a result, this decision opened up opportunities for citizens to kill wolves. For example, owners of livestock that were threatened by a wolf would have the legal right to kill the wolf.

The decision of the U.S. Fish and Wildlife Service was based on the fact that they considered two distinct populations of wolves to be thriving. This was the requirement needed to change the listing. The one population is in Minnesota and the other one is in Wisconsin-Michigan.

A group of 19 conservation and wildlife-protection groups filed a lawsuit against the U.S. Fish and Wildlife Service. The groups argued that reclassifying wolves was against the Endangered Species Act. In February of 2005, a judge ruled that the government could not reduce federal protection for the gray wolf. In his judgment, U.S. District Court Judge Robert Jones of Oregon argued that the decision to delist the wolf was not based on valid scientific data.

While supporters of the wolf celebrated a victory, opponents quickly opposed the ruling. The U.S. Fish and Wildlife Service released a statement disagreeing with the decision. Similarly, private groups with interests in the wool and cattle industries argued that their rights were not being protected.

A growing portion of society is now in favor of saving wolf populations and reintroducing wolves to their natural habitats. Scientists are beginning to see the effect of the loss of wolves on their natural habitats. Some ecologists fear that the complete effect of such a loss may not be realized for generations to come.

Ecologists point out that wolves help bring balance to their ecosystems. They keep populations of large herbivores, such as deer and moose, from growing too large for the resources. In this way, wolves preserve the biodiversity of the region.

Supporters argue that restoring wolves to their natural habitats can also bring economic benefits to nearby communities. A significant increase in tourist dollars has been generated in locations where wolves live. The reintroduction of wolves to North Carolina and to Yellowstone National Park has resulted in millions of dollars for local communities. Supporters predict that returning the wolf to New England would have a similar effect.

Many supporters of the wolf want an active plan to restore the wolf to New England. A similar program was carried out at Yellowstone National Park. However, restoration of the wolf is a subject of controversy wherever it occurs. It is proving to be controversial in New England as well.

Some supporters of the wolf advocate a passive recovery. This program would allow wolves to move down, naturally, from Canada. However, many scientists think it is unlikely that wolves will return to the region on their own. The wolves would have to pass through farmlands, roadways, and waterways. Even if a stray wolf did make the journey, it would not have other wolves to breed with.

A large a portion of the population is against the reintroduction of the wolf. They do not want wolves in New England under any circumstances; and their concerns are equally valid. Their opposition is based on the fact that as the population of wolves increases, conflicts with humans will also increase. These opponents fear that wolves will attack people, prey on livestock, and kill pets.

Deer hunting in northern New England is another concern. This pursuit is a strong tradition and hunters often equate reintroduction of the wolf with the loss of the animals they hunt. Others fear that protection of wolves and their habitat will cause a loss of logging jobs.

Unlike the program in Yellowstone, there is no federal mandate to restore wolves to the Northeast. As a result, public support is required for restoration to succeed. Supporters are working on plans to find an acceptable compromise.

For example, paying back owners for livestock losses may decrease some anger toward wolves. Another plan involves developing zones for separate populations of wolves that can exist without posing a threat to livestock.

Many people on both sides of the argument have misinformation about wolves. Some may have an unnecessary fear of the wolf; others may have an unrealistically innocent image of the wolf. Perhaps the most important thing needed for successful restoration of wolves is the education of people. A management plan that includes some form of wolf population control may be a good compromise, too. Only time—and our reaction to the sound of wolves howling—will tell.

Questions

1. Why is it that wolves do not inhabit New England today?

2. What does it mean for the wolf to be classified as endangered?

3. What are some arguments for and against the restoration of the wolf to New England?

Nanotubes: Can Tiny Tubes Deliver Drugs to Cells?

A new material is showing great promise in industries from communication to electronics to medicine. The material consists of tiny tubes called *nanotubes* (a nanometer is 1×10^{-9} m). Each tube is so tiny that it is about 10,000 times thinner than a human hair.

A nanotube is made from a network of carbon atoms. There are two basic types of nanotubes. A single-walled nanotube is a sheet of carbon atoms rolled up in a tube. A multiwalled nanotube is a single tube placed inside a series of larger tubes.

The excitement about nanotubes stems from their unusual combination of properties. The tubes have proved to be extremely strong and can conduct heat and electricity well. In fact, they are stronger than steel but as flexible as plastic; and they conduct energy better than any other known material.

Nanotubes have the potential to revolutionize many industries. Perhaps one of the most desirable properties of nanotubes is that they can be made from common materials, such as methane gas.

Some nanotube researchers hope that nanotubes can replace silicon chips in many items. One of the reasons is that the carbon atoms from which nanotubes are formed bond tightly to each other. They shift to form stable, hexagonal rings. As a result, nanotubes basically repair themselves by moving around to replace any atoms that are moved.

Unlike silicon, which may not function with slight defects, carbon nanotubes can work with defects. Large, expensive machines are required to manufacture silicon chips. Replacing silicon chips with nanotubes that assemble themselves and correct for defects will eliminate the need for these machines.

The properties of the bonds also make the tubes resistant to damage. A single-walled nanotube can be bent at a 120-degree angle and bounce back to its original form without damage.

Another advantage of nanotubes is that they can be relatively long. Researchers have created nanotubes as long as 4 microns. Although that may not be long compared to the objects around you, it is 40 times as long as average-sized features on silicon chips. In theory, nanotubes could be used to replace wires on such devices as airplanes. The tubes would add strength while decreasing weight.

Some of the most exciting uses for nanotubes may be right under your nose—in the human body. In one research study, scientists at the University of Florida used nanotubes to find and retrieve DNA dissolved in a solution. This task is commonly used in medical research as well as in criminal forensics. This is the first time that a nanotube filter, based on DNA, was used to perform the task. Scientists formed strands of DNA molecules inside nanotubes within a membrane. The strands in the nanotubes captured complementary strands of DNA molecules in a solution on one side of the membrane and carried them through the nanotubes to a solution on the other side of the membrane.

Researchers hope that this technique will someday be used for technologies that involve gene separation. Separating genes is a process that may become essential to health care.

Another team of researchers from Rice University, the University of Texas Health Science Center, and Houston's Texas Heart Institute made medical breakthroughs with nanotubes. They incubated white blood cells in dilute solutions of nanotubes. Their research suggests that the cells treat nanotubes like other types of particles. The cells ingest the nanotubes and seal them off inside phagosomes. (Phagosomes are membrane-bound vesicles formed when the cell membrane folds inward to hold matter taken into the cell.)

The first important observation they made was that the nanotubes did not have any negative effects on the cells. The second observation was that the nanotubes retain their unique properties. Each type of single-walled carbon nanotube gives off a unique fluorescent signature. Scientists were able to use specialized microscopes tuned to the near-infrared region to detect the

nanotubes. In this way they were able to identify their specific locations within cells.

This research may lead to the use of nanotubes as imaging agents. Because they give off light in wavelengths not normally emitted by biological tissues, scientists hope to be able to selectively detect nanotubes within the body. This may be particularly useful in situations where traditional materials used to trace processes in the body cannot be used.

In other studies, French and Italian scientists used nanotubes to inject drugs into targeted cells. The nanotubes can drill through cell membranes without damaging the cell. If substances such as proteins or nucleic acids are attached to the nanotubes, they can also go through the membrane. The researchers were trying to find out if small molecules such as antibiotics or drugs could be carried directly into the cells as well.

The researchers also wanted to find out if they could attach a marker to the nanotube. In this way, they could not only deliver drugs directly to a cell, but they could trace the movement of the drug throughout the cell. In fact, the scientists were successful in delivering an antifungal agent into a cell and they developed a method of attaching a marker in a controlled fashion.

One of the main drawbacks of using nanotubes on a large scale is how to produce enough nanotubes to make them useful. The difficult and slow process of manufacturing nanotubes still makes them too costly for widespread use.

One of the challenges to the manufacturing process is to remove impurities. Another challenge has to do with how the hexagons are arranged. If the carbon hexagons run in parallel lines, the atoms will act like a metal. If the rows of tubes are slightly swirled, they will act like semiconductors.

A third challenge is to find a reliable method of arranging nanotubes in products. For many devices, the placement is not a problem because the nanotubes are painted into place. In chips, however, individual nanotubes must be placed in specific locations. This placement is more difficult to control.

Questions

1. What are nanotubes made up of?
2. How do nanotubes repair themselves?
3. How is this self-repair related to their structure?
4. Why are nanotubes so useful for delivering drugs to specific cells?

The New Food Pyramids: Building a Better Diet?

Do you know how many servings of fruits and vegetables you should eat each day? What about meats, grain, or dairy products? The Food Guide Pyramid is an educational tool in which the United States Department of Agriculture (the USDA) makes recommendations about the types and amounts of foods that should be eaten each day. In theory, if a person follows a diet that sticks to these recommendations, he or she will obtain the nutrients needed to maintain good health.

The original pyramid, which was introduced decades ago, listed the foods that should be eaten in the greatest quantities at the bottom. Those foods that should be eaten in the smallest amounts were listed at the top.

Since the original pyramid was developed, researchers have formed a better understanding of the relationship between diet and health. In 2005, the USDA replaced the original food pyramid with a new one called *MyPyramid.*

The new dietary guidelines upon which MyPyramid is based reflect an increased focus on physical activity and recognize the fact that some fats are beneficial.

Source: http://www.mypyramid.gov/

Rather than the layers of the old pyramid, the new pyramid is divided into six sections. The pyramid is color coded. Orange represents grains, green is for vegetables, red is for fruits, yellow is for oils, blue is for dairy, and purple is for meat and beans. Although each stripe of color begins as the same size at the top of the pyramid, they expand to different widths at the bottom. The width is related to the amount of each food type that should be consumed. A figure shown running up a set of stairs along the edge of the pyramid represents physical activity.

The new food pyramid aims to correct misconceptions people have based on the old pyramid. There are no words or labels on the new pyramid; it is designed to be quick and simple to use. By clicking on each section at the USDA's web site, you can find information about the foods represented in that section.

The USDA did not put recommended servings because they can vary with each person's diet. At the web site, people can develop a personalized guide. However, millions of Americans without access to the Internet will not be able to get this important information.

Those who do read the information in MyPyramid will learn that obtaining half their supply of grains from refined starches is acceptable. Unfortunately, refined starches behave like sugars. They provide empty calories and increase the risks of diabetes and heart disease.

In addition, readers will find that MyPyramid does not differentiate among the different types of fat in red meat, poultry, fish, and beans.

Researchers at the Harvard School of Public Health, led by Walter Willett, have posed a challenge to the USDA guidelines in a different pyramid known as the *Healthy Eating Pyramid.*

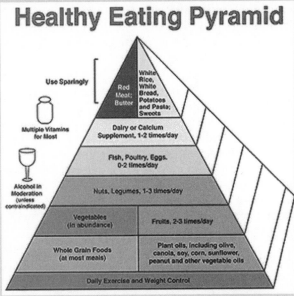

source: http://whyfiles.org/179food_pyramid/2.html

At first glance MyPyramid and the Healthy Eating Pyramid appear similar because they have the same shape; and both pyramids recommend plenty of fruits and vegetables. However, that is where the similarities end. Willett's pyramid has exercise at the bottom. This stresses that while a good diet is important, diet alone is not enough to maintain good health.

Willett's pyramid also reduces the suggested intake of carbohydrates and encourages people to identify the healthier options in each food group. The pyramid is based on the premise that not all fat is bad fat, a high intake of starch from refined grains is not healthy, and red meat increases the risk of diabetes and heart disease.

The pyramid emphasizes whole-grain, high-fiber carbohydrates and healthy fats. At the tip of the pyramid are foods with little nutritional value such as butter, refined starches, and sugars.

Willett's pyramid is not without scientific research. As part of the Nurse's Health Study, which began in 1976, Willett's group followed the eating habits of 140,000 people. Willett's group scored the people according to both the USDA's system and the similar scoring system his team developed according to the Healthy Eating Pyramid guidelines.

At the end of 2002, Willett's team concluded that people who followed the guidelines of the Healthy Eating Pyramid had a much smaller risk of chronic disease than the people who adhered to the USDA's guidelines.

Both the USDA and Willett's group stand by their food pyramids. It is up to society to weigh the validity of each and to use the research to develop a healthful diet.

Questions

1. Why should people include a variety of food types in their diets every day?
2. How is the USDA's new MyPyramid different from the original Food Guide Pyramid?
3. What shortcomings in MyPyramid led the researchers at Harvard University to develop the Healthy Eating Pyramid?

Obesity: Is It An American Epidemic?

Over 60 percent of Americans over the age of 20 are overweight. About half of those people are considered to be obese. Obesity is not limited to adults. A nationwide study conducted by the Centers for Disease Control and Prevention (CDC) showed that more than 16 percent of high school students were overweight and nearly 10 percent were obese. This recent increase in obesity in the United States has led some people to call it an epidemic.

To determine whether a person is obese, doctors use the body mass index, or BMI. A person's BMI can be found by multiplying his/her weight in pounds by 703 and then dividing the result by his/her height in inches squared. A BMI of less than 25 is considered healthy. A BMI of more than 30 is considered obese. Measurements between 25 and 30 indicate that a person is overweight.

While genetics does play a role in obesity, the major increase in obesity during the last several decades is much greater than can be explained by possible genetic changes. Instead, society must look to behaviors and lifestyles to find the immediate cause for the increase.

Perhaps the greatest underlying cause of obesity is a sedentary lifestyle and lack of exercise. With the increased accessibility to electronic devices, such as televisions, computers, and video games, people are spending more time sitting in front of a screen. In previous years, this time might have been spent performing some physical activity.

Processed foods have become more available and affordable. Advertising exposes children and adults alike to the convenience of high-fat (and high fructose) foods. As a result, many people consume too many high-calorie foods.

Obesity is much more than a cosmetic issue. Every year in the United States, almost 300,000 adult deaths are related to obesity. Obesity is about to surpass smoking as the number-one preventable cause of death in the United States. Obese people are at risk for many health problems, including heart disease, type-2 diabetes, high blood pressure, stroke, and some forms of cancer. Other health problems linked to obesity include gallbladder disease, liver disease, osteoarthritis, breathing problems, and reproductive problems in women.

Aside from the physical risks of obesity, emotional stress may be one of the most painful aspects. Obese people are often ridiculed and criticized. They may suffer discrimination at school, in the workplace, and in social situations. This can lead to severe depression.

Treating obesity is a lifelong goal. The way that an obese person is treated depends on their level of obesity and their overall health. The three main aspects of treatment include modifying nutrition, physical activity, and behavior.

Nutrition. People are taught to analyze their food intake in terms of nutrition, calories, and fat. Simply reducing the amount of food or following fad or binge diets is not the solution.

Physical Activity. Under the care of a physician, obese people are guided through an exercise regimen aimed at burning calories and losing weight.

Behavior Modification. Improving health and losing weight requires a complete change in behavior. Some people who start to lose weight soon revert back to old habits. As part of a treatment program, people are taught that the changes they make must continue for life. During the process, it is suggested that people rely on family, friends, or healthcare professionals for support.

In some cases of severe obesity, surgery may be recommended. Some types of surgery, known as restrictive operations, limit the amount of food that can be taken in without altering the normal operation of the digestive system.

In one type of surgery, a band made of silicone is placed around the upper end of the stomach. This creates a small pouch at the top of the stomach and a narrow tube leading to the rest of the stomach. The pouch can hold between 1 and 3 ounces of food. The narrow tube slows the passage of food into the stomach. This makes the person feel a sensation of fullness for a longer period of time.

The band contains a salt solution and is attached to a tube that leads to a port just under the skin. By increasing or decreasing the amount of salt solution in the band, a doctor can adjust the size of the band from outside the body.

In another type of surgery, known as a *malabsorptive operation,* the process of digestion is altered. In the digestive system, food passes from the mouth, through the esophagus, and into the stomach. Form there it passes into the small intestine and then to the large intestine. The small intestine is where most of the nutrients and calories from food are absorbed. In a malabsorptive operation, part of the small intestine is taken out of the food pathway. As a result, fewer calories and nutrients are taken into the body. This means that a smaller amount of material is available to be converted into fat.

Malabsorptive operations are rarely recommended on their own because they have a high risk of causing nutritional deficiencies. Although the person will lose weight, they will also miss out on nutrients the body needs to survive. Combined operations that restrict the stomach and bypass just part of the small intestine are more common.

In 2002, the Internal Revenue Service (IRS) declared that obesity is a medical disease. This means that treatments for obesity can be claimed as medical deductions on a person's taxes. Obese people may also be entitled to disability payments from the Social Security Administration. A recent government study indicates that medical costs related to obesity reach $75 billion per year. Roughly half that cost is funded by taxpayers through Medicare and Medicaid programs.

Questions

1. Why is obesity sometimes called an "epidemic"? Do you think it is one?

2. What factors are considered more important than genetics in causing the recent trend toward obesity?

3. Besides surgery, what are three ways in which obesity can be treated?

4. Do you think taxpayer money should be used to fund medical treatments for obesity? Present your argument.

Vaccinations: Is the Medical Miracle a Mercury Risk?

In the early part of the 1900s, in the United States, as many as 20,000 people became paralyzed from polio every year; an average of 450 children died from illnesses related to measles; and 5 out of every 1000 children died from whooping cough.

Fortunately, people today are not faced with these same statistics. What changed these gruesome data? It was the introduction of vaccines. Not many scientific discoveries have had as great an effect on health as have vaccines.

A vaccine is a substance that helps the body fight infections that cause disease. It works by triggering the body's own immune system. Once the immune system has fought a particular agent of disease, it produces antibodies to it. If the person is exposed to the same agent again, those antibodies detect and inactivate them.

Most vaccines work by exposing a person to a weakened or killed form of a disease agent, such as a microbe, to cause their immune system to produce antibodies. The vaccine does not cause the disease, but does "teach" the immune system to recognize and destroy the agent that causes it.

The development of the modern vaccine began more than 200 years ago with the work of an English doctor named Edward Jenner. In 1796, Jenner tested the first vaccine by injecting a young boy with live cowpox virus to develop immunity to the smallpox virus.

Almost 100 years later, the French chemist Louis Pasteur investigated bacteria. He showed that people could be protected against disease by first being exposed to the weakened germs. In 1885, Pasteur administered the first rabies vaccine.

Scientists learned that vaccines could be made using killed germs. By the early 1900s, vaccines were developed to prevent typhoid, cholera, and plague. Two other vaccines produced since then are the inactivated polio vaccine by Dr. Jonas Salk and the live polio vaccine by Dr. Albert Sabin. Vaccines have also been developed to prevent measles, mumps, rubella, and hepatitis B.

Most healthcare providers would agree that routinely immunizing children saves millions of lives every year. Yet over two million children around the world continue to die from preventable diseases. More lives could be saved if vaccines were produced faster and less expensively. New vaccines to prevent diseases such as malaria and new forms of the flu are also needed.

Recent vaccine research has shifted away from vaccines made up of whole organisms. Instead, scientists are focusing on using only those parts needed to cause the immune system to respond. A useful tool in this endeavor is genetic engineering. Using this process, scientists can take all or part of the DNA from one organism (the disease agent) and insert it into the cells of another. The researchers would use microorganisms to produce proteins from infectious agents, such as the hepatitis virus, so they can be used in vaccines.

However, no single process can be used for every vaccine. Each disease presents its specific characteristics and challenges. Some microbes are more likely to vary and mutate than others. In addition, once a possible vaccine is developed in the laboratory, it must be tested through clinical trials. These trials can take many years as researchers track the effectiveness of the vaccine.

Meanwhile, researchers are working toward developing a single vaccine that would protect a child from all of the major infectious diseases for life. The vaccine might have a time-release property by which some contents would be released right away, while others would be delivered over time.

Vaccines, like other technologies, have always had opponents. From the very start, people were upset by the idea of injecting a disease into a healthy individual. Only over time did people come to recognize the value of vaccines.

Recent controversy has centered on a possible connection between childhood immunizations and autism. Most concerns are based on the use of mercury as an ingredient in vaccines. Mercury is a chemical element that is toxic to humans at high levels. It can result in disorders of the nervous system.

Mercury has been used in vaccines in a compound called *thimerosal*, which is a preservative used to prevent bacterial contamination. Although thimerosal has been used since the 1930s, vaccine manufacturers began reducing the amount of thimerosal in childhood vaccines in 1999. Now all recommended vaccines manufactured for children in the United States contain no thimerosal or only trace amounts. The flu vaccine, however, still contains thimerosal.

A growing number of parents and members of the medical community are looking at the possible link between mercury and autism. Autism is a developmental disability for which there is no known cure.

Over the last two decades, the number of cases of autism has been increasing at an alarming rate. Some states have reported a complete doubling in cases in as little as four years. While once considered rare, autism is now surpassing all other disabilities in children. In California, for example, the number of children diagnosed with autism has increased from 1 in 10,000 to 1 in 150 in just 20 years.

People are looking for answers and although other possibilities have been suggested, the number one source of blame has been childhood vaccines. In addition to thimerosal, vaccines may even contain such toxic chemicals as formaldehyde,

aluminum phosphate, and methanol. As a result, some parents are refusing to even have their children vaccinated.

In addition, the required vaccination schedule has increased from 23 doses of seven vaccines to 38 doses of 12 vaccines in the last two decades as well. Some argue that multiple vaccines at the same time may be too much for a young child's immune system to handle.

At this point, studies have not proved either side of the argument. Further studies are being conducted in the hopes of stopping this unfortunate trend.

Questions

1. How does a vaccine protect a person from getting a disease?
2. What is genetic engineering and how might it be useful in vaccine production?
3. Why are some people refusing to have their children vaccinated?
4. What are the possible benefits and risks of these parents' decision?

Fetal Surgery: Do the Benefits Outweigh the Risks?

Each day thousands of people undergo operations to correct or repair some health problem. Occasionally, health problems are identified in fetuses while they are still developing in the uterus. Although most health problems diagnosed before birth are best treated after the baby is born, a few conditions benefit from immediate treatment. A surgical procedure conducted on a fetus before it is born is called *fetal surgery*.

Fetal surgery has been performed for more than two decades in the United States. Only a few facilities around the world perform such surgery. It requires a combination of many specialists (such as pediatric surgeons and obstetricians), as well as specialists to monitor heart rate and blood pressure and to take sonogram images and echocardiograms.

Although no two procedures are identical, fetal surgery basically involves surgically opening the mother's uterus, correcting the problem for the fetus, returning the fetus to the uterus, and closing the uterus. The fetus may be reached through a Cesarean section or through a series of smaller incisions called fetoscopic port injections. Doctors insert small scopes and use video equipment during this type of procedure.

As with all types of surgery, fetal surgery is not without risks. In this case, however, there are two sets of risks to consider. Each situation requires doctors to weigh the benefits of the procedure against the risks to the mother and fetus.

Risks to the mother include developing an infection, losing blood, gaining weight due to bed rest, and forming gestational diabetes. This is a form of diabetes that can be developed by pregnant women.

The major risk to the fetus is that the surgery might lead to premature labor and delivery. If the baby is born too early, its organs will not be mature enough to work properly and the baby will be unable to survive. In addition, the baby may require a blood transfusion, bleed from the brain, or even die.

At most facilities where fetal surgery is done, it is performed to correct one of the following conditions:

- Urinary tract obstruction
- Congenital diaphragmatic hernia
- Congenital cystic adenomatoid malformation
- Twin-twin transfusion syndrome

Urinary Tract Obstruction (UTO). Problems with the fetal urinary tract can lead to a deficiency of amniotic fluid, incomplete development of the lungs, and death. During fetal surgery, doctors attempt to clear any valves that may be blocked and return amniotic fluid to normal levels. Amniotic fluid is the liquid in which the fetus floats during development.

Congenital Diaphragmatic Hernia. This condition results from a defect that forms as the fetal diaphragm develops. The diaphragm is a large muscle at the bottom of the chest cavity. When a hole forms in the muscle, organs such as the liver move up into the chest cavity. They compress the lungs, preventing them from growing properly. During surgery, organs are moved to where they belong and the opening in the diaphragm is repaired. After the baby is born, it is usually placed on support immediately to give the lungs a chance to recover and expand. Infants who survive may still suffer from long-term complications such as recurrent lung infections, intestinal problems, and hypertension.

Congenital Cystic Adenomatoid Malformation. This condition involves a mass of tissue that forms in the lung. The abnormal lung tissue does not work like normal lung tissue.

Several different events can occur. The mass may grow with the fetus, but not prevent the lungs from developing properly. The mass may remain the same size and become relatively small as the fetus grows. The mass may even shrink or disappear before birth. In a few cases, however, the mass grows so large that it threatens the life of the fetus. The mass can also push on the heart

and the esophagus. In these cases, fetal surgery may be an option used to cut out the mass in the lung.

Twin-Twin Transfusion Syndrome. This is a serious progressive disorder that can occur in twin fetuses that share a placenta. Blood transfuses from one fetus, known as the donor twin, to the other. The donor fetus shuts down many of its internal organs. The other fetus, known as the recipient, produces extra amounts of urine and is surrounded by a large amount of amniotic fluid. The recipient's blood becomes thick and difficult to pump. Without treatment, roughly 80 percent of the twins will die. The most common treatment is to drain amniotic fluid from the sac of the recipient fetus. In some cases, a thin fiber-optic scope is inserted into the sac of the recipient. Doctors use the scope to direct a laser to destroy the abnormal vessel connecting the twins.

Spina Bifida. Fetal surgery was once performed only to correct conditions that threatened the life of a fetus, such as those just described. However, fetal surgery has in addition been successfully used for fetuses diagnosed with spina bifida. Spina bifida is not necessarily life threatening. It is a debilitating birth defect in which babies experience lifelong disabilities such as weakness or paralysis of the lower limbs. People with spina bifida may also have deformed skeletons and mental impairment.

The spinal tissue of a fetus with spina bifida is exposed in the womb. Movement and amniotic fluid damage the tissue. Through surgery, doctors can protect the spinal tissue. They cannot repair neurological functions that have already been damaged, but they can prevent further loss.

Doctors have also discovered an added benefit of this type of fetal surgery. Children with spina bifida usually need a device called a shunt to drain fluid off the brain. The average child needs three or more shunts throughout his or her lifetime. The surgery for spina bifida somehow affects development of the brain, causing some defects to repair themselves. As a result, children who have had the fetal surgery have a reduced need for shunts later on in life.

Questions

1. What is fetal surgery?

2. In what types of situations is fetal surgery considered?

3. Most facilities that perform fetal surgery have an ethics committee. What might be the role of this committee?

Introducing Snuppy: Give the Dog a Clone?

Look around and you will find that people have unique combinations of traits. With the exception of identical twins, even siblings in the same families have physical variations. The reason is that humans and other animals inherit one set of chromosomes from each parent. The way in which these chromosomes separate and recombine during sexual reproduction determines the traits that are passed down.

In bacteria and other organisms that can reproduce asexually, an exact copy of the genetic information of the parent is passed down to the offspring. As a result, offspring are genetically identical to each other and to the parent.

In 2005, a team of South Korean scientists, led by Hwang Woo-suk of Seoul National University's College of Veterinary Medicine, produced a dog that was genetically identical to its one parent. The dog was not the result of sexual reproduction as normally occurs in dogs. It was also not the result of asexual reproduction. Instead, the dog was produced through cloning. A *clone* is generally defined as an organism that is produced from a single cell of a parent organism; that is why the clone is genetically identical to its parent.

The dog, named Snuppy, was not the first cloned animal. In fact, he comes at the end of a long list of cloned frogs, sheep, cats, goats, cows, mice, pigs, and others. What makes Snuppy so special is that he is the first canine to be cloned. Differences in the reproductive system of dogs make them much harder to clone than other animals.

The science of cloning is not new. Scientists were able to clone frogs in the 1950s and mice in the 1980s. In the 1990s, scientists cloned a sheep. In 1996, the technology was revolutionized when researchers cloned a sheep named Dolly by using a body cell of an adult. Up until that time, only reproductive cells were used as parent cells for cloning. The new procedure allowed scientists to use body, or *somatic*, cells.

In this procedure, the genetic material from the nucleus of a donor adult cell is transferred to an egg cell from which the nucleus has been removed. The egg cell holding the genetic material from the donor cell is treated with chemicals or an electric current to stimulate cell division. The cloned embryo is then placed in the uterus of a surrogate mother, where it continues to grow until birth.

Snuppy was produced from a single ear cell taken from a three-year-old male Afghan hound. His embryo was placed in the uterus of a yellow Labrador retriever to develop. After finishing development, Snuppy was born by means of Caesarean section.

The announcement of Snuppy's existence has led to ongoing scientific, political, and ethical debates about the uses of cloning technology. While some people welcome the achievement, others fear it is just one more step toward the eventual cloning of humans. In fact, that same research team had claimed they created the first cloned human embryos but, in January 2006, their research claims were discredited.

Hwang's research team argues that the canine clone is an important research tool. They suggest positive outcomes, such as possibly preserving endangered species. By using cells from endangered adult animals, scientists may be able to increase their population at a faster rate than would otherwise be possible. In addition, they say it will provide an opportunity to learn how genes affect traits in different breeds of dogs.

Many animal activists criticize the research as not only inhumane, but wasteful. First, cloning Snuppy was not an easy task. The researchers succeeded only after conducting numerous surgical procedures on more than 100 dogs to implant over 1000 embryos. In the end, three dogs became pregnant. One dog had a miscarriage and lost the embryo. Of the two cloned puppies that were born, one died after just three weeks. Opponents argue that too many failures occur before just one clone is produced.

Second, animal activists point out that there already is an overabundance of unwanted dogs throughout the world. It seems irresponsible to deliberately invest such a tremendous amount of effort and money into producing yet more dogs.

So why continue the cloning? Let's consider the stem cell issue. During the early stages of an embryo's growth, the cells are all basically the same. These cells, known as *stem cells*, are not yet specialized for their specific functions. As an embryo develops, its cells become specialized into all the cells of its body.

Many researchers think they can use stem cells to grow tissues and organs that can be implanted in people who need them. Researchers also hope to learn how to control the development of cells in order to cure some diseases.

A great deal of controversy surrounds stem cells because the cells used in research must be taken from human embryos. Some supporters of the canine clone suggest that developing cloned dog embryos could parallel studies in human stem cells. They point to the possibility of testing human stem cell therapies on dogs. In the process, this may also give them an opportunity to find cures for some dog diseases.

Many companies hope to profit from clones by tapping the market of people grieving over the loss of a favorite pet. In 2004, the first cloned pet was produced and sold to fill an order in the United States. The kitten, named Little Nicky, was created for a woman who was saddened by the loss of her cat.

While Little Nicky's owner says she is happy with her kitten, scientists are quick to point out that clones are not exact replacements. A cloned animal looks like the parent, but it does not necessarily have the same personality. Furthermore, it does not have the same experiences and is not born with the same bond formed by the original pet.

In addition, scientists warn that cloned animals are known for having short lives and suffering from more health problems than the parent animals. Although the exact reason is unknown, scientists think it is related to damage caused in the genetic information during the cloning process.

The cloned sheep Dolly gained worldwide fame. What many people do not know is that Dolly had to be put to sleep just halfway through her normal lifespan after developing degenerative lung disease and arthritis. It is sad to think that a similar fate may await Snuppy and other cloned animals.

Questions

1. What is a clone?
2. How was the cloning technique that produced Dolly different from previous cloning procedures?
3. Why was the cloning of a dog (Snuppy) so significant?
4. What are some arguments for and against the cloning of dogs?

Modern Dinosaurs: Are They Flying Overhead?

Watch a hawk soaring toward its prey or an ostrich speeding across a field and you may be getting a glimpse into the ancient lives of dinosaurs. A recent theory supported by the majority of paleontologists is that modern birds evolved from dinosaurs.

The paleontologists base their theory on fossil evidence. Over the years, many scientists have noted the surprising number of characteristics shared by birds and dinosaurs. Using a classification system known as *cladistics*, researchers compare the shapes and structures of bird and dinosaur skeletons to determine evolutionary relationships.

From such studies, scientists can draw a branching diagram, called a *cladogram*, in which they organize the organisms based on their shared characteristics. Each branch of the cladogram breaks into two parts to represent two species diverging from a common ancestor.

In early paleontology, scientists compared all of the characteristics of organisms before deciding to group them together. In the 1960s, a German entomologist named Willy Hennig suggested that, instead of looking at all the similar characteristics, one should look mainly at the unique characteristics that are shared by different organisms.

For example, a scientist studying the evolutionary history of a rodent, a lizard, and a fish can ignore similarities, such as the fact that all three have a backbone. This basic feature does not provide any new information about the relationships among them. However, the fact that only the rodent and the lizard have four legs indicates that they share a trait that the fish does not have; therefore they must share a more recent common ancestor.

While many researchers agree that birds evolved from dinosaurs, they do not agree about which group of dinosaurs gave rise to ancestral birds. Much of the original evidence for the connection comes from an unusual fossil discovered in a quarry in Germany in 1860. This fossil was *Archaeopteryx lithographica*, which is generally accepted to be the oldest known bird. It is a 145-million-year-old, crow-sized skeleton. The skeleton had some distinct bird features, such as a furcula (the so-called wishbone) and birdlike feet. In addition, impressions of its feathers were preserved along with the skeleton. However, unlike modern birds, the skeleton had a long bony tail, teeth instead of a beak, and claws on its wings. If feathers had not been found with the skeleton, *Archaeopteryx* would probably have been classified as a small dinosaur.

In 1969, Dr. J. H. Ostrom of Yale University presented a description of the dromaeosaur dinosaur *Deinonychus antirrhopus* and its similarity to *Archaeopteryx*. He described 22 features that were common in birds and meat-eating dinosaurs, but did not appear in other organisms. For example, both have very similar shoulders, hips, thighs, and ankles, as well as light, hollow bones. Although scientists had noticed some similarities before, Ostrom's work started a true shift in their thinking.

In the mid-1980s, Dr. Jacques Gauthier, also of Yale University, developed a cladogram to show the connection of birds to dinosaurs. His conclusion was that modern birds share a common evolutionary ancestor with ancient dinosaurs. Other scientists have independently agreed with his results.

Archaeopteryx was not the only fossil used to develop this theory. Many other specimens have been discovered. A *Unenlagia*, or half bird, was found in Argentina. It is significant because it was found with its arms folded in the very same way that birds' wings fold.

The remains of oviraptors have been found in Mongolia crouching over nests. It was once thought that these dinosaurs were stealing the eggs. Upon further analysis, however, scientists concluded that they were actually tending their own nests much like birds do.

The skeleton of *Rahonavis* found in Madagascar is important because it has a mix of bird and dinosaur features. This animal is more like a bird than *Archaeopteryx* is, but it has significant dinosaur features as well, such as a slashing claw.

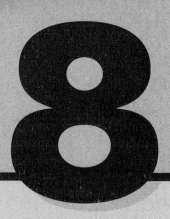

Perhaps the most significant fossil, found in China, represents what may be a feathered dinosaur. This species, named *Incisivosaurus gauthieri* in honor of Dr. Gauthier, came from an organism that lived almost 130 million years ago. It has the body of a dinosaur, but hairlike projections that appear to be an early form of feathers. The existence of its feathers is interesting because it is unlikely that such a complex structure would have evolved twice.

Three more discoveries from the same region of China also appear to be feathered dinosaurs. These specimens were originally thought to be birds, based on the presence of feathers, but were later reclassified as dinosaurs.

All of this evidence does not mean that modern birds can be considered dinosaurs. What it does mean is that, even though they look and behave differently from dinosaurs, modern birds appear to be linked to dinosaurs through a common evolutionary history.

As with most theories, the theory that birds evolved from dinosaurs has met some opposition. Arguments against the "Dino-Bird" theory take a logical, rather than a cladistic approach.

Major opponents reason that flight would have likely begun with a tree-climbing (arboreal) organism. They argue that because dinosaurs were ground-dwellers, they are unlikely candidates for ancestors. Those who support the "Dino-Bird" theory respond by saying that the ancestor of birds may have been arboreal or may have developed flight from the ground, as a result of chasing or leaping after insects.

Opponents further argue that there are important differences between birds and dinosaurs. The dinosaur ribcage, for example, is compressed from side to side. A bird's ribcage is compressed from back to belly. Supporters argue that similarities provide more information than differences. Because organisms change over time, only the features that remain the same can be used to trace evolutionary relationships.

Another study suggests that birds followed a different evolutionary path than thought. This idea, which has yet to be proved, investigates the development of hands and feet in turtles, alligators, lizards, and mice. It then attempts to compare these data with data on birds.

Questions

1. How are similarities and differences in traits used to establish a cladogram?

2. What important conclusion did the scientists make about the oviraptors?

3. What are some of the fossil traits that provide support for the theory that birds evolved from dinosaurs?

4. Why do some scientists think that dinosaurs are "unlikely candidates for ancestors" of birds?

Rain Forests: Does Cutting Trees Cut Our Chances for Survival?

Many kilometers away, in places such as Brazil and Madagascar, about 1000 m^2 of rain forest have been destroyed while you opened your book and read this sentence!

Rain forests used to cover about 14 percent of Earth's land surface. Now they are down to only 6 percent and shrinking rapidly. At this rate, some scientists estimate that the remaining natural rain forests will disappear within 40 years. How does this affect you? The answer might surprise you.

A tropical rain forest is characterized by dense vegetation of several layers, topped by a canopy of tall trees. As the name suggests, rainfall is plentiful and the dry season is limited to a few months. Temperatures remain high throughout the year. These warm, moist conditions allow for abundant vegetation and wildlife.

The main reason that rain forests are being destroyed is the cutting of trees by logging companies. Hardwoods such as teak, mahogany, and rosewood are removed from rain forests and sold for great profits around the world. Timber for furniture, building materials, paper, charcoal, and other wood products are also being taken. In addition, rain forests are cleared to raise cattle, build dams, and construct highways.

Most of the world's rain forests are in developing nations that have large national debts. Selling the rights to use rainforest land to commercial companies is a large economic resource. In many cases, it is the main source of money to pay off a poor country's national debt.

Large areas of rain forest must be destroyed to harvest just a small number of logs. The machinery used to obtain the logs causes extensive damage to the land and makes it very difficult, if not impossible, for the forest to regenerate.

To make matters worse, the roads constructed for logging provide access to farmers who need land. They enter the rain forest, settle there, and clear the land to grow food for their families. Farmers usually practice this form of subsistence farming when they are forced off their own lands by the government or by large corporations.

Unlike the fertile soil of farmland, the nutrients of a rain forest are mostly in the vegetation. Within a short time, rainforest soil is not fertile enough to support crops. This leads farmers to move inward, destroying additional rain forest to obtain more land.

Despite covering so little of Earth's surface, rain forests are home to roughly 50 percent of its species. This incredible diversity of organisms means that many species are adapted to specific niches within the rainforest ecosystem. The species depend on one another in an intricate food web. When humans disrupt even one part of a rain forest, many organisms are not able to survive. Over long periods of time, most species naturally become extinct. However, species are disappearing up to 1000 times faster than the natural rate. Much of this is due to rainforest destruction.

Losing so many species is not just a matter of concern due to loss of natural beauty. The rain forest is the source of many commonly used medicines. In fact, nearly half of the medicines used today have come from rainforest plants. With so many species undiscovered, the rain forests may still hold the ingredients to medicines that can fight or cure other diseases. As the rain forests are being destroyed, sources of possible medicines are disappearing as well.

The loss of rainforest species is of concern for another reason. Some domesticated species are so inbred that they are susceptible to disease. Wild plant breeds can often be crossed with domestic breeds to obtain genes for traits that protect the crops better against pests and disease. By destroying wild species we put our food supply at risk.

It is worthwhile to explore medicinal and nutritional treasures of the rain forests; but it is also important to stop the ongoing destruction of these forests. Because governments in control of rain forests need funds, the only way to protect the rain forests is to give these governments a valid economic reason not to destroy them.

Such a solution may be possible. Financial experts have determined that the income earned from clearing rain forests in order to raise cattle and

Facts About Rain Forests

- Madagascar has 10,000 species of plants. Of those, 80% are not found anywhere else on Earth.

- In a 10-hectare plot of Malaysian rain forest, 780 tree species have been found. That is more than the total number of tree species native to the United States.

- Over 2000 rainforest plants are thought to have anti-cancer properties. To date, scientists have tested fewer than 10% of rainforest plants for these properties.

- Roughly 90% of all non-human primates on Earth are found in rain forests.

- One pond in Brazil can support a greater variety of fish than is found in all of Europe's rivers.

- A single rainforest reserve in Peru contains far more species of birds than are found in all of the United States.

- More species of fish are found in the Amazon ecosystem than in the entire Atlantic Ocean.

harvest timber is about $460 per acre. If the same land is left undisturbed instead, and renewable resources such as nuts, fruits, oil-producing plants, and medicinal plants are harvested, an owner could earn about $2400 per acre. Unlike clearing the land, which yields a one-time income, harvesting renewable resources provides a sustainable income year after year. It is these sustainable resources that are the real treasures of the rain forest.

Questions

1. What are the general characteristics of a rain forest?

2. What are some products people use that come from rainforest resources?

3. Biologists Paul and Anne Ehrlich have compared the loss of individual species to the loss of the rivets that hold an airplane together. How might this statement apply to tropical rain forests?

Glossary

abiotic: Describes the nonliving parts of an organism's environment.

acidity: Describes a low pH level due to dissolved acids, such as in acid rain.

acquired immunity: Type of immunity a person gets when he/she develops antibodies (i.e., active immunity) or receives antibodies (i.e., passive immunity) to a particular antigen/pathogen.

active immunity: Type of immunity a person gets when exposed to an antigen/pathogen directly or in a vaccine, and then develops antibodies to it.

active transport: Movement of substances across a membrane from an area of lower concentration to an area of higher concentration; requires energy.

adaptations: Special characteristics that make an organism well suited for a particular environment.

AIDS (acquired immunodeficiency syndrome): An immunodeficiency disease, caused by HIV in humans.

algae: Plantlike organisms, often single-celled, that carry out photosynthesis.

alleles: The two alternate versions of a gene for a particular trait, located on homologous chromosomes.

allergic reactions: Conditions caused by an overreaction of the immune system.

amino acids: Organic compounds that are the building blocks of proteins.

antibiotics: Chemicals that kill specific microorganisms; frequently used to combat infectious diseases.

antibodies: Molecules that individuals produce as a defense against foreign objects in the body; antibodies bind to specific antigens.

antigens: Proteins on a foreign object that stimulate the immune system to produce antibodies.

asexually: Describes reproduction that requires only one parent to pass on genetic information; e.g., budding and fission.

atmosphere: The blanket of gases that covers Earth; commonly called "air".

atoms: The smallest units of an element that can combine with other elements.

ATP (adenosine triphosphate): The substance used by cells as an immediate source of chemical energy for the cell.

autotrophic: Describes a self-feeding organism that obtains its energy from inorganic sources; e.g., plants and algae (producers).

bacteria: Single-celled organisms that have no nuclear membrane to surround and contain their DNA molecule.

biodiversity: The variety of different species in an ecosystem or in the world.

biome: A very large area characterized by a certain climate and types of plants and animals.

biotechnological: Describes new procedures that utilize discoveries in biology; usually refers to recombinant DNA technology (i.e., genesplicing).

biotic: Describes the living parts of an organism's environment.

cancer: A disease that results from uncontrolled cell division, which damages normal tissues.

carbon: One of the six most important chemical elements for living things; carbon atoms form the backbone of nearly all organic compounds.

carbon dioxide: The inorganic molecule from which plants get carbon for photosynthesis; waste product of cellular respiration; a greenhouse gas.

carnivores: Animals that obtain their energy by eating other animals; see also **consumers**, **predators**, and **heterotrophic**.

carrying capacity: The size of a particular population that an ecosystem can support.

catalysts: Substances that increase the rate of a chemical reaction, but are not changed during the reaction.

cell membrane: A selectively permeable plasma membrane that separates and regulates substances that pass between the inside and the outside of a cell.

cells: The smallest functioning units of an organism; all living things are made up of at least one cell.

cellular respiration: The process that uses oxygen to create ATP for energy use.

chloroplasts: The organelles within plant cells that contain the pigment chlorophyll and carry out photosynthesis.

chromosomes: Structures composed of DNA that contain the genetic material.

circulation: The movement of blood throughout the body of an animal.

cloning: The production of identical individuals (i.e., clones) from the cell of another individual.

community: Populations of different species that interact within an area.

competition: The struggle between organisms for limited resources such as food and space.

consumers: Organisms that obtain their energy by feeding on other organisms; heterotrophic life forms.

coordination: The means by which body systems work together to maintain homeostasis; a property of living things.

cytoplasm: The watery fluid that fills a cell, surrounding its organelles.

decomposers: Heterotrophic organisms that obtain their energy by feeding on decaying organisms.

deforestation: The cutting down and clearing away of forests; clear-cutting.

depleted: Describes natural resources, e.g., topsoil, that have been used up and cannot be replaced within our lifetimes.

development: The changes in an organism that occur from fertilization until death.

deviations: Changes in the body's normal functions that are detected by control mechanisms, which maintain a balanced internal environment.

differentiation: The creation of specialized cells from less specialized parent cells through controlled gene expression.

diffusion: The movement of molecules from an area of higher concentration to an area of lower concentration.

digestion: The process of breaking down food particles into molecules small enough to be absorbed by cells.

diversity: The variety of different traits in a species or different species in an ecosystem.

DNA (deoxyribonucleic acid): The hereditary material of all organisms, which contains the instructions for all cellular activities.

dynamic equilibrium: In the body, a state of homeostasis in which conditions fluctuate yet always stay within certain limits.

ecology: The study of the interactions of living things with their environment.

ecosystem: An area that contains all living and nonliving parts that interact.

egg: The female gamete that supplies half the genetic information to the zygote.

embryo: An organism in an early stage of development before it is hatched, born, or germinated.

energy pyramid: Describes the flow of energy through an ecosystem; most energy is at the base (producers) and decreases at each higher level (consumers).

enzymes: Protein compounds that act as catalysts for biological reactions.

equilibrium: In ecosystems, an overall stability in spite of cyclic changes.

estrogen: In females, along with progesterone, a major sex hormone that affects secondary sex characteristics and reproduction.

evolution: The change in organisms over time due to natural selection acting on genetic variations that enable them to adapt to changing environments.

excretion: The removal of metabolic wastes from the body.

expression: The use of genetic information in a gene to produce a particular characteristic, which can be modified by interactions with the environment.

extinction: The death of all living members of a species.

feedback mechanisms: Systems that reverse an original response that was triggered by a stimulus; also, *negative feedback mechanisms* or *feedback loops*.

fertilization: In sexual reproduction, process by which an egg cell and a sperm cell unite to form a zygote.

fetus: In humans, a developing embryo after the first three months of development.

food chain: The direct transfer of energy from one organism to the next.

food web: The complex, interconnecting food chains in a community.

fossils: The traces or remains of dead organisms, preserved by natural processes.

fungi (singular, **fungus**): Heterotrophic organisms that obtain their energy by feeding on decaying organisms; e.g., yeast and mushrooms.

gametes: The male and female sex cells that combine to form a zygote during fertilization.

gene expression: See **expression**.

genes: The segments of DNA that contain the genetic information for a given trait or protein.

genetic engineering: Recombinant DNA technology, i.e., the insertion of genes from one organism into the genetic material of another.

genetic variation: See **variability**.

genetics: The branch of biology concerned with the study of patterns of inheritance.

geologic time: Earth's history divided into vast units of time by which scientists mark important changes in Earth's climate, surface, and life forms.

global warming: An increase in average global atmospheric temperature, caused by the greenhouse effect, which is due to more heat-trapping CO_2 in the air.

glucose: A simple sugar that has six carbon atoms bonded together; a subunit of complex carbohydrates.

habitat: The place in which an organism lives; a specific environment that has an interacting community of organisms.

herbivores: Animals that obtain their energy by eating plants; see also **consumers** and **heterotrophic**.

hereditary: Describes the genetic information that is passed from parents to offspring.

heterotrophic: Describes an organism that obtains its energy by feeding on other living things; e.g., animals (consumers).

homeostasis: In the body, the maintenance of a constant (i.e., stable) internal environment.

hormones: Chemical messengers secreted directly into the bloodstream, which bind with receptor proteins to affect gene activity; result in long-lasting changes in the body.

host: The organism that a parasite uses for food and shelter by living in it or on it.

hydrogen: One of the six most important chemical elements for living things.

immune system: Recognizes and attacks specific invaders, such as bacteria, to protect the body against infection and disease.

immunity: The ability to resist or prevent infection by a particular microbe.

inheritance: The process by which traits are passed from one generation to the next.

innate immunity: Type of immunity that is present at birth and has nothing to do with exposure to pathogens or antigens.

inorganic: In cells, substances that allow chemical reactions to take place; in ecosystems, substances that are cycled between living things and the environment.

insulin: A hormone secreted by the pancreas that maintains normal blood sugar levels.

internal development: Describes an embryo's development within the female's body.

internal fertilization: Describes the sperm fertilizing the egg cell within the female's body.

kingdoms: The major groupings into which scientists categorize all living things.

lipids: The group of organic compounds that includes fats, oils, and waxes.

malfunction: When an organ or body system stops functioning properly, which may lead to disease or death.

meiosis: The division of one parent cell into four daughter cells; reduces the number of chromosomes to one-half the normal number.

membrane: See **cell membrane**.

metabolic: Describes the chemical reactions (building up and breaking down) that take place in an organism.

microbes: Microscopic organisms that may cause disease when they invade another organism's body; e.g., microorganisms such as bacteria and viruses.

mitochondria: The organelles at which the cell's energy is released.

mitosis: The division of one cell's nucleus into two identical daughter cell nuclei.

molecules: The smallest unit of a compound, made up of atoms.

movement: The flow of materials between the cell and its environment; a property of living things, i.e., locomotion.

multicellular: Describes organisms that are made up of more than one cell.

mutations: Errors in the linear sequence (gene) of a DNA molecule that can affect gene expression.

natural selection: The process by which organisms having the most adaptive traits for an environment are more likely to survive and reproduce.

nerve cells: In animals, the cells that transmit nerve impulses to other nerve cells and to other types of cells; also called *neurons*.

niche: An organism's role in, or interaction with, its habitat and ecosystem.

nitrogen: One of the six most important chemical elements for living things.

nucleotides: The building blocks, or subunits, of DNA; they include four types of nitrogen bases, which occur in two pairs.

nucleus: The dense region of a (eukaryotic) cell that contains the genetic material.

nutrients: Important molecules in food, such as lipids, proteins, and vitamins.

nutrition: The life process by which organisms take in and utilize nutrients.

omnivores: Animals that eat both plants and animals; see also **consumers** and **heterotrophic**.

organ: Describes a level of organization in living things, i.e., a structure made up of similar tissues that work together to perform the same task; e.g., the liver.

organelles: Structures within a cell that perform a particular task; e.g., the vacuole.

organic: Relating to compounds that contain carbon and hydrogen (in living things).

organisms: Living things; life forms.

organ system: A group of organs that works together to perform a major task; e.g., the respiratory system.

ovaries: The female reproductive organs that produce the mature egg cells.

oxygen: One of the six most important chemical elements for living things; released as a result of photosynthesis; essential to cellular (aerobic) respiration.

ozone shield: The layer of ozone gas (O_3) that surrounds Earth high in the atmosphere and blocks out harmful ultraviolet (UV) radiation.

pancreas: Gland that secretes pancreatic juice (containing enzymes that aid digestion) and insulin (maintains normal blood sugar levels).

parasite: An organism that lives in or on another organism (the host), deriving nutrients from it and usually causing it harm.

passive immunity: Type of immunity a person gets when antibodies are passed from one person to another, e.g., through the placenta to a fetus or by breast-feeding.

passive transport: Movement of substances across a membrane; requires no use of energy.

pathogens: Microscopic organisms that cause diseases, such as certain bacteria and viruses; see also **microbes**.

pesticides: Chemicals used to kill agricultural pests, mainly insects, some of which have evolved resistance to the chemicals.

pH: A measurement (on a scale of 0 to 14) of how acidic or basic a solution is.

photosynthesis: The process that, in the presence of light energy, produces chemical energy (glucose) and water, and releases oxygen.

placenta: The organ that forms in the uterus of mammals to nourish a developing embryo and remove its waste products.

population: All the individuals of the same species that live in the same area.

predators: Organisms that feed on other living organisms (the prey); see also **carnivores**, **consumers**, and **heterotrophic**.

pregnancy: In animals, the condition of having a developing embryo within the body.

prey: An organism that is eaten by another organism (the predator).

producers: Organisms on the first trophic level, which obtain their energy from inorganic sources (i.e., by photosynthesis); autotrophic life forms such as plants and algae.

progesterone: In females, along with estrogen, a major sex hormone; see also **estrogen**.

proteins: The group of organic compounds made up of chains of amino acids.

radiation: A form of energy that can cause genetic mutations in sex cells and body cells.

receptors: Molecules that play an important role in the interactions between cells; e.g., molecules that bind with hormones.

recombination: The formation of new combinations of genetic material due to crossing-over during meiosis or due to genetic engineering.

recombining: During meiosis, the process that causes an increase in genetic variability due to the exchange of material between homologous chromosomes.

replicate: The process by which DNA makes a copy of itself during cell division and protein synthesis.

reproduction: The production of offspring (i.e., the passing on of hereditary information), by either sexual or asexual means.

residue: The remains of dead organisms, which are recycled in ecosystems by decomposers; also, the remains of harmful chemicals in the soil, air, and water.

respiration: In the lungs, the process of exchanging gases; in cells, the process that releases the chemical energy stored in food; see also **cellular respiration**.

response: An organism's reaction to an internal or external stimulus.

RNA (ribonucleic acid): The material that assists DNA in carrying out protein synthesis; the genetic material of some viruses.

ribosomes: The organelles at which protein synthesis occurs, and which contain RNA.

scavengers: Animals that eat the remains of a kill, rather than hunt the living animals.

selective breeding: The process by which humans encourage the development of specific traits by breeding only the plants or animals that have those traits.

sex cell: The male or female gamete; it has one-half the normal chromosome number as a result of meiosis.

sexually: Describes reproduction that requires two parents to pass on genetic information.

simple sugars: Single sugars that have six carbon atoms; e.g., glucose.

solar energy: Radiant energy from the sun that is a renewable resource.

species: A group of related organisms that can breed and produce fertile offspring.

sperm: The male gamete that supplies half the genetic information to the zygote.

stability: The ability of an ecosystem to continue and to remain healthy; usually, the greater the species diversity, the more stable the ecosystem; see also **homeostasis**.

starch: A complex carbohydrate made up of many glucose molecules; used for energy storage in plants.

stimulus (plural, **stimuli**): Any event, change, or condition in the environment that causes an organism to make a response.

subunits: The four types of nucleotide bases that make up the DNA molecule; also, the components (i.e., amino acids) that make up a protein molecule.

succession: The gradual replacement of one ecological community by another until it reaches a point of stability.

symbiosis: A close relationship between two or more different organisms that live together, which is often but not always beneficial.

synthesis: The building of compounds that are essential to life; e.g., protein synthesis.

system: Describes a level of organization in living things; e.g., in an ecosystem, the living and nonliving parts that function together; and, in a body, the groups of organs that work together to perform the same task; see also **organ system**.

template: In DNA replication, the original molecule that is used to make copies.

testes (singular, **testis**): The pair of male reproductive organs that produces the sperm cells.

testosterone: In males, the main sex hormone that influences secondary sex characteristics and reproduction.

tissues: Describes a level of organization in living things, i.e., groups of similar cells that work together to perform the same function; e.g., muscle tissue.

toxins: Chemicals that can harm a developing fetus if taken in by the mother during pregnancy; harmful chemicals that may get passed from one energy level to the next as they move up the food chain.

uterus: In female mammals, the reproductive organ that holds the developing embryo; also called *womb*.

vaccination: A medical substance (usually an injection) that prepares the immune system to better fight a specific disease in the future.

vacuoles: Organelles that store materials, e.g., food or wastes, for plant and animal cells.

variability: The differences that exist among offspring in their genetic makeup.

virus: A particle of genetic material that can replicate only within a living host cell, where it usually causes harm.

white blood cells: Several types of cells that work to protect the body from disease-causing microbes and foreign substances.

zygote: The fertilized egg cell that is formed when the nuclei of two gametes (a male and a female) fuse.

Index

Chemosynthesis, 2
Chlorophylls, 32
Chloroplasts, 20, 32, 34
Chromatids, 79
Chromatin, 79
Chromosomal alterations, 103
Chromosome number, 83–84
Chromosomes, 20
  changes in structure of, 103
Cilia, 36
Circulation, 18, 38
  intercellular, 38
  intracellular, 38
  pathways of, 56
Cleavage, 86–87
Climate, 14
Climax communities, 12
Clones, 79
Cloning, 110
Codon, 109
Coenzyme, 27
Color blindness, 102
Commensalism, 2
Communities, 1
Comparative anatomy, 117
Comparative biochemistry, 117–118
Comparative cytology, 117
Comparative embryology, 117
Competition, 8–9, 120
Compound light microscope, 22
  identifying cell parts with, 141
  using, 141
Compounds, 25
  inorganic, 25
  organic, 25
Concentration gradient, 37
Condensation, 10
Connections, making, 145
Conservation of resources, 131
Constipation, 51
Consumers, 4
  primary, 4
  secondary, 4
Contractile vacuoles, 44
Coordination, 18
Coronary circulation, 56
Coronary thrombosis, 56
Corpus luteum, 90
Cotyledons, 94
Counseling, genetic, 107
Covalent bond, 25
Crossing-over, 100
Cross-pollination, 94
Cuticle, 34
Cyclosis, 38
Cytology, comparative, 117
Cyton, 45
Cytoplasm, 20
  division of, 79
Cytosine, 108

D

Darwin's theory of natural selection, 120–121
Decomposers, 4–5
Deforestation, 129
Denaturation, 27
Dendrites, 45
Denitrifying bacteria, 11
Deoxyribose, 108
Depletion, 129
Development, 86
  external, 87
  fertilization and, 90–91
  human, 89–91
  internal, 87–88
  stages of, 86–87

Deviations, 18
Diabetes, 67
Diarrhea, 51
Diet, balanced, 51
Differentiation, 87
Diffusion, 37
  through membrane, 146
Digestion, 18, 35
  extracellular, 35
  intracellular, 35
Digestive system, disorders of, 51
Dipeptide, 25
Diploid, or $2n$, chromosome number, 84, 85
Disaccharides, 25
Disease
  causes of, 72–73
  protection against, 74–75
  symptoms of, 73
Disjunction, 103
Dissecting microscope, 22
Diversity, 119
DNA (deoxyribonucleic acid), 19, 20, 25–26
  replication of, 108
  structure of, 108
Dominance, 97, 98
Down syndrome, 103
Dynamic equilibrium, 72

E

Ecological organization, levels of, 1
Ecological succession, 12
Ecology, 1
Ecosystems, 1
  abiotic factors of, 2
  biotic factors of, 2–3
  characteristics of, 1–3
  formation of, 11–12
  requirements of, 1–2
Ectoderm, 87
Effectors, 45, 63
Egestion, 18
Egg, 67
Egg cells, 83, 84
  production of, 89
Electron microscope, 22
Electrophoresis, 111
Elements, 24
Embryo, 67, 83
Embryology, comparative, 117
Emphysema, 59
Emulsification, 50
Endocrine glands, 46, 66–67
Endocrine system, 66–67
  disorders of, 67
Endoderm, 87
Endoskeleton, 47
Energy, solar, 4
Energy flow relationships, 4
Energy pyramid, 5
Environment, 32
  effect of people on, 128–129
  heredity and, 105
  role of people in protecting, 131–132
Enzymes, 26–28
  factors influencing action, 27
  models of action of, 27
  restriction, 111
  role of, 26–27
  structure, 27
  substrate concentrations and, 27–28
Enzyme-substrate complex, 27
Epidermis, 34
Epigenetic modification, 114
Epiglottis, 58
Equilibrium, 37

Esophagus, 49
Essential amino acids, 51
Estrogen, 67, 96
Evaporation, 10
Evolution, 116
  evidence of, 116–118
  by natural selection, 120–121
  theories of, 119
  time frame for, 123–124
Excretion, 18, 43–44, 61–62
  adaptations for, 43
Exhalation, 59
External development, 87
External fertilization, 86
Extinction, 117
Extracellular digestion, 35

F

Fallopian tube, 89
Fats, 25, 51
  saturated, 51
  unsaturated, 51
Feces, 50
Feedback mechanisms, 59, 72
Female reproductive system, 89
Fermentation, 41, 120
Fertilization, 83, 86–87, 94
  development and, 90–91
  external, 86
  internal, 86
Fetus, 91
Fibrin, 54
Fibrinogen, 54
Filament, 93
Filtrate, 61
Filtration, 61
Finches, beaks of, 145–146
First filial (F_1) generation, 98
Flagella, 47
Flowering plants, sexual reproduction in, 93–95
Flowers, structure of, 93
Follicles, 89
Follicle-stimulating hormone (FSH), 66, 90
Food chains, 4–6
Food webs, 5
Fossils, 116–117
Founder effect, 123
Fraternal twins, 91
Freshwater biomes, 16
Fruits, 94
Fungi, 35

G

Gallbladder, 50
Gallstones, 51
Gametes, 83
Gametogenesis, 84
Gas exchange, 59
Gastritis, 71
Gastrointestinal (GI) tract, 49
Gastrula, 87
Gastrulation, 87, 91
Gel, 111
Gel electrophoresis, 111
Gene–chromosome theory, 97
Gene expression, 97–98
Gene mutations, 103–104, 110
Gene regulation, 114–115
Genes, 97
  control of cellular activities, 108–109
  sex-linked, 102
Gene therapy, for Alzheimer's disease, 47–48
Genetic counseling, 107
Genetic disorders, detection of, 107

Muscles, 69–70
Mutagenic agents, 104, 122
Mutations, 103
  genetic, 103–104, 110
  silent, 122
Mutualism, 3

N
Nasal cavity, 58
Natural selection, 121
  evolution by, 120–121
  genetic variation and, 122–123
  impact of humans on, 124
Negative feedback, 67, 72, 90
Nephron, 61
Nerve cells, 63
  structure of, 45
Nerve control, 45
  adaptations for, 45–46
Nerve growth factor (NGF), 47–48
Nerves, 63
Nervous system, 63–65
  behavior and, 64
  central, 46
  disorders of, 64–65
  peripheral, 46
Neurons, 63
Neurotransmitters, 45
Niche, 9
Nitrates, 11
Nitrifying bacteria, 11
Nitrogen, 24
Nitrogen cycle, 10–11
Nitrogen-fixing bacteria, 3, 11
Nitrogenous base, 108
Nondisjunction, 103
Nucleic acids, 25–26
Nucleotides, 108
Nucleus, 20
Nutrients, 32
Nutrition, 32–33, 49–51
  autotrophic, 32
  heterotrophic, 32, 35–36
  relationships in, 2

O
Oils, 25
Omnivores, 2
One gene–one polypeptide hypothesis, 109–110
Oogenesis, 84–85
Oral cavity, 49
Oral groove, 36
Organic compounds, 25
  synthesis of, 119
Organisms, 18
Osmosis, 37
Ovaries, 67, 89, 93
Overpopulation, 120
Ovulation, 89
Ovules, 93
Oxygen, 24
Ozone shield, 129

P
Palisade layer, 34
Pancreas, 50
Parasite, 3
Parasitism, 3
Parathormone, 66
Parathyroid glands, 66
Parent generation, 98
Passive immunity, 54, 75
Passive transport, 37
Pathogens, 73
Penis, 89
Peptide bond, 25

Peripheral nervous system, 46, 64
Peristalsis, 49
Perspiration, 61
Pesticides, 17, 124
Petals, 93
Petrifaction, 116
pH, 28
Phagocytes, 53
Phagocytosis, 36, 38
Pharynx, 58
Phase-contrast microscope, 22
Phenotype, 98
Phenylketonuria (PKU), 107
Phloem, 40
Phosphate group, 108
Photosynthesis, 10, 20, 32, 120
  adaptations for, 34
  chemistry of, 32–33
Photosynthetic pigments, 32
pH paper, 142
Pigments, 32
  accessory, 32
Pinocytosis, 38
Pioneer organisms, 12
Pistils, 93
Pituitary gland, 66, 72
Pivot joint, 69
Placenta, 88
Placental mammals, 87–88
Plant cells, 79
Plant hormones, 46
Plants
  breeding, 105
  dissecting specimens, 143
  excretion in, 44
  growth of, 95
  respiration in, 41–42
  transport in, 39–40
Plasma, 53
Plasma membrane, 20
Plasmid, 110
Platelets, 53
Polar body, 84
Polio, 64
Pollen grains, 93, 94
Pollination, 94
Pollution
  control of, 131
  fighting, with bacteria, 133–134
  thermal, 129
Polypeptide, 25
Polyploidy, 103
Polysaccharides, 25
Population, 1
  control of, 131
  growth of human, 128
Postnatal development, 91
Pregnancy, 90
Premenstrual dysphoric disorder, 96
Prey, 2
Primary consumers, 4
Prions, 19, 73
Producers, 4
Progesterone, 67, 96
Proteins, 25, 49, 51
Protein synthesis, 109
Protists, 36, 41
  excretion in, 43–44
  locomotion among, 46–47
  transport in, 40
Pseudomonas aeruginosa, 31
Pseudopods, 36
Puberty, 90
Pulmonary circulation, 56
Punctuated equilibrium, 123–124
Punnett square, 100

Pyramid
  biomass, 5–6
  energy, 5

R
Radiation, 73, 104, 119
Radioactive dating, 116
Receptors, 45
Receptor sites, 37
Recombinant DNA technology, 67
Recombination, 97, 99, 110
Red blood cells, 53
Reduction division, 84
Reflex arc, 64
Reflexes, 64
Regeneration, 81–82
Regulation, 45–46
Relationships and biodiversity, 145
Replicates, 79
Replicating, 26
Reproduction, 18, 121
  asexual, 81–82
  human, 89–91
  sexual, 83–85
Reproductive isolation, 123
Resolution, 22
Resources, conservation of, 131
Respiration, 18, 41–42, 58–59
  adaptations for, 41–42
  aerobic, 10, 41, 58
  anaerobic, 41, 58
  cellular, 41, 58
Respiratory system, disorders of, 59
Responses, 45
Restriction enzymes, 110, 111
Rheumatoid arthritis (RA), 76
Ribonucleic acid (RNA), 26, 109
Ribosomal RNA (rRNA), 109
Ribosomes, 20, 109
Ribozyme, 27
Rock, sedimentary, 116
Root hairs, 40
Roots, 40
Roughage, 51
RUNX3, 71

S
Safety in laboratory, 140
Saliva, 49
Salivary glands, 49
Saturated fats, 51
Scavengers, 2
Schleiden, Matthias, 19
Schwann, Theodor, 19
Scientific method
  collecting, organizing, and graphing data, 136
  defining problem and developing hypothesis, 135
  designing and conducting an experiment, 136
  making generalizations and drawing conclusions, 136, 138
  making predictions based on experimental data, 136
  skills using, 135
Scrapie, 73
Scrotum, 89
Secondary consumers, 4
Second filial (F_2) generation, 98
Sedimentary rock, 116
Seed coat, 94
Seeds
  germination of, 95
  structure of, 94
Segregation, 97, 99
Selective breeding, 105

Sample Examinations

LIVING ENVIRONMENT
JUNE 2010

Part A

Answer all questions in this part. [30]

Directions (1–30): For *each* statement or question, write on your separate answer sheet the *number* of the word or expression that, of those given, best completes the statement or answers the question.

1 Why is a mushroom considered a heterotroph?
(1) It manufactures its own food. *autotrophs recieve from sunlight.*
(2) It divides by mitosis.
(3) It transforms light energy into chemical energy.
(4) It obtains nutrients from its environment.

2 Three days after an organism eats some meat, many of the organic molecules originally contained in the meat would be found in newly formed molecules of

(1) glucose (3) starch
(2) protein (4) oxygen

3 Which body system is correctly paired with its function?

(1) excretory—produces antibodies to fight disease-causing organisms
(2) digestive—produces hormones for storage and insulation
(3) circulatory—transports materials for energy release in body cells
(4) respiratory—collects waste material for digestion

4 Which statement best explains why some cells in the reproductive system only respond to certain hormones?

(1) These cells have different DNA than the cells in other body systems.
(2) These cells have specific types of receptors on their membranes.
(3) Reproductive system cells could be harmed if they made contact with hormones from other body systems.
(4) Cells associated with the female reproductive system only respond to the hormone testosterone.

5 In the cell shown below, which lettered structure is responsible for the excretion of most cellular wastes?

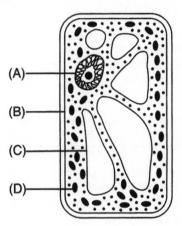

(1) A (3) C
(2) B (4) D

6 What is the main function of a vacuole in a cell?
(1) storage
(2) coordination
(3) synthesis of molecules
(4) release of energy

7 If 15% of a DNA sample is made up of thymine, T, what percentage of the sample is made up of cytosine, C?

(1) 15% (3) 70%
(2) 35% (4) 85%

8 Global warming has been linked to a *decrease* in the
(1) size of the polar ice caps
(2) temperature of Earth
(3) rate of species extinction
(4) rate of carbon dioxide production

9 Several structures are labeled in the diagram of a puppy shown below.

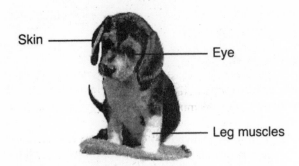

Every cell in each of these structures contains

(1) equal amounts of ATP
(2) identical genetic information
(3) proteins that are all identical
(4) organelles for the synthesis of glucose

10 A characteristic that an organism exhibits during its lifetime will only affect the evolution of its species if the characteristic

(1) results from isolation of the organism from the rest of the population
(2) is due to a genetic code that is present in the gametes of the organism
(3) decreases the number of genes in the body cells of the organism
(4) causes a change in the environment surrounding the organism

11 Agriculturists have developed some varieties of vegetables from common wild mustard plants, which reproduce sexually. Which statement best explains the development of these different varieties of vegetables?

(1) Different varieties can develop from a single species as a result of the recombination of genetic information.
(2) Different species can develop from a single species as a result of the effect of similar environmental conditions.
(3) Mutations will occur in the genes of a species only if the environment changes.
(4) Variations in a species will increase when the rate of mitosis is decreased.

12 The diagram below represents a technique used in some molecular biology laboratories.

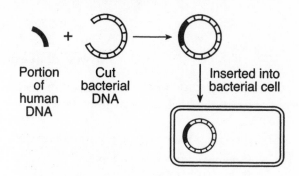

This technique is a type of

(1) chromatography
(2) gel electrophoresis
(3) direct harvesting
(4) genetic engineering

13 A species of bird known as Bird of Paradise has been observed in the jungles of New Guinea. The males shake their bodies and sometimes hang upside down to show off their bright colors and long feathers to attract females. Females usually mate with the "flashiest" males. These observations can be used to support the concept that

(1) unusual courtship behaviors lead to extinction
(2) some organisms are better adapted for asexual reproduction
(3) homeostasis in an organism is influenced by physical characteristics
(4) behaviors that lead to reproductive success have evolved

14 Which statement concerning the evolution of species *A, B, C, D,* and *E* is supported by the diagram below?

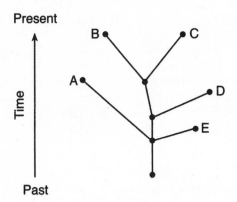

(1) Species *B* and *C* can be found in today's environments.
(2) Species *A* and *D* evolved from *E.*
(3) Species *A* and *C* can still interbreed.
(4) Species *A, B,* and *E* all evolved from a common ancestor and all are successful today.

15 The diagram below represents a process that occurs during human reproduction.

(Not drawn to scale)

The process represented by the arrow will ensure that the

(1) zygote contains a complete set of genetic information
(2) gametes contain a complete set of genetic information
(3) zygote contains half of the genetic information
(4) gametes contain half of the genetic information

16 Even though identical twins have the same genetic material, they may develop slightly different characteristics because

(1) each twin receives different chromosomes from the egg
(2) one twin may only have genes from the father
(3) gene expression may be influenced by factors that switch genes on and off
(4) a gene mutation may have occurred before the zygote divided

17 What normally happens immediately after fertilization in sexual reproduction?

(1) specialization of cells to form a fetus from an egg
(2) production of daughter cells having twice the number of chromosomes as the parent cell
(3) production of daughter cells having half the number of chromosomes as the parent cell
(4) division of cells resulting in the development of an embryo from a zygote

18 The human female reproductive system is represented in the diagram below.

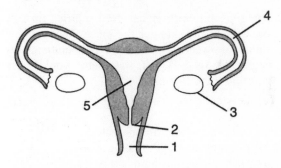

Production of gametes and support of the fetus normally occur in structures

(1) 1 and 2   (3) 3 and 5
(2) 2 and 4   (4) 4 and 5

19 Essential materials needed for development are transported to a human fetus through the

(1) reproductive hormones
(2) egg cell
(3) placenta
(4) ovaries

20 The failure to regulate the pH of the blood can affect the activity of

(1) enzymes that clot blood
(2) red blood cells that make antibodies
(3) chlorophyll that carries oxygen in the blood
(4) DNA that controls starch digestion in the blood

21 Young birds that have been raised in isolation from members of their species build nests characteristic of their species. This suggests that the nest-building behavior is

(1) genetically inherited from parents
(2) learned by watching members of their species
(3) a disadvantage to the survival of the species
(4) a direct result of the type of food the bird eats

22 Some people with spinal cord injuries do not sweat below the area of the injury. Without the ability to sweat, the human body temperature begins to rise. Which statement would best describe this situation?

(1) Feedback mechanisms regulate blood sugar levels.
(2) Gene mutations are increased.
(3) Energy from ATP is not available.
(4) Dynamic equilibrium is disrupted.

23 Decomposers are necessary in an ecosystem because they

(1) produce food for plants by the process of photosynthesis
(2) provide energy for plants by the process of decay
(3) can rapidly reproduce and evolve
(4) make inorganic materials available to plants

24 A manatee is a water-dwelling herbivore on the list of endangered species. If manatees were to become extinct, what would be the most likely result in the areas where they had lived?

(1) The biodiversity of these areas would not be affected.
(2) Certain producer organisms would become more abundant in these areas.
(3) Other manatees would move into these areas and restore the population.
(4) Predators in these areas would occupy higher levels on the energy pyramid.

25 A serious threat to biodiversity is

(1) habitat destruction
(2) maintenance of food chains
(3) competition within a species
(4) a stable population size

26 Which action will result in the greatest *decrease* in rain forest stability?

(1) removing one species of plant for medicine
(2) harvesting nuts from some trees
(3) cutting down all the trees for lumber
(4) powering all homes with wind energy

27 One way that humans could have a positive impact on local environments is to

(1) generate waste products as a result of technological advances
(2) use resources that are renewable
(3) increase planting large areas of one crop
(4) increase the use of pesticides

28 Which statement provides evidence that evolution is still occurring at the present time?

(1) The extinction rate of species has decreased in the last 50 years.
(2) Many bird species and some butterfly species make annual migrations.
(3) New varieties of plant species appear more frequently in regions undergoing climatic change.
(4) Through cloning, the genetic makeup of organisms can be predicted.

29 The diagram below represents the various stages of ecological succession in New York State.

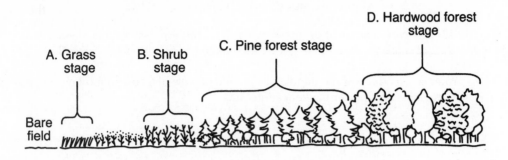

If the ecosystem is not altered, which stage would be the most stable?

(1) grass

(2) shrub

(3) pine forest

(4) hardwood forest

30 Because of an attractive tax rebate, a homeowner decides to replace an oil furnace heating system with expensive solar panels. The trade-offs involved in making this decision include

(1) high cost of solar panels, reduced fuel costs, and lower taxes

(2) low cost of solar panels, increased fuel costs, and higher taxes

(3) increased use of fuel, more stable ecosystems, and less availability of solar radiation

(4) more air pollution, increased use of solar energy, and greater production of oil

Part B–1

Answer all questions in this part. [13]

Directions (31–43): For *each* statement or question, write on the separate answer sheet the *number* of the word or expression that, of those given, best completes the statement or answers the question.

31 A clear plastic ruler is placed across the middle of the field of view of a compound light microscope. A row of cells can be seen under low-power magnification (100×).

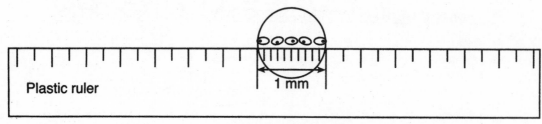

1 mm = 1000 μm

What is the average length of a single cell in micrometers (μm)?

(1) 10 μm   (3) 200 μm

(2) 100 μm   (4) 2000 μm

32 The graph below represents the populations of two different species in an ecosystem over a period of several years.

Population Changes in an Ecosystem

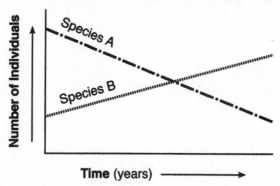

Which statement is a possible explanation for the changes shown?

(1) Species *A* is better adapted to this environment.

(2) Species *A* is a predator of species *B*.

(3) Species *B* is better adapted to this environment.

(4) Species *B* is a parasite that has benefited species *A*.

33 A mineral supplement designed to prevent the flu was given to two groups of people during a scientific study. Dosages of the supplement were measured in milligrams per day, as shown in the table below.

Supplement Dosages

Group	Dosage (mg/day)
A	100
B	200

After 10 weeks, neither group reported a case of the flu. Which procedure would have made the outcome of this study more valid?

(1) test only one group with 200 mg of the supplement
(2) test the supplement on both groups for 5 weeks instead of 10 weeks
(3) test a third group that receives 150 mg of the supplement
(4) test a third group that does not receive the supplement

34 The diagram below shows a normal gene sequence and three mutated sequences of a segment of DNA.

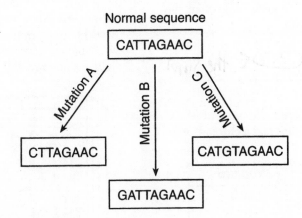

Which row in the chart below correctly identifies the cause of each type of mutation?

Row	Mutation A	Mutation B	Mutation C
(1)	deletion	substitution	insertion
(2)	insertion	substitution	deletion
(3)	insertion	deletion	substitution
(4)	deletion	insertion	substitution

Base your answers to questions 35 and 36 on the energy pyramid below and on your knowledge of biology.

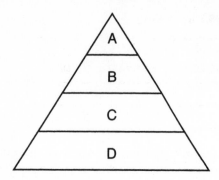

35 Which level includes organisms that receive their energy from level *B*?

(1) A   (3) C
(2) B   (4) D

36 Which level includes organisms that get their energy exclusively from a source other than the organisms in this ecosystem?

(1) A   (3) C
(2) B   (4) D ← the Sun!

37 The chart below compares the size of three structures: a gene, a nucleus, and a chromosome.

Size	Structure
smallest in size	A
↓	B
greatest in size	C

Based on this information, structure *A* would most likely be a

(1) chromosome that is part of structure C
(2) chromosome that contains structures B and C
(3) nucleus that contains both structure B and structure A
(4) gene that is part of structure B

38 The diagram below shows molecules represented by *X* both outside and inside of a cell.

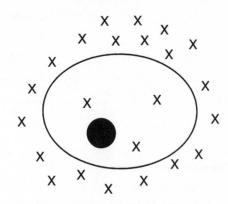

A process that would result in the movement of these molecules out of the cell requires the use of

(1) DNA   (3) antigens
(2) ATP   (4) antibodies

39 Which statement most accurately predicts what would happen in the aquarium shown below if it were tightly covered and maintained in natural light for one month?

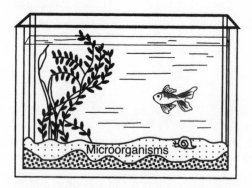

(1) The water temperature would rapidly decrease.
(2) The process of respiration in the snail would decrease.
(3) The rate of reproduction of the fish would be affected.
(4) The organisms would probably survive because materials would cycle.

40 The data table below shows an effect of secondhand smoke on the birth weight of babies born to husbands and wives living together during pregnancy.

Effect of Secondhand Smoke on Birth Weight

	Wife: Nonsmoker Husband: Nonsmoker	Wife: Nonsmoker Husband: Smoker
Number of Couples	837	529
Average Weight of Baby at Birth	3.2 kg	2.9 kg

Based on these data, a reasonable conclusion that can be drawn about secondhand smoke during pregnancy is that secondhand smoke

(1) is unable to pass from the mother to the fetus
(2) slows the growth of the fetus
(3) causes mutations in cells of the ovaries
(4) blocks the receptors on antibody cells

41 A limiting factor unique to a field planted with corn year after year is most likely

(1) temperature
(2) sunlight
(3) water
(4) soil nutrients

Base your answers to questions 42 and 43 on the information below and on your knowledge of biology.

After the Aswan High Dam was built on the Nile River, the rate of parasitic blood-fluke infection doubled in the human population near the dam. As a result of building the dam, the flow of the Nile changed. This changed the habitat, which resulted in an increase in its population of a certain aquatic snail. The snails, which were infected, released larvae of the fluke. These larvae then infected humans.

42 This situation best illustrates that

(1) the influence of humans on a natural system is always negative in the long term
(2) the influence of humans on a natural system can have unpredictable negative impacts
(3) human alteration of an ecosystem does not need to be studied to avoid ecological disaster
(4) human alteration of an ecosystem will cause pollution and loss of finite resources

43 The role of the snail may be described as a

(1) host
(2) parasite
(3) producer
(4) decomposer

Part B–2

Answer all questions in this part. [12]

Directions (44–55): For those questions that are followed by four choices, circle the *number* preceding the choice that, of those given, best completes the statement or answers the question. For all other questions in this part, follow the directions given in the question and record your answers in the spaces provided.

44 The table below shows the abundance of some greenhouse gases in the atmosphere.

Abundance of Some Atmospheric Greenhouse Gases

Greenhouse Gases	Abundance (%)
carbon dioxide (CO_2)	99.438
methane (CH_4)	0.471
nitrous oxide (N_2O)	0.084
other gases (CFCs, etc.)	0.007
Total	**100.000**

Identify the most abundant greenhouse gas and state *one* human activity that is a source of this gas. [1]

Greenhouse gas: _____

_____

_____

45 The United States government does not allow travelers from foreign countries to bring plants, fruits, vegetables, animals, or other living organisms into this country. State *one* biological reason for keeping these out of the United States. [1]

_____

_____

For Teacher Use Only

44 ☐

45 ☐

Base your answers to questions 46 through 49 on the information and data table below and on your knowledge of biology.

Birds colliding with aircraft either on the ground or in the air create problems for the Air Force. An organization known as BASH (Bird Aircraft Strike Hazard) studied the impact of birds colliding with aircraft. In 2001, there were 3854 bird collisions reported at a total cost to the Air Force of over 31 million dollars in damage—approximately eight thousand dollars per collision. August, September, and October were the busiest months with 1442 collisions. Nearly 50% of all these collisions occurred in the airfield environment, an environment that can most easily be controlled.

The top five species of birds involved in these collisions are listed in the data table below.

Top Five Bird Species Involved in Collisions in 2001

Type of Bird	Number of Collisions
American mourning dove (species A)	123
horned lark (species B)	100
barn swallow (species C)	83
American cliff swallow (species D)	55
American robin (species E)	55

Source of data: Bird Aircraft Strike Hazard by Matt Granger, http://www.find.articles.com

Directions (46–47): Using the information in the data table, construct a bar graph on the grid, following the directions below.

46 Mark an appropriate scale on the axis labeled "Number of Collisions." [1]

47 Construct vertical bars to represent the data. Shade in each bar. [1]

**Top Five Bird Species
Involved in Collisions in 2001**

Number of Collisions (vertical axis)

A   B   C   D   E

Bird Species

48 Is the problem with birds and aircraft limited to birds living on or near airport grounds? Support your answer using information from the passage. [1]

_____

_____

49 State *one* possible reason that the greatest number of bird collisions occurs during August, September, and October. [1]

_____

_____

_____

Base your answers to questions 50 through 53 on the information below and on your knowledge of biology. •

The Control of Transpiration

Plants normally lose water from openings (stomates) in their leaves. The water loss typically occurs during daylight hours when plants are exposed to the Sun. This water loss, known as transpiration, is both beneficial and harmful to plants.

Scientists believe wind and high temperatures increase the rate of transpiration, but the size of each stomate opening can be regulated. Reducing the size of the openings during drought conditions may help reduce the dehydration and wilting that would otherwise occur.

A leaf may lose more than its own weight in water each day. Transpiration also lowers the internal temperature of the leaf as water evaporates. On hot days, temperatures in the leaves may be from 3° to 15°C cooler than the outside air. With stomates open, vital gases may be exchanged between the leaf tissues and the outside environment.

Researchers have also found many plants that use another response when leaf temperatures rise. Special molecules known as heat shock proteins are produced by plant cells and help to hold enzymes in their functional shapes.

50 State *one* way transpiration is beneficial to plants. [1]

_____

_____

50 □

51 Identify *two* of the "vital gases" that are exchanged between leaf tissues and the outside environment. [1]

_____ and _____

51 □

52 Identify the specific leaf structures that regulate the opening and closing of stomates. [1]

_____

52 □

53 Explain why it is important for plants to "hold enzymes in their functional shapes." [1]

_____

_____

53 □

54 The graph below shows the growth of a population of coyotes in a wilderness area.

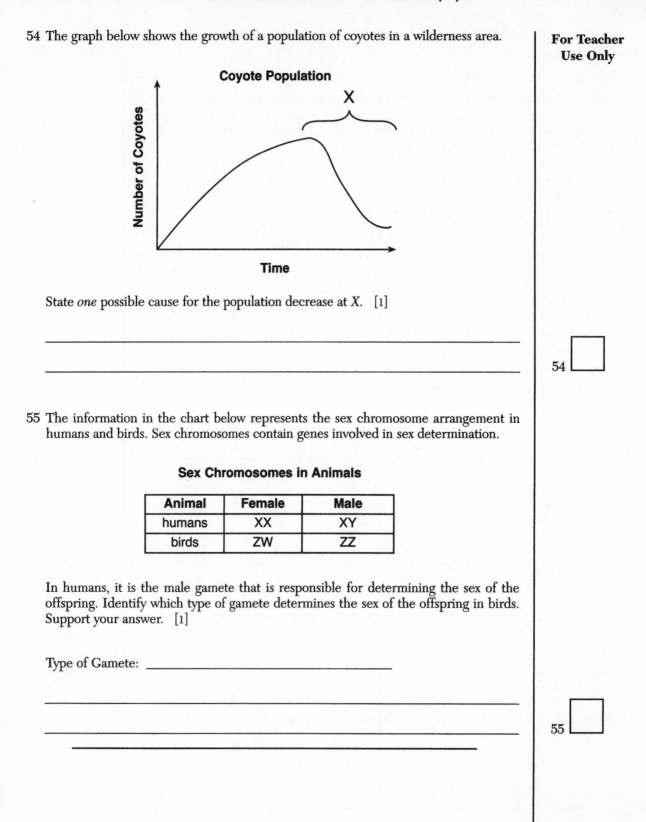

Coyote Population

X

Number of Coyotes

Time

State *one* possible cause for the population decrease at *X*.   [1]

_____

_____

54 ☐

55 The information in the chart below represents the sex chromosome arrangement in humans and birds. Sex chromosomes contain genes involved in sex determination.

Sex Chromosomes in Animals

Animal	Female	Male
humans	XX	XY
birds	ZW	ZZ

In humans, it is the male gamete that is responsible for determining the sex of the offspring. Identify which type of gamete determines the sex of the offspring in birds. Support your answer.   [1]

Type of Gamete: _____

_____

_____

55 ☐

Part C

Answer all questions in this part.   [17]

Directions (56–64): Record your answers in the spaces provided in this examination booklet.

Base your answer to question 56 on the information below and on your knowledge of biology.

For Teacher Use Only

Help for Aging Memories

As aging occurs, the ability to form memories begins to decrease. Research has shown that an increase in the production of a certain molecule, BDNF, seems to restore the processes involved in storing memories. BDNF is found in the central nervous system and seems to be important in maintaining nerve cell health. Researchers are testing a new drug that seems to increase the production of BDNF.

56 Design an experiment to test the effectiveness of the new drug to increase the production of BDNF in the brains of rats. In your answer, be sure to:

- state the hypothesis your experiment will test   [1]
- describe how the control group will be treated differently from the experimental group   [1]
- identify *two* factors that must be kept the same in both the experimental and control groups   [1]
- identify the dependent variable in your experiment   [1]

56

For Teacher Use Only

Base your answers to questions 57 through 59 on the information below and on your knowledge of biology.

Rabbits eat plants and in turn are eaten by predators such as foxes and wolves. A population of rabbits is found in which a few have a genetic trait that gives them much better than average leg strength.

57 Predict how the frequency of the trait for above average leg strength would be expected to change in the population over time. Explain your prediction. [1]

____

____

57 ☐

58 State what is likely to happen to the rabbits in the population that do *not* have the trait for above average leg strength. [1]

____

____

58 ☐

59 It was later discovered that the rabbits born with the trait for above average leg strength also inherited the trait for poor eyesight. Taking into account this new information, explain how your predictions would change. Support your answer. [1]

____

____

____

59 ☐

Base your answer to question 60 on the information below and on your knowledge of biology.

Bacterial resistance to antibiotic treatment is becoming an increasing problem for the medical community. It is estimated that 70% of bacteria that cause infections in hospitals are resistant to at least one of the drugs used for treatment. Dangerous strains of tuberculosis (TB) have emerged that are resistant to several major antibiotic drugs. While drug-resistant TB is generally treatable, it requires much longer treatments with several antibiotics that are very expensive.

60 Explain the loss of effectiveness of antibiotic drugs. In your explanation, be sure to:
- identify the genetic event that resulted in the original antibiotic resistance in some strains of bacteria   [1]
- explain how the overuse of antibiotics can increase bacterial resistance   [1]

_____

_____

_____

Base your answers to questions 61 and 62 on the information below and on your knowledge of biology.

The average life expectancy of humans in the United States increased from 63.3 years in 1943 to 77.6 years in 2003. This, combined with other factors, has led to an increase in population.

61 State *one* factor that contributed to the increase in life expectancy in the United States. [1]

_____

_____

62 State *one* way the increase in population affects other species.   [1]

_____

_____

63 The diagram below represents a cell found in some complex organisms. The enlarged section represents an organelle, labeled *X*, found in this cell.

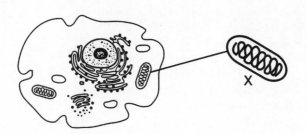

Describe the function of organelle *X* and explain how it is important to the survival of the cell. In your answer, be sure to:

- identify organelle *X*   [1]
- state the process that this organelle performs   [1]
- identify the *two* raw materials that are needed for this process to occur   [1]
- identify *one* molecule produced by this organelle and explain why it is important to the organism   [2]

Cellular
respiration
Glucose + O₂
enzymes
CO₂ + H₂O +
36 ATP

Organelle X is the mitochondria. The process this
Organelle performs, is to make energy into a
more usable form for the rest of the cell.
The two raw materials that are needed
for this process to occur are glucose and
oxygen.

63 ☐

Base your answer to question 64 on the passage below and on your knowledge of biology.

The Arctic National Wildlife Refuge

The Arctic National Wildlife Refuge (ANWR) in Alaska is the last great wilderness in America. Many migratory animals stop there to feed and rest. This region also supports an abundance of wildlife, including various types of vegetation, herbivores such as musk oxen and reindeer, and carnivores such as polar bears and wolves.

64 Wolves often hunt reindeer for food. State the effect on the size of the wolf population if the amount of vegetation were to drop suddenly. Support your answer. [1]

_____

_____

_____

Part D

Answer all questions in this part. [13]

Directions (65–77): For those questions that are followed by four choices, circle the *number* of the choice that, of those given, best completes the statement or answers the question. For all other questions in this part, follow the directions given in the question and record your answers in the spaces provided.

65 The amino acid sequences of three species shown below were determined in an investigation of evolutionary relationships.

Species A:	Val	His	Leu	Ser	Pro	Val	Glu
Species B:	Val	His	Leu	Cys	Pro	Val	Glu
Species C:	Val	His	Thr	Ser	Pro	Glu	Glu

Based on these data, which *two* species are most closely related? Support your answer. [1]

_____

_____

65 ☐

66 A student carried out a lab activity where she was asked to squeeze a clothespin as many times as she could in one minute and record that number. She immediately tried the same activity again, thinking she could do better the second time, but the number was lower. She immediately tried again, but the number was lower still.

State *one* reason why she continued to get lower numbers, even though she tried to increase the number of squeezes several times. [1]

_____

_____

66 ☐

67 A laboratory setup using an artificial cell made from dialysis tubing is shown in the diagram below.

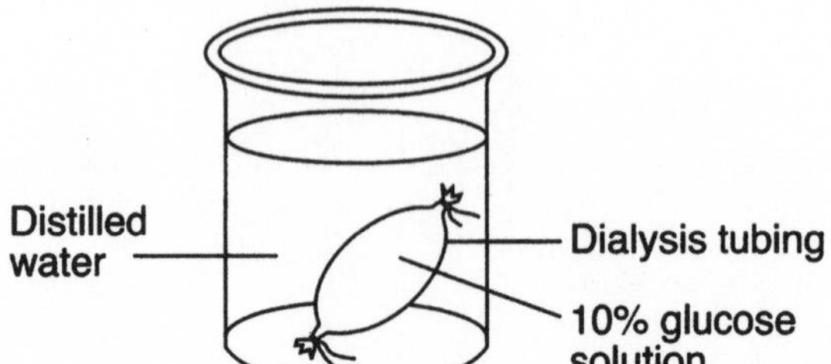

Distilled water — Dialysis tubing — 10% glucose solution

Identify the process that would most likely be responsible for the movement of glucose from inside the artificial cell to the solution outside of the cell. [1]

Diffusion Active transport

going IN!

67 ☐

For Teacher Use Only

Base your answers to questions 68 through 71 on the information below and on your knowledge of biology.

Scientists attempted to determine the evolutionary relationships between three different plant species, *A*, *B*, and *C*. In order to do this, they examined the stems and DNA of these species. Diagram 1 represents a microscopic view of the cross sections of the stems of these three species. DNA was extracted from all three species and analyzed using gel electrophoresis. The results are shown in diagram 2. Based on the data they collected, they drew diagram 3 to represent the possible evolutionary relationships.

Diagram 1

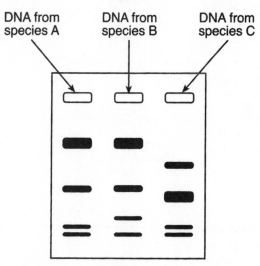

Species A   Species B   Species C

Diagram 2

DNA from   DNA from   DNA from
species A   species B   species C

Diagram 3

**Possible Evolutionary Relationships
Between Species A, B, and C**

A   B   C

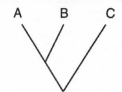

Base your answers to questions 72 through 74 on the information below and on your knowledge of biology.

A wet-mount slide of red onion cells is studied using a compound light microscope. A drawing of one of the cells as seen under high power is shown below.

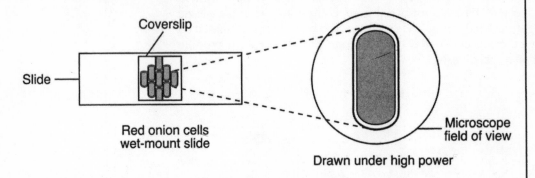

72 On the diagram below, label the location of each of the cell structures listed. [1]

cell wall
cytoplasm
cell membrane

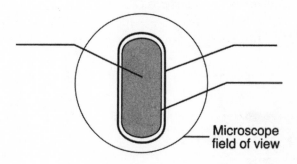

72 ☐

73 Describe the proper way to add a saltwater solution to the cells without removing the coverslip. [1]

_____

_____

_____

73 ☐

74 In the space below, sketch how the cell would look after the saltwater solution is added to it. [1]

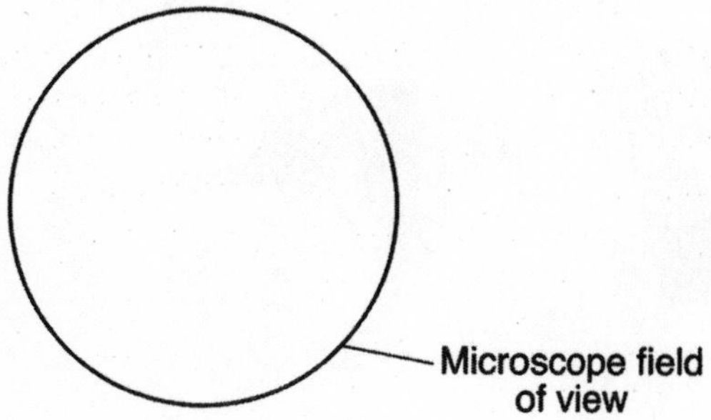

Microscope field of view

74 ☐

75 A student added an enzyme to a test tube containing a sample of DNA. After a period of time, analysis of the DNA sample indicated it was now broken into three segments. The purpose of the enzyme was most likely to

(1) cut the DNA at a specific location

(2) move the DNA to a different organism

(3) copy the DNA for protein synthesis

(4) alter the DNA sequence in the segment

75 ☐

Base your answers to questions 76 and 77 on the diagram below and on your knowledge of biology. The diagram shows the heads of four different species of Galapagos Islands finches.

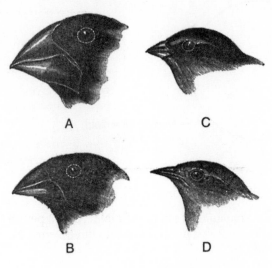

A

C

B

D

Source: http://Darwin-online.org

76 The four different types of beaks shown are most likely the result of

(1) gene manipulation

(2) natural selection

(3) unchanging environmental conditions

(4) patterns of behavior learned from parents

76 ☐

77 Scientists observed that when two closely related species of predatory birds live in different areas, they seek prey early in the morning. However, when their territories overlap, one species hunts at night and the other hunts in the morning. When these two species live in the same area, they apparently modify their

(1) habitat

(2) niche

(3) ecosystem

(4) biodiversity

77 ☐

LIVING ENVIRONMENT
JUNE 2010

ANSWER SHEET

Student ....

Teacher ....

School .... Grade ....

Part	Maximum Score	Student's Score
A	30	
B–1	13	
B–2	12	
C	17	
D	13	
Total Raw Score (maximum Raw Score: 85)		
Final Score (from conversion chart)		

Raters' Initials

Rater 1 .... Rater 2 ....

Record your answers to Part A and Part B–1 on this answer sheet.

Part A

1	11	21 ....
2	12	22 ....
3	13	23 ....
4	14	24 ....
5	15	25 ....
6	16	26 ....
7	17	27 ....
8	18	28 ....
9	19	29 ....
10	20	30 ....

Part A Score

Part B–1

31	38 ....
32	39 ....
33	40 ....
34	41 ....
35	42 ....
36	43 ....
37	

Part B–1 Score

LIVING ENVIRONMENT
AUGUST 2010

Part A

Answer all questions in this part. [30]

Directions (1–30): For *each* statement or question, write on your separate answer sheet the *number* of the word or expression that, of those given, best completes the statement or answers the question.

1 An ecosystem that has almost the same number and type of organisms for many years is exhibiting

(1) feedback
(2) global instability
(3) environmental change
(4) equilibrium

2 Which system is correctly paired with its function?

(1) immune system—intake and distribution of oxygen to cells of the body
(2) excretory system—remove potentially dangerous materials from the body
(3) digestive system—transport energy-rich molecules to cells
(4) circulatory system—produce building blocks of complex compounds

3 Which statement concerning the reproductive cells in the diagram below is correct?

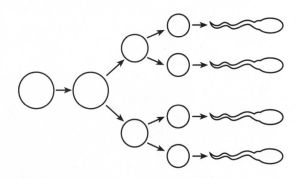

(1) The cells are produced by mitosis and contain all the genetic information of the father.
(2) If one of these cells fertilizes an egg, the offspring will be identical to the father.
(3) Each of these cells contains only half the genetic information necessary for the formation of an offspring.
(4) An egg fertilized by one of these cells will develop into a female with the same characteristics as the mother.

4 Which set of functions is directly controlled by the cell membrane?

(1) protein synthesis, respiration, digestion of food molecules
(2) active transport, recognition of chemical messages, protection
(3) enzyme production, elimination of large molecules, duplication of DNA codes
(4) release of ATP molecules, regulation of cell reproduction, food production

glucose+oxygen ⊃ ATP

5 When a new viral infection appears in a population, scientists usually try to develop a vaccine against the virus. Which substances would most likely be contained in the new vaccine?

(1) live bacteria that ingest viruses
(2) white blood cells from an infected individual
(3) weakened viruses associated with the infection
(4) a variety of microbes that will attack the virus

6 The human heart and lungs contain cells that

(1) produce a hormone involved in respiration
(2) have the same genetic information but perform different specialized functions
(3) use one part of the genetic code to synthesize all enzymes needed by the cell
(4) contain different numbers of DNA molecules

7 The diversity of organisms present on Earth is the result of

(1) ecosystem stability
(2) homeostasis
(3) natural selection
(4) direct harvesting

8 Some stages in the development of an organism are represented in the diagram below.

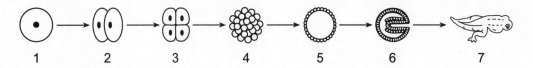

Which levels of biological organization do stages 2 and 7 have in common?

(1) cells and organs

(2) cells and tissues

(3) tissues and organelles

(4) organelles and cells

9 Plants in species *A* cannot fight most fungal infections. Plants in species *B* make a protein that kills many fungi. One possible way for humans to produce species *A* plants with the ability to synthesize this protein would be to

(1) mutate fungal DNA and introduce the mutated DNA into species *B* using a virus

(2) add DNA from species *B* into the soil around species *A*

(3) insert the gene for the protein from species *B* into a chromosome in species *A*

(4) cross species *A* and a fungus to stimulate the synthesis of this protein

10 The diagram below shows a process that affects chromosomes during meiosis.

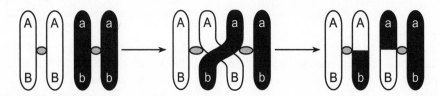

This process can be used to explain

(1) why some offspring are genetically identical to their parents

(2) the process of differentiation in offspring

(3) why some offspring physically resemble their parents

(4) the origin of new combinations of traits in offspring

11 Which phrase, if placed in box *X*, would correctly complete the flowchart shown below?

| Exposed to sunlight | ? | Increased use of CO_2 | ? | X |

(1) Increased use of starch in root cells

(2) Increased concentration of glucose in leaf cells

(3) Decreased ATP in root cells

(4) Decreased concentration of oxygen in leaf cells

12 The diagram below represents levels of organization within a cell of a multicellular organism.

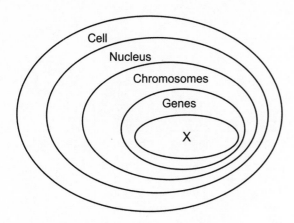

The level represented by X is composed of

(1) four types of base subunits
(2) folded chains of glucose molecules
(3) twenty different kinds of amino acids
(4) complex, energy-rich inorganic molecules

13 Scientists have discovered that the Oklahoma salamander, *Eurycea tynerensis*, develops into its adult form in streams where the streambeds are made of fine, tightly packed gravel. Salamanders living in streams with streambeds made of large, loosely packed gravel remain immature. This situation is an example of

(1) the production of gametes
(2) faulty genes found in aquatic organisms
(3) development influenced by the environment
(4) the production of new organisms by environmental engineering

14 Which statement is best supported by the theory of evolution?

(1) Genetic alterations occur every time cell reproduction occurs.
(2) The fossil record provides samples of every organism that ever lived.
(3) Populations that have advantageous characteristics will increase in number.
(4) Few organisms survive when the environment remains the same.

15 A chemical known as 5-bromouracil causes a mutation that results in the mismatching of molecular bases in DNA. The offspring of organisms exposed to 5-bromouracil can have mismatched DNA if the mutation occurs in

(1) the skin cells of the mother
(2) the gametes of either parent
(3) all the body cells of both parents
(4) only the nerve cells of the father

16 A species that lacks the variation necessary to adapt to a changing environment is more likely to

(1) develop many mutated cells
(2) become extinct over time
(3) begin to reproduce sexually
(4) develop resistance to diseases

17 A particular species of shark normally reproduces sexually. In captivity, it was found that a female could also reproduce asexually. One *negative* result from asexual reproduction is

(1) increased gene recombinations
(2) increased number of males produced
(3) decreased number of eggs used
(4) decreased biodiversity within the species

18 Which situation involves a risk to a fetus due to the mother smoking during pregnancy?

(1) decreased digestive activity in the stomach of the fetus
(2) a decrease in the amount of oxygen in the ovary of the mother
(3) inhalation of secondhand smoke by the fetus
(4) toxins in the bloodstream of the mother

19 Drugs to reduce the risk of rejection are given to organ transplant patients because the donated organ contains

(1) foreign antigens
(2) foreign antibodies
(3) DNA molecules
(4) pathogenic microbes

20 A reproductive system is represented in the diagram below.

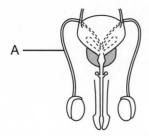

If an injury occurred to the structure labeled A, the most likely result would be a problem with

(1) delivery of sperm
(2) production of gametes
(3) production of hormones
(4) excretion of urine

21 The leaves of a plant are dotted with openings known as stomata. When open, stomata allow the plant to exchange gases and allow moisture to evaporate, helping to draw water from the roots up into the plant. These activities help the plant to

(1) produce light energy
(2) maintain homeostasis
(3) decompose organic matter
(4) synthesize minerals

22 A stable ecosystem is characterized by having

(1) predators that outnumber their prey
(2) a continual input of energy
(3) limited autotrophic nutrition
(4) no competition between species

23 The pedigree of Seattle Slew, a racehorse considered by some to be one of the fastest horses that ever lived, includes very fast horses on both his mother's side and his father's side. Seattle Slew most likely was a result of

(1) environmental selection
(2) alteration of DNA molecules
(3) selective breeding
(4) a sudden mutation

24 Changes in an ecosystem over a long period of time are shown in the diagram below.

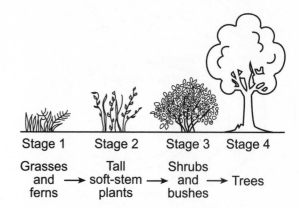

| Stage 1 | Stage 2 | Stage 3 | Stage 4 |

Grasses and ferns → Tall soft-stem plants → Shrubs and bushes → Trees

These changes will most likely lead to a

(1) stable ecosystem that can last for many years
(2) loss of heterotrophs that cannot be recovered
(3) long-term rise in environmental temperatures
(4) forest consisting of only producers and decomposers

25 Which situation indicates a serious organ system malfunction?

(1) The ovary releases estrogen, which quickly binds to cell receptors.
(2) Blood flow throughout the entire body is suddenly reduced.
(3) White blood cells release enzymes in response to the proteins on inhaled pollen.
(4) Mitochondria stop functioning in a unicellular organism exposed to pollutants.

26 Which pair of organisms would most likely compete for the same ecological niche?

(1) bacteria and fungi
(2) deer and wolf
(3) tree and fungi
(4) deer and bacteria

27 Rabbits introduced into Australia over one hundred years ago have become a serious pest. Rabbit populations have increased so much that they have displaced many native species of herbivores. Which statement best explains the reason for their increased numbers?

(1) Rabbits have a high metabolic rate.
(2) There are few native predators of rabbits.
(3) Additional rabbit species have been introduced.
(4) There is an increase in rabbit competitors.

28 Which human activity would preserve finite resources?

(1) deforestation
(2) removing carnivores from a forest
(3) recycling aluminum
(4) heating homes with fossil fuels

29 Abandoned railroad tracks are overgrown with weeds. Ten years later there are small aspen trees growing in the middle of the tracks. This change is an example of

(1) ecological succession
(2) biological evolution
(3) genetic variation
(4) heterotrophic nutrition

30 Which action would be *least* likely to harm endangered species?

(1) releasing more carbon dioxide into the atmosphere
(2) reducing the human population
(3) decreasing the amount of dissolved oxygen in the oceans
(4) reducing the thickness of the ozone layer

Part B–1

Answer all questions in this part. [13]

Directions (31–43): For *each* statement or question, write on the separate answer sheet the *number* of the word or expression that, of those given, best completes the statement or answers the question.

31 The graph below represents the growth of bacteria cultured at three different temperatures over a period of 24 hours.

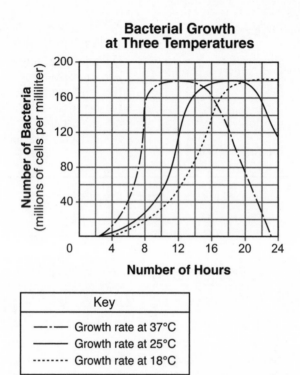

**Bacterial Growth
at Three Temperatures**

Key

—·— Growth rate at 37°C
——— Growth rate at 25°C
········· Growth rate at 18°C

Which statement concerning the rate of cell division in the bacteria culture is correct?

(1) Cell division is most rapid at 37°C between 6 and 8 hours after it began.
(2) Cell division is most rapid at 25°C between 20 and 24 hours after it began.
(3) Cell division is most rapid at 18°C between 4 and 8 hours after it began.
(4) Cell division occurs at the same rate no matter what the temperature.

32 A wet-mount slide preparation of a specimen is stained in order to

(1) eliminate some organelles
(2) make cell structures more visible
(3) use the high-power lens
(4) remove water from the slide

33 Which set of terms correctly identifies the procedure shown in the diagram below and a substance produced by this procedure?

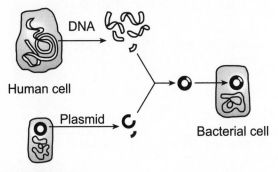

(1) selective breeding—growth hormone
(2) cloning—antibiotics
(3) genetic engineering—insulin
(4) replicating—glucose

34 The diagram below represents structures found in a human female.

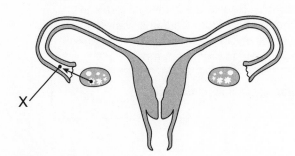

Which process results in the formation of structure X?

(1) mitosis   (3) recombination
(2) meiosis   (4) cloning

35 The graph below shows changes in the stability of an ecosystem over a period of time.

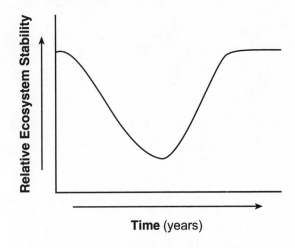

Which statement best describes the change in ecosystem stability shown in the graph?

(1) A stable ecosystem can be altered, then it can recover to a point of stability.
(2) An ecosystem remains unchanged as its stability decreases.
(3) The stability of an ecosystem remains unchanged but its biodiversity decreases.
(4) A stable ecosystem cannot recover after it is altered.

36 Two interactions between organisms are shown in the table below. X and Y do *not* represent the same organisms in the two interactions.

	Organism X	Organism Y
Interaction 1	predator	prey
Interaction 2	parasite	host

Which statement best describes the relationship between organism X and organism Y in each interaction?

(1) Organism X is positively affected by the relationship and organism Y is negatively affected.
(2) Organism X is negatively affected by the relationship and organism Y is positively affected.
(3) Both organisms are positively affected by the relationship.
(4) Both organisms are negatively affected by the relationship.

37 The diagrams below illustrate types of asexual reproduction.

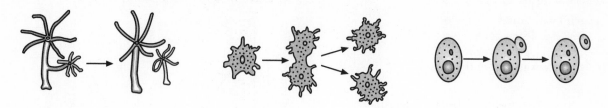

Which statement correctly describes the offspring?

(1) They vary genetically from the parent.

(2) They are produced by the union of gametes.

(3) They obtain nourishment from a placenta.

(4) They result without the union of gametes.

Base your answer to question 38 on the information below and on your knowledge of biology.

A reporter conducted a number of "on-the-street" interviews with people selected at random. The reporter found that many people gave responses similar to those of the person quoted below.

Question	Response of Person Interviewed
Would you be concerned if winters in this area became more severe and the cost of plowing and sanding snowy roads increased?	Of course I would be concerned. I can?t afford higher taxes!
Would you be willing to pay more for a car that has better fuel economy if it would benefit the environment?	No! Cars that would use less gasoline would have to be much smaller. I like my big car—and besides that, it?s safer.
If droughts became more common, would you be upset if you had to pay more for your food at the grocery store or a restaurant?	Definitely. My weekly food bill is too high already!
Would it bother you if the sea level increased a foot or two, causing many lowland areas to flood?	Not really. People could always move to higher ground. But I wouldn?t want to see my taxes go up because we have to spend more on aid to help them move.
Are you concerned about global warming?	Not really. It doesn?t affect me.

38 Which statement is best supported by these interviews?

(1) Many people are very aware of the possible effects of global warming.

(2) Many people care more about their personal comforts than about the possible effects of global warming.

(3) Many people are willing to sacrifice to reduce the possible effects of global warming.

(4) Many people are now taking action to reduce the possible effects of global warming.

Base your answers to questions 39 and 40 on the diagram below, which represents a pond food web, and on your knowledge of biology.

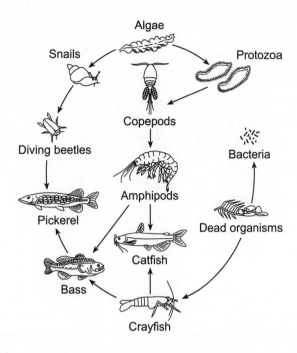

39 Which energy pyramid most accurately shows the energy relationships between three organisms in this food web?

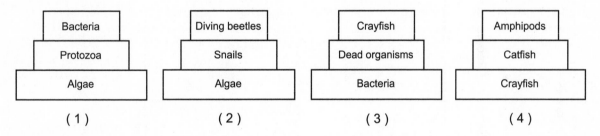

Bacteria
Protozoa
Algae

(1)

Diving beetles
Snails
Algae

(2)

Crayfish
Dead organisms
Bacteria

(3)

Amphipods
Catfish
Crayfish

(4)

40 Which statement best describes what will most likely happen if the amphipod population is removed from this food web?

(1) Population sizes of species at feeding levels both before and after amphipods will decrease.

(2) Population sizes of species at feeding levels both before and after amphipods will increase.

(3) Population sizes of species at feeding levels after amphipods will increase and before amphipods will decrease.

(4) Population sizes of species at feeding levels after amphipods will decrease and before amphipods will increase.

41 A biological process that occurs in both plants and animals is shown below.

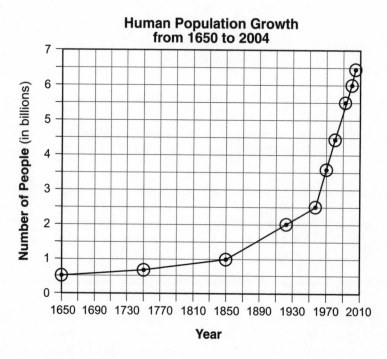

Which row in the chart below identifies the lettered substances in this process?

Row	A	B	C	D
(1)	O_2	CO_2	glucose	enzymes
(2)	glucose	O_2	enzymes	CO_2
(3)	enzymes	O_2	CO_2	glucose
(4)	glucose	CO_2	enzymes	O_2

42 The graph below shows data on human population growth.

**Human Population Growth
from 1650 to 2004**

The trend shown on the graph would most likely result in

(1) a decreased demand for deforestation
(2) an increase in available freshwater
(3) a decrease in air pollution
(4) an increased demand for land use

43 The diagram below represents the varying biodiversity in three ecosystems.

The level of biodiversity in ecosystem A is high because it has the

(1) least variety of energy levels
(2) greatest variety of genetic material
(3) greatest number of decomposers
(4) least number of ecological niches

Part B–2

Answer all questions in this part. [12]

Directions (44–55): For those questions that are followed by four choices, circle the *number* preceding the choice that, of those given, best completes the statement or answers the question. For all other questions in this part, follow the directions given in the question and record your answers in the spaces provided.

44 The graph below shows the growth of a field mouse population in an ecosystem over time.

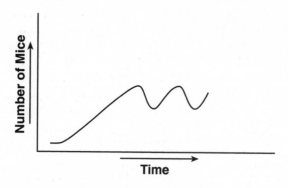

The dashed line indicating the carrying capacity for the mouse population is correctly shown on which graph?

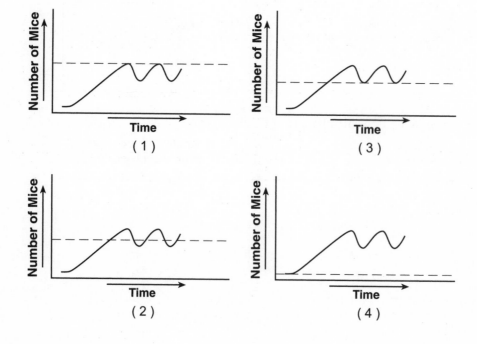

44 ☐

Base your answers to questions 45 through 48 on the passage below and on your knowledge of biology.

Sudden Death from a Marine Predator

Members of the Conidae family (cone snails) have been collected for centuries for their beautiful and elaborately detailed shells. Cone snails are marine mollusks found in reef environments throughout the world.

Cone snails feed on organisms such as fish, worms, and other mollusks. They are very slow moving but capture their prey by paralyzing them using venom. The venom contains some of the most deadly neurotoxins known. The neurotoxins work by attaching to receptor molecules on nerves, blocking the transmission of nerve impulses. The neurotoxins are injected into the prey by way of a hollow, spearlike tooth and the effects are usually immediate. One species, a fish-eating cone snail, can paralyze the prey in about two seconds.

The venom produced by each species is prey specific. It may contain two or more different types of neurotoxins, each composed of long chains of amino acids.

45 Explain how a neurotoxin present in the venom can paralyze one type of prey but not another. [1]

_____

_____

45 ☐

46 State *one* way the neurotoxin protein in the venom of cone snails can be different. [1]

_____

_____

46 ☐

47 Explain why paralyzing its prey in only two seconds is an advantage to fish-eating cone snails. [1]

_____

_____

47 ☐

48 Cone snails of the same species often exhibit variations in the patterns of their shells. State *one* possible cause for these variations appearing in the shell pattern within the population of the cone snails. [1]

_____

_____

48 ☐

Base your answers to questions 49 and 50 on the information below and on your knowledge of biology.

Due to the negative effects on the environment of burning coal and oil, society is looking for alternate energy resources that are renewable.

49 Identify *one* renewable resource that can be used to generate energy. [1]

_____

50 State *one* benefit, other than the fact that it is renewable, of using this resource. [1]

_____

_____

_____

51 A student, using a metric ruler, measured a larva as represented in the diagram below.

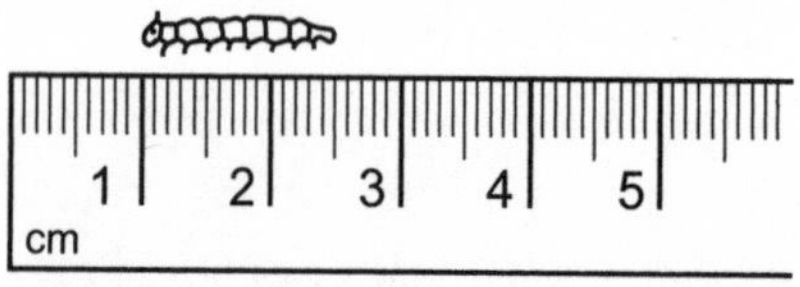

What is the length of the larva, in millimeters? [1]

_____ **mm**

49 ☐

50 ☐

51 ☐

Base your answers to questions 52 through 55 on the information and data table below and on your knowledge of biology.

An investigation was carried out over a five-year period to measure the effect of color on the survival of trout in a stream. The stream contained many brightly colored stones and food was plentiful. At the start of the investigation (year 0), 100 bright-colored trout and 100 drab-colored trout were placed into a section of the stream that had been blocked with netting. Investigators monitored the trout populations for five years and recorded the water condition each time a count was done. The data collected are shown in the table below.

Trout Population Over Five Years

Year	Bright-Colored Trout	Drab-Colored Trout	Condition of Water
0	100	100	clear
1	64	36	clear
2	86	25	clear
3	25	77	cloudy
4	14	86	cloudy
5	90	9	clear

Directions (52–54): Using the information in the data table, construct a line graph on the grid on the following page, following the directions below.

52 Mark an appropriate scale on each labeled axis. [1]

53 Plot the data for the bright-colored trout on the grid. Surround each point with a small circle and connect the points. [1]

Example:

54 Plot the data for the drab-colored trout on the grid. Surround each point with a small triangle and connect the points. [1]

Example:

Trout Population Over Five Years

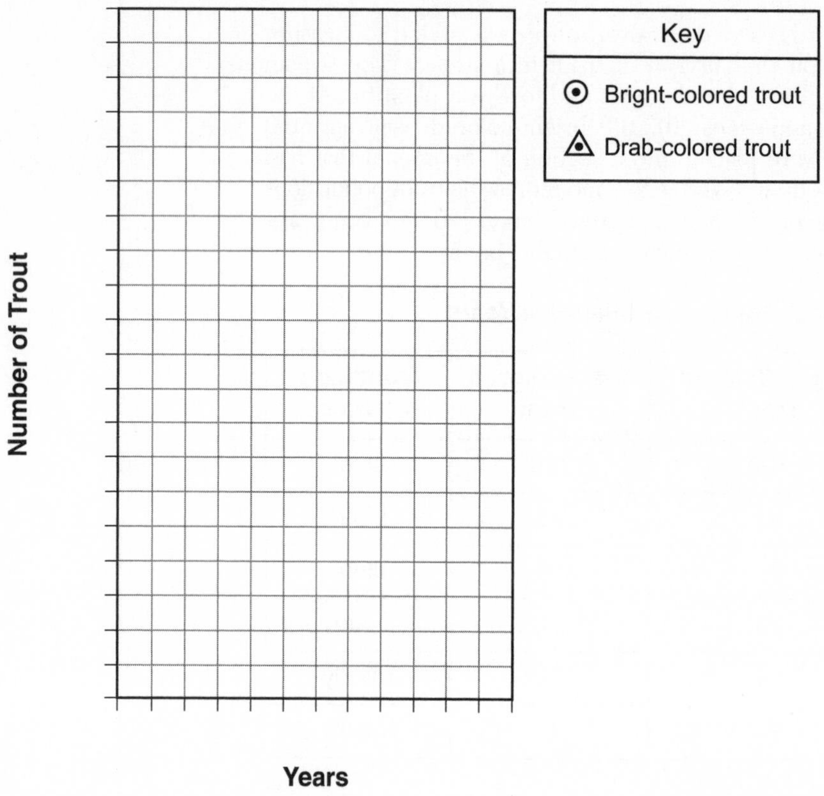

Key

⊙ Bright-colored trout

△ Drab-colored trout

Number of Trout

Years

52 ☐

53 ☐

54 ☐

55 Explain how trout survival is related to the color of trout and the environmental condition of the stream. [1]

_____

_____

55 ☐

Part C

Answer all questions in this part. [17]

Directions (56–62): Record your answers in the spaces provided in this examination booklet.

Base your answer to question 56 on the diagram below and on your knowledge of biology.

In a cell, a variety of structures perform specific functions and interact to maintain homeostasis. The diagram below represents a typical cell with three cell structures labeled 1, 2, and 3.

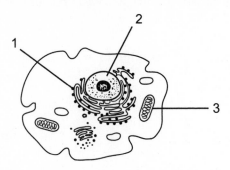

56 Select *one* cell structure labeled in the diagram and write its number in the space below. Explain how the cell structure you selected helps maintain homeostasis in a cell. In your answer, be sure to:

- identify the cell structure you selected [1]
- state *one* function of this cell structure [1]
- identify *one* substance that is often associated with the cell structure you selected and state how that substance is associated with the cell structure [1]
- identify *one* other cell structure and explain how it interacts with the cell structure you selected to maintain homeostasis in the cell [1]

Cell structure number: 3)

The cell structure I selected is the mitochondria. The function of this structure is to provide energy for the rest of the cell.

56 ☐

57 The table below lists enzymes that function in different locations in the human body, and the temperature and pH ranges of these locations.

Enzyme	Location	Temperature (°C)	pH
ptyalin	mouth	36.7–37.0	6.5–7.0
pepsin	stomach	37.3–37.6	1.0–3.0
trypsin	small intestine	37.3–37.6	7.5–9.0

Different enzymes are secreted in each of the three locations. Ptyalin digests carbohydrates. Pepsin and trypsin both digest proteins. Discuss the activity of these enzymes. In your answer, be sure to:

• state how the activity of pepsin will most likely change after it moves with the food from the stomach to the small intestine   [1]
• support your answer using data from the table   [1]
• state how a fever of 40°C would most likely affect the activity of these enzymes and support your answer   [1]
• identify the characteristic of enzymes that prevents ptyalin and trypsin from digesting the same type of food   [1]

57

58 Consuming large volumes of soft drinks containing sugar during the day can disrupt homeostasis. Describe how the human body responds to restore sugar balance. In your answer, be sure to:

- identify the hormone responsible for restoring homeostasis [1]
- identify the organ that releases this hormone [1]
- state *one* possible reason why sugar levels may remain high even though this hormone has been released [1]

_____

_____

_____

_____

_____

_____

58 ☐

59 Each of the environmental problems listed below has had an impact on ecosystems.

increased ultraviolet radiation
global warming

Select *one* of these problems and write it in the space below. Explain how this problem has affected an ecosystem. In your answer, be sure to:

- identify *one* specific cause of the environmental problem [1]
- identify *one* organism that has been affected by the problem and state *one* way that organism has been affected [1]
- state *one* action that can be taken to lessen the impact of this problem on the environment [1]

Problem: _____

_____

_____

_____

_____

_____

59 ☐

Base your answers to questions 60 through 62 on the information below and on your knowledge of biology.

In the abyssal zones (deepest zones) of oceans, organisms live in an ecosystem that lacks sunlight. Other environmental conditions include temperatures of 4°C and extremely high water pressure. Dead material from upper ocean zones sinks and settles in the abyssal zone.

60 State *one* possible way that some organisms living permanently in the abyssal zone could obtain energy. [1]

_____

_____

60 ☐

61 Many of the animals in the abyssal zone possess light-producing cells in specific parts of their bodies. State *one* possible use for these lights. [1]

_____

_____

61 ☐

62 Animals from the abyssal zone can *not* survive in upper ocean zones. State *one* possible reason for this. [1]

_____

_____

62 ☐

Part D

Answer all questions in this part. [13]

Directions (63–75): For those questions that are followed by four choices, circle the *number* of the choice that, of those given, best completes the statement or answers the question. For all other questions in this part, follow the directions given in the question and record your answers in the spaces provided.

Base your answers to questions 63 through 66 on the information and diagram below and on your knowledge of biology.

For Teacher Use Only

DNA samples were collected from four children. The diagram below represents the results of a procedure that separated the DNA in each sample.

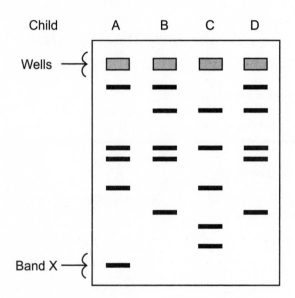

63 Identify the procedure used to obtain these results. [1]

_____

63 ☐

64 Band *X* represents the

(1) largest fragment of DNA that traveled the fastest

(2) smallest fragment of DNA that traveled the fastest

(3) largest fragment of DNA that traveled the slowest

(4) smallest fragment of DNA that traveled the slowest

64 ☐

65 The DNA is most similar in which two children? Support your answer.   [1]

_____

_____

65 ☐

66 State *one* way information obtained from this procedure can be used.   [1]

_____

_____

_____

66 ☐

67 A technique used to analyze pigments in spinach leaves is shown in the diagram below.

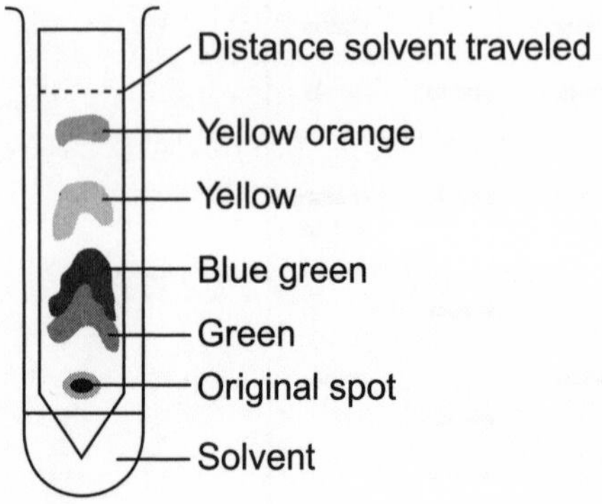

Distance solvent traveled
Yellow orange
Yellow
Blue green
Green
Original spot
Solvent

This technique is known as

(1) paper chromatography

(2) gene manipulation

(3) dissection

(4) staining

67 ☐

68 A student conducted an experiment to determine if listening to different types of music would affect pulse rate. She thought that pulse rate would change with different types of music. Each person participating in her experiment listened to seven different selections of music for 30 seconds each. The pulse rates were taken after each 30-second interval of music. Based on her experiment, the student concluded that a person's pulse rate changed when listening to different types of music.

The component missing from this experiment is a

(1) prediction

(2) hypothesis

(3) control group

(4) research plan

68 ☐

69 An experiment was carried out to determine whether drinking caffeinated soda increases pulse rate. The pulse rates of two groups of people at rest were measured. Group *A* was then given caffeinated soda and group *B* was given caffeine-free soda. One hour after drinking the soda, the pulse rates were measured. The participants in the experiment were all the same age, and they were all given the same amount of soda.

The dependent variable in this experiment is the

(1) type of soda given to each group

(2) amount of soda given to each group

(3) pulse rate of each group

(4) age of participants in each group

69 ☐

Base your answer to question 70 on the information below and on your knowledge of biology.

A student states that exercise will affect the number of times a person can squeeze a clothespin in a certain amount of time. An experiment is carried out to test this hypothesis. One group of ten students sits quietly before squeezing a clothespin as many times as possible during a one-minute interval. A second group of ten students does 25 jumping jacks before squeezing a clothespin as many times as possible during a one-minute interval.

70 State *one* way the experiment could be improved in order to increase the validity of the results. [1]

_____

_____

_____

70 ☐

Base your answer to questions 71 and 72 on the information in the diagram below and on your knowledge of biology.

Finch Diversity

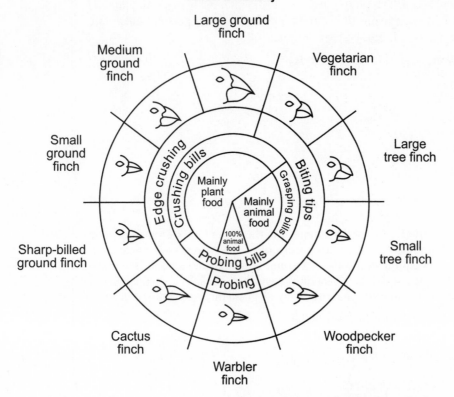

—From *Galapagos: A Natural History Guide*

71 Small ground finches and medium ground finches live on an island with abundant plant and animal food. Predict how the small ground finch and the medium ground finch would be affected if warbler finches migrated to the island where these finches live. Support your answer. [1]

_____

_____

71 ☐

72 The differences observed in the bird beaks are most likely due to

(1) asexual reproduction of these finch species

(2) the selection for different shaped beaks that best suit different niches

(3) the genetic recombination associated with mitotic cell division

(4) the genetic engineering of the DNA of each of these species

72 ☐

Base your answers to questions 73 through 75 on the laboratory setup illustrated below and on your knowledge of biology.

73 Identify the color of the contents of the artificial cell after two hours. [1]

_____

73 ☐

74 After two hours, the color of the liquid in the beaker did *not* change. This shows that

(1) glucose moved from the artificial cell into the beaker

(2) starch did not pass out of the artificial cell

(3) starch was digested to glucose in the artificial cell

(4) glucose molecules combined to produce starch in the artificial cell

74 ☐

75 This laboratory setup would most likely be used to demonstrate

(1) carbohydrate synthesis

(2) active transport

(3) diffusion

(4) dehydration

75 ☐

LIVING ENVIRONMENT
AUGUST 2010

ANSWER SHEET

Student . . . .

Teacher . . . .

School . . . . Grade . . . .

Part	Maximum Score	Student's Score
A	30	
B–1	13	
B–2	12	
C	17	
D	13	
Total Raw Score (maximum Raw Score: 85)		
Final Score (from conversion chart)		

Raters' Initials

Rater 1 . . . .   Rater 2 . . . .

Record your answers to Part A and Part B–1 on this answer sheet.

Part A

1	11	21 . . . .
2	12	22 . . . .
3	13	23 . . . .
4	14	24 . . . .
5	15	25 . . . .
6	16	26 . . . .
7	17	27 . . . .
8	18	28 . . . .
9	19	29 . . . .
10	20	30 . . . .

Part A Score

Part B–1

31	38 . . . .
32	39 . . . .
33	40 . . . .
34	41 . . . .
35	42 . . . .
36	43 . . . .
37	**Part B–1 Score**

LIVING ENVIRONMENT
JANUARY 2011

Part A

Answer all questions in this part. [30]

Directions (1–30): For *each* statement or question, write on your separate answer sheet the *number* of the word or expression that, of those given, best completes the statement or answers the question.

1 The greatest number of relationships between the organisms in an ecosystem is best shown in
(1) a food chain
(2) an energy pyramid
(3) a food web
(4) an ecological succession diagram

2 The diagram below shows stages of human reproduction.

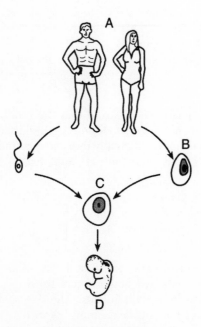

The direct result of fertilization is represented at
(1) A
(2) B
(3) C
(4) D

3 Certain organisms are able to store energy from the Sun in energy-rich compounds. Which event best illustrates this activity?
(1) A fox captures and eats a young rabbit.
(2) A caterpillar is eaten by a blackbird.
(3) Lettuce produces organic substances.
(4) Bacteria change organic material into simple nutrients.

4 The diagram below shows how a chemical message produced by one cell is received by other cells.

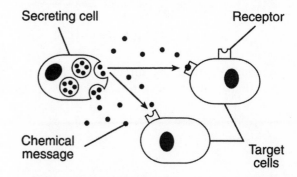

If these chemical messages are destroyed, the target cells will
(1) produce their own chemical messages
(2) not respond with appropriate actions
(3) develop different receptors
(4) no longer be produced in the organism

5 Coded instructions that are passed from one generation to the next can be most directly changed by the processes of
(1) passive transport, natural selection, and synthesis
(2) selective breeding, replication, and absorption
(3) recombination, mutation, and genetic engineering
(4) evolution, reproduction, and digestion

6 The diagram below represents a cross section of a leaf of a green plant, showing an opening (stomate) in the lower surface.

Structure of a Leaf

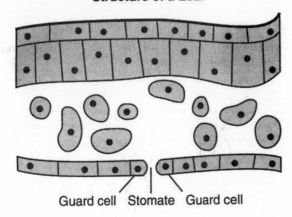

Guard cell   Stomate   Guard cell

A stomate in the lower surface of the leaf has a function most similar to the function of which cell structure?

(1) cell membrane   (3) ribosome
(2) vacuole   (4) nucleus

7 When *S. marcescens*, a bacterium, is grown in a refrigerator, it produces red-colored colonies. However, if the bacterium is grown at room temperature, the colonies are white. The best explanation for this situation is that

(1) refrigeration changes the structure of genes
(2) room temperature stimulates the synthesis of a red pigment
(3) temperature has an effect on the expression of genes
(4) only temperature is responsible for the expression of a trait

8 In sexually reproducing organisms, mutations can be inherited if they occur in

(1) the egg, only
(2) the sperm, only
(3) any body cell of either the mother or the father
(4) either the egg or the sperm

9 The diagram below represents a structure found in most cells.

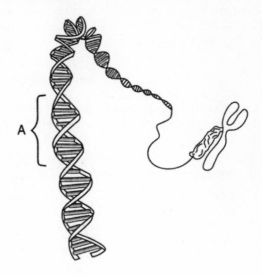

A

The section labeled *A* in the diagram is most likely a

(1) protein composed of folded chains of base subunits
(2) biological catalyst
(3) part of a gene for a particular trait
(4) chromosome undergoing a mutation

10 Researchers have reported that the number of different species of fish found in certain areas of the ocean has been greatly reduced over the past 50 years. This situation is an example of

(1) a loss of biodiversity
(2) an increase in ecological succession
(3) a lack of differentiation
(4) an increased carrying capacity

11 Large rebates and low-cost loans have been made available to homeowners to install solar panels to heat their homes. The use of these incentives benefits ecosystems because it

(1) encourages conservation of resources
(2) reduces the need for recycling
(3) promotes the use of nonrenewable resources
(4) discourages the use of alternative energy

12 Which sequence represents the correct order of events for the production of necessary complex molecules after food is taken in by a multicellular animal?

    (1) diffusion → synthesis → absorption → digestion → circulation
    (2) circulation → diffusion → synthesis → absorption → digestion
    (3) digestion → absorption → circulation → diffusion → synthesis
    (4) synthesis → digestion → absorption → diffusion → circulation

13 The number in each circle below represents the chromosome number of the cell. Which diagram represents the production of offspring by an asexually reproducing organism?

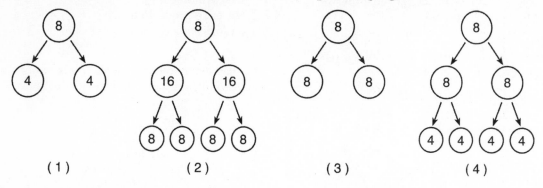

   (1)   (2)   (3)   (4)

14 The arrows in the diagram below indicate the development of four different varieties of vegetable plants from wild mustard.

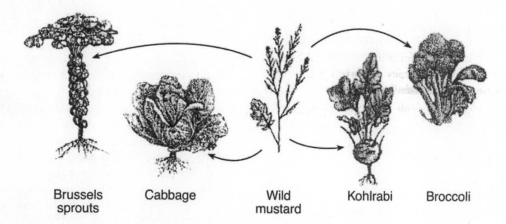

   Brussels   Cabbage   Wild   Kohlrabi   Broccoli
   sprouts   mustard

   Each of these varieties was most likely produced as a result of

   (1) asexual reproduction in the wild for many years   (3) competition between plants
   (2) changes in light availability   (4) selective breeding over many generations

15 The sorting and recombination of genes during reproduction is important to evolution because these processes

(1) decrease variation and help maintain a stable population

(2) increase variation that enables species to adapt to change

(3) decrease the chances of producing offspring that are adapted to the environment

(4) increase the ability of all the offspring to adapt to the environment

16 A diagram of evolutionary pathways of various animal species is shown below.

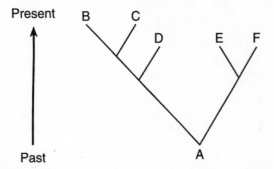

The pattern of these evolutionary pathways is most likely the result of alterations within which structure?

(1) vacuole   (3) nucleus

(2) cell membrane   (4) ribosome

17 Which situation is *least* likely to result in new inherited characteristics?

(1) altering genetic information

(2) changes in the structure of genes

(3) producing new individuals by means of cloning

(4) changes in the structure of individual chromosomes

18 In most mammals, the placenta is essential to the embryo for the processes of

(1) meiosis and excretion

(2) nutrition and excretion

(3) milk production and digestion

(4) blood exchange and digestion

19 Ancestors of the giant panda had rounded paws with five very short toes. Today, the giant panda has a sixth toe, often referred to as a thumb, even though it develops from a wrist bone. This unique thumb is an adaptation that allows the panda to easily hold and eat bamboo shoots. The presence of the giant panda's thumb is most likely the result of

(1) natural selection

(2) selective breeding

(3) asexual reproduction

(4) ecological succession

20 The diagram below represents levels of organization within a cell of a multicellular organism.

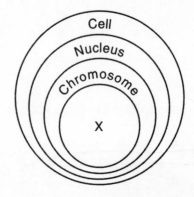

Which statement is correct regarding the structure represented by X?

(1) Structure X is composed of many different amino acids that determine the type of cell it will become in the organism.

(2) Structure X has the same base sequence in all the body cells of the organism.

(3) Structure X is a folded chain arrangement of carbohydrate found in all the body cells of the organism.

(4) Structure X contains 20 different kinds of subunits that are present in all the cells of the organism.

21 A pathogen passing from a mother to her fetus could cause

(1) a decrease in the chromosome number of the fetus
(2) an increase in milk production in the mother
(3) gamete production to increase
(4) an infection in the fetus

22 The diagram below represents the human male reproductive system.

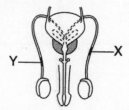

Which activity would be prevented by blockages at X and Y?

(1) transport of urine out of the body
(2) passage of testosterone to the female to stimulate egg production
(3) movement of sperm out of the body
(4) movement of testosterone to the testes to stimulate sperm production

23 One environmental problem caused by the use of nuclear power as an energy source is the

(1) destruction of the ozone shield
(2) disposal of wastes
(3) production of acid rain
(4) accumulation of CO_2 in the atmosphere

24 Which method of protecting members of an endangered species is most ecologically sound?

(1) protecting the habitats where these animals live from human development
(2) capturing these animals and putting them in wildlife parks
(3) feeding and constructing shelters for these organisms
(4) passing laws that encourage hunting of the predators of these species

25 The interaction of which two systems provides the molecules needed for the metabolic activity that takes place at ribosomes?

(1) digestive and circulatory
(2) reproductive and excretory
(3) immune and nervous
(4) respiratory and muscular

26 The swordfish contains a heat generating organ that warms its brain and eyes up to 14°C above the surrounding ocean water temperature. Which structures are most likely to be found at relatively high concentrations within the cells of this heat generating organ?

(1) nuclei
(2) chloroplasts
(3) chromosomes
(4) mitochondria

27 Two species of animals with a similar appearance live in the same habitat but do *not* compete for food. This is because they most likely

(1) reproduce at different times of the year
(2) are the same size
(3) occupy different ecological niches
(4) are active at night

28 During its annual migration, the red knot, a medium-size shorebird, flies the entire length of North and South America. During one critical stop to feed on the eggs of horseshoe crabs, the birds nearly double their body mass. The relationship between the red knot and the horseshoe crab is that of

(1) parasite–host
(2) consumer–producer
(3) scavenger–producer
(4) predator–prey

29 It is recommended that people at risk for serious flu complications be vaccinated so that their bodies will produce

(1) antigens to fight the flu virus
(2) antibodies against the flu virus
(3) toxins to fight the infection caused by the flu virus
(4) antibiotics to reduce symptoms caused by the flu virus

30 The diagram below represents a process that occurs during normal human development.

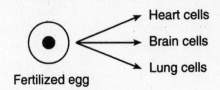

Fertilized egg

Heart cells

Brain cells

Lung cells

Which statement is correct regarding the cells and DNA?
(1) All the cells have identical DNA.
(2) The DNA of the fertilized egg differs from the DNA of all the other cells.
(3) The DNA of the fertilized egg differs from some, but not all, of the other cells.
(4) Only the fertilized egg contains DNA.

Part B–1

Answer all questions in this part. [15]

Directions (31–45): For *each* statement or question, write on the separate answer sheet the *number* of the word or expression that, of those given, best completes the statement or answers the question.

31 Activities in the human body are represented in the diagram below.

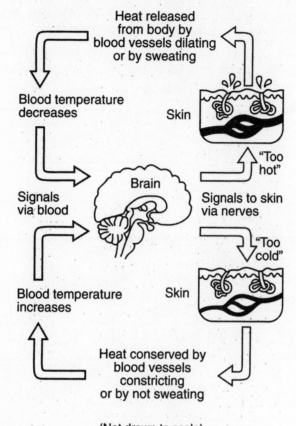

(Not drawn to scale)

Source: Campbell and Reece,
Biology, 6th edition (adapted)

Which title would be appropriate for the diagram?
(1) Rate of Excretion Varies in Response to Amount of Water Taken In
(2) Feedback Mechanisms Help to Maintain Homeostasis
(3) Respiratory Rate Responds to an Increase in Muscle Activity
(4) The Nervous System Responds to Changes in Blood Sugar Levels

32 A company that manufactures a popular multivitamin wanted to determine whether their multivitamin had any side effects. For its initial study, the company chose 2000 individuals to take one of their multivitamin tablets per day for one year. Scientists from the company surveyed the participants to determine whether they had experienced any side effects. The greatest problem with this procedure is that

(1) only one brand of vitamin was tested

(2) the study lasted only one year

(3) the sample size was not large enough

(4) no control group was used

33 In a particular ecosystem, squirrels make up a large portion of the diet of coyotes. A fatal disease in the squirrel population begins to reduce their population over a period of months. Which graph best represents the expected changes in population size of the coyotes and the squirrels?

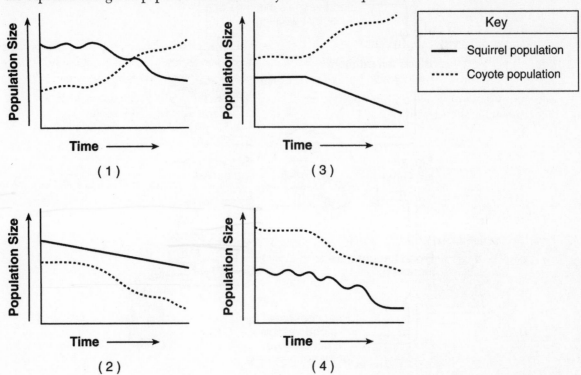

34 Which statement would most likely be used to describe the procedure represented in the diagram below?

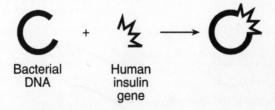

Bacterial DNA   Human insulin gene

(1) Enzymes are used to assemble an insulin gene, which is then attached to bacterial DNA.

(2) Bacterial DNA is cut from a human DNA strand and inserted into a human cell to form an insulin gene.

(3) The insulin gene is cut out of a human DNA strand using an enzyme and inserted into bacterial DNA, resulting in a combination of different DNA segments.

(4) A gene is deleted from bacterial DNA to produce an insulin gene, which is then inserted into human DNA.

35 Part of a molecule found in cells is represented below.

Which process is most directly affected by the arrangement of components 1 through 4?

(1) diffusion through cell membranes
(2) fertilization of a sex cell
(3) sequencing of amino acids in cells
(4) increasing the number of cells in an organism

36 What is the volume of water represented in the graduated cylinder shown below?

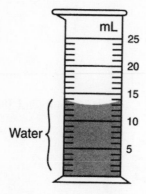

(1) 10.3 mL   (3) 14.0 mL
(2) 13.0 mL   (4) 15.0 mL

37 A student prepared a test tube containing yeast, glucose, and water. After 24 hours, the test tube was analyzed for the presence of several substances.

What substance would the student expect to find if respiration occurred in the test tube?

(1) a hormone   (3) nitrogen
(2) starch   (4) carbon dioxide

38 A student used the low-power objective of a compound light microscope and observed a single-celled organism as shown in the diagram below.

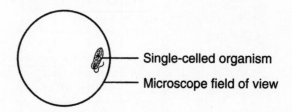

Single-celled organism
Microscope field of view

When he switched to high power, the organism was no longer visible. This most likely happened because switching to high power made the

(1) field too bright to see the organism
(2) image too small to be seen
(3) area of the slide being viewed smaller
(4) fine-adjustment knob no longer functional

39 The daphnia shown below has produced three egg cells, eats live single-celled organisms, lives in freshwater, and is caught and eaten by animals known as hydra.

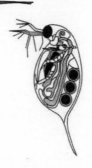

Which terms would most likely be used in a description of this organism?

(1) asexual reproduction, herbivore, prey, aquatic, heterotrophic
(2) sexual reproduction, predator, aquatic, heterotrophic, prey
(3) asexual reproduction, autotrophic, predator, terrestrial, scavenger
(4) sexual reproduction, carnivore, aquatic, autotrophic, prey

40 Changes in a deer population are shown in the graph below.

Changes in a Deer Population

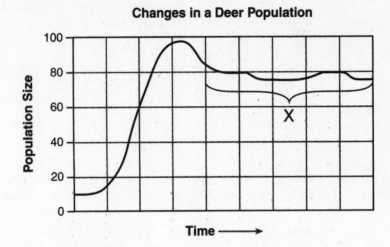

Which statement best explains section *X*?

(1) The population has reached the carrying capacity of its environment.
(2) Energy is used for interbreeding between members of different species.
(3) A predator recycles the remains of dead organisms.
(4) Competition does not occur between members of different species in the same habitat.

41 The diagram below shows various ecological communities that occupied an area over a period of 300 years.

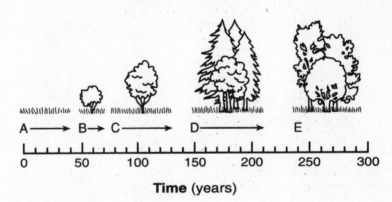

Time (years)

Which statement best describes the diagram?

(1) Community *A* is the most stable community.
(2) Community *B* replaced community *C* after a period of 100 years.
(3) Community *C* developed into community *A* after a period of 75 years.
(4) Community *D* modified the environment, making it more suitable for community *E*.

Base your answers to questions 42 and 43 on the food web below and on your knowledge of biology.

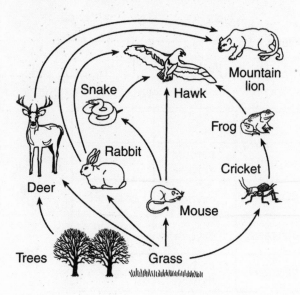

42 Which organisms are carnivores?

    (1) grass and trees   (3) deer and mountain lion

    (2) mouse, rabbit, and cricket   (4) frog, snake, and hawk

43 A *decrease* in the grass population will most immediately *decrease* the available energy for the

    (1) mouse   (3) snake

    (2) hawk   (4) frog

---

44 The diagram below shows two different kinds of substances, *A* and *B*, entering a cell.

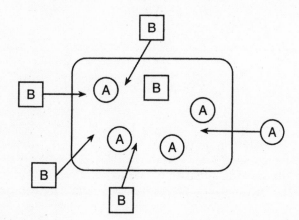

    ATP is most likely being used for

    (1) substance *A* to enter the cell   (3) both substances to enter the cell

    (2) substance *B* to enter the cell   (4) neither substance to enter the cell

45 A biological process that occurs in plants is represented below.

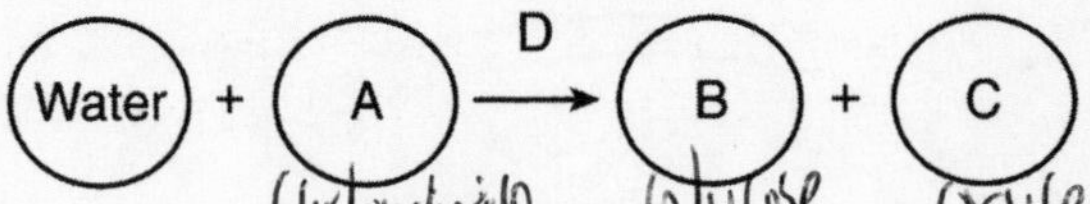

Water + A $\xrightarrow{\text{D}}$ B + C

carbon dioxide   glucose   oxygen

Which row in the chart below identifies the lettered substances in this process?

Row	A	B	C	D
(1)	enzymes	oxygen	carbon dioxide	glucose
(2)	carbon dioxide	glucose	oxygen	enzymes
(3)	glucose	enzymes	oxygen	carbon dioxide
(4)	oxygen	glucose	carbon dioxide	enzymes

Photosynthesis:

$H_2O + CO_2 \xrightarrow{\text{enzymes}} Glucose + O_2$

Answer all questions in this part. [10]

Directions (46–55): For those questions that are followed by four choices, circle the *number* preceding the choice that, of those given, best completes the statement or answers the question. For all other questions in this part, follow the directions given in the question and record your answers in the spaces provided.

Base your answers to questions 46 through 50 on the data table below and on your knowledge of biology. The data table shows the concentrations of oxygen in parts per million (ppm) present in freshwater and seawater at various temperatures.

Concentration of Oxygen in Water

Temperature (°C)	Oxygen Concentration in Freshwater (ppm)	Oxygen Concentration in Seawater (ppm)
1	14.0	11.0
10	11.5	9.0
15	10.0	8.0
20	9.0	7.5
25	8.0	7.0
30	7.5	6.0

Directions (46–48): Using the information in the data table, construct a line graph on the grid on the next page, following the directions below.

46 Mark an appropriate scale on each labeled axis. [1]

47 Plot the data for freshwater oxygen concentration on the grid. Surround each point with a small circle and connect the points. [1]

Example:

48 Plot the data for seawater oxygen concentration on the grid. Surround each point with a small triangle and connect the points. [1]

Example:

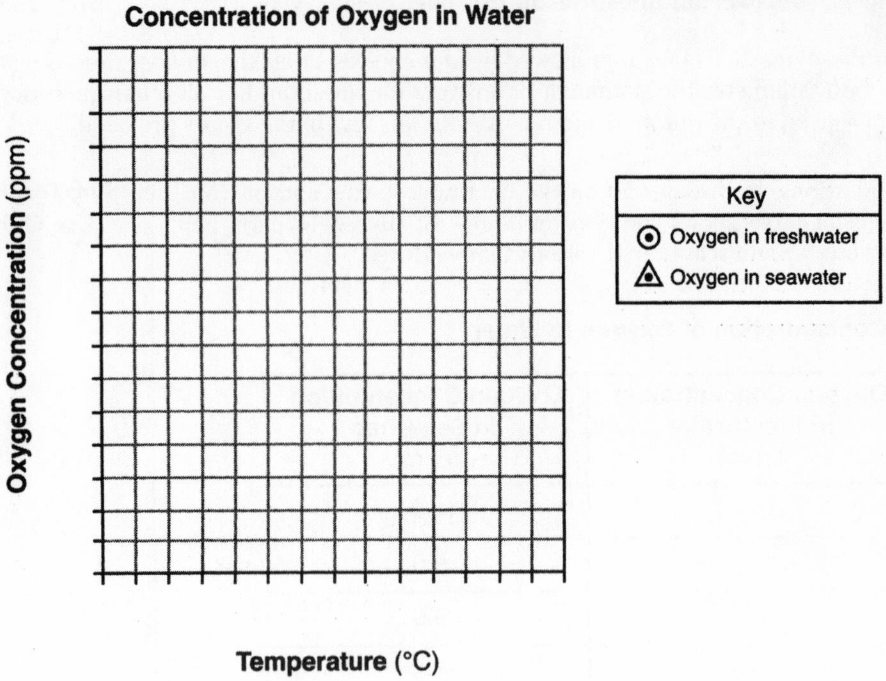

Concentration of Oxygen in Water

Oxygen Concentration (ppm)

Temperature (°C)

Key
⊙ Oxygen in freshwater
△ Oxygen in seawater

49 Predict the oxygen concentration in freshwater at 35°C. [1]

_____ ppm

50 State *one* relationship between temperature and dissolved oxygen concentration in water. [1]

_____

_____

46

47

48

49

50

Base your answers to questions 51 through 53 on the passage below and on your knowledge of biology.

A New Theory on Malaria Transmission

Plasmodium falciparum, one parasite that causes malaria, spreads rapidly, infecting up to 500 million people every year. Malaria spreads when an infected mosquito bites an uninfected human, who then becomes infected. This infected human is bitten by an uninfected mosquito, which then becomes infected. This infected mosquito then bites and infects an uninfected human. Malaria transmission is illustrated below.

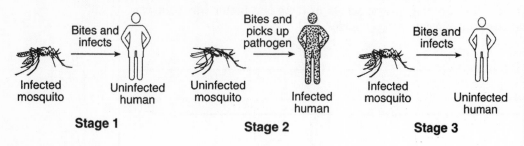

(Not drawn to scale)

Scientists have a new idea about how this disease spreads. When the malaria parasite is passed to humans through the bite of an infected mosquito, there is a great possibility that this action alters the chemical scent of the human. The altered human body scent then attracts more uninfected mosquitoes, which bite the infected person, thus spreading the disease.

To test this hypothesis, an experiment was conducted on humans infected with malaria. The results indicate that malaria gametocytes (a stage of the parasite) may trigger the production of chemicals that change the scent of the human. The change in human scent makes humans more appealing to mosquitoes.

Scientists are now studying ways to copy this chemical scent for use in traps that would attract mosquitoes.

51 Which statement best describes the role of gametocytes in the spread of malaria?

(1) They give off a scent that attracts infected mosquitoes.

(2) They absorb human body scents that attract mosquitoes.

(3) They release a scent into the human body.

(4) They cause a chemical reaction that alters human scent.

51

52 Malaria is easily spread because uninfected mosquitoes are attracted to

  (1) humans without malaria

  (2) humans infected with gametocytes

  (3) gametocytes in other mosquitoes

  (4) mosquitoes that are uninfected

52 ☐

53 State *one* reason why the use of synthetic scents in traps is a better way to lower mosquito populations than spraying with pesticides. [1]

_____

_____

53 ☐

Base your answers to questions 54 and 55 on the information below and on your knowledge of biology.

The graph below shows the effect of substrate concentration on the action of enzyme X. This enzyme is functioning at its optimal temperature, 36°C, and at its optimal pH, 5.5.

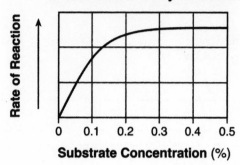

Effect of Substrate Concentration on the Rate of Enzyme Action

54 When the substrate concentration increases from 0.4% to 0.5%, the rate of the reaction

(1) decreases

(2) increases

(3) remains the same

(4) increases, then decreases

54 ☐

55 State what would most likely happen to the rate of enzyme action if the temperature were reduced by 10 degrees. Support your answer. [1]

_____

_____

_____

55 ☐

Part C

Answer all questions in this part. [17]

Directions (56–67): Record your answers in the spaces provided in this examination booklet.

Base your answers to questions 56 and 57 on the experimental setup shown below. The tubing connected to both flask setups used in the experiment provides oxygen to the solution.

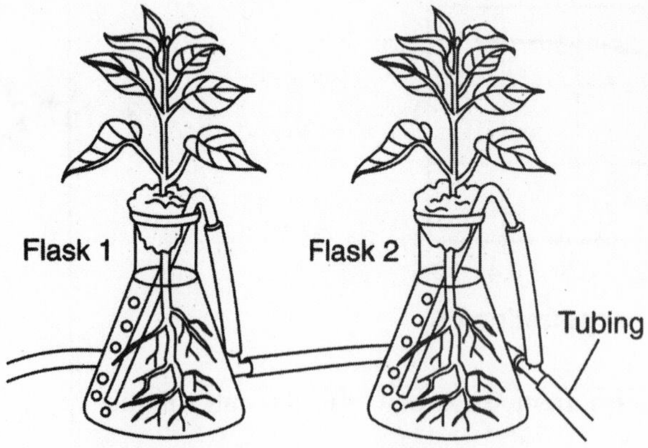

For Teacher Use Only

Source: Campbell and Reece, *Biology*, 6th edition (adapted)

56 State *one* possible hypothesis for the experiment. [1]

_____

_____

56 ☐

57 State what type of data would most likely be collected during the experiment to support or refute your hypothesis. [1]

_____

_____

_____

57 ☐

Base your answer to question 58 on the information below and on your knowledge of biology.

Stem cells present in an embryo are responsible for the formation of various tissues and organs. Recent research suggests that it may be possible to replicate stem cells from sections of skin taken from adult mice, rather than having to use stem cells from the embryos of mice. In the future, human stem cells may be used to replace human tissue damaged by diseases such as Parkinson's disease and multiple sclerosis.

58 Discuss why the use of stem cells taken from a patient to replace damaged tissues and organs may decrease the potential risk to a patient. In your answer, be sure to:

- identify the major problem that may occur when tissues and organs donated by another individual are used [1]
- explain why this problem may occur [1]
- explain why this problem will *not* occur if tissues and organs produced by stem cells from the patient are used [1]

_____

_____

_____

_____

_____

_____

58

59 *Staphylococcus aureus* is a type of bacterium that lives on the skin and in the nostrils of most people. Generally, it is controlled by the immune system of the body. Occasionally, the antibiotic penicillin is needed to control the bacterium. However, some strains of *S. aureus* have a resistance to penicillin, which makes them hard to kill and infections difficult to cure.

Explain how the resistance to penicillin affects the *S. aureus* population. In your answer, be sure to include an explanation of:

- how the exposure to penicillin affects the survival of some bacteria in the population [1]
- why the frequency of penicillin-resistant bacteria can change over time within the population [1]
- how it is still possible to cure patients who are infected with penicillin-resistant bacteria [1]

59

Base your answers to questions 60 through 63 on the information below and on your knowledge of biology.

An ecology class is trying to help reduce the problem of global warming by asking their school district to change all of their old lightbulbs to compact fluorescent lightbulbs that use less electricity.

60 Identify *one* specific gas that contributes to the problem of global warming. [1]

_____

60 ☐

61 State *one* activity of humans that increases the concentration of this gas. [1]

_____

61 ☐

62 Describe *one negative* effect of global warming on humans or ecosystems. [1]

_____

_____

62 ☐

63 Explain why switching to more efficient lightbulbs will help reduce the school's contribution to global warming. [1]

_____

_____

_____

63 ☐

Base your answers to questions 64 and 65 on the information below and on your knowledge of biology.

There's No Place Like Home!

Some pets need expensive food, or grow to large sizes, or have nasty, dangerous behavior. Because of this, some people realize that they can no longer care for their pets. A pet twist-neck turtle in a state of near starvation was found by rescuers at the Brooklyn Botanic Garden. The food that this species eats is not commonly found in New York State. In Florida and other warm states, people have released pet snakes such as pythons and anacondas into local lakes and swamps, where they have become a threat to other animals and even to humans. Those released pets that survive in their new environment can eventually breed and multiply, causing even more problems!

64 Identify *one* abiotic factor that might affect the survival of a released pet and explain why that factor would affect survival. [1]

_____

_____

64

65 State *one* reason released pets that survive in a new environment may be able to form a large population. [1]

_____

_____

65

Base your answers to questions 66 and 67 on the information and diagrams below and on your knowledge of biology.

There are over 40 different species of butterfly fish found in tropical reefs throughout the world. Three different species of butterfly fish are shown below.

Fish A   Fish B   Fish C

The fish fin diagram and dichotomous key shown below can be used to determine the species of each of these fish.

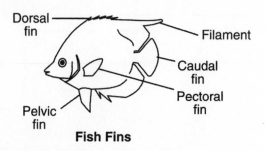

Fish Fins

Dichotomous Key to Butterfly Fish	
1. a. Pelvic fin dark	2
b. Pelvin fin light	4
2. a. Two large white spots below dorsal fin	*C. quadrimacul*
b. Lacks two large white spots below dorsal fin	3
3. a. Caudal fin with two dark bars at tip	*C. reticulatus*
b. Caudal fin with one dark bar at tip	*C. kleinii*
4. a. Dorsal fin has long filament extension	5
b. Filament extension lacking from dorsal fin	6
5. a. Large dark spot on body near filament	*C. ephippium*
b. Small dark spot on body near filament	*C. auriga*
6. a ...	

Directions (66–67): Using the information and dichotomous key, complete the table following the directions below.

For Teacher Use Only

66 Use the dichotomous key and fish fin diagram to identify fish *A*, *B*, and *C* and write the name of *each* fish in the column labeled "Scientific Name" in the table below. [1]

67 Select *two* characteristics from the dichotomous key that are useful for determining the identity of fish *A*, *B*, and *C*. Using these characteristics, label the headings for the last two columns in the table and complete the last two columns in the table. [2]

Fish	Scientific Name	Subgroup		
A		Rabdophorus		
B		Lepidochaetodon		
C		Rabdophorus		

66

67

Part D

Answer all questions in this part. [13]

Directions (68–78): For those questions that are followed by four choices, circle the *number* of the choice that, of those given, best completes the statement or answers the question. For all other questions in this part, follow the directions given in the question and record your answers in the spaces provided.

Base your answer to question 68 on the chart below and on your knowledge of biology.

Universal Genetic Code Chart
Messenger RNA and the Amino Acids for Which They Code

	U	C	A	G	
U	UUU } PHE UUC UUA } LEU UUG	UCU UCC UCA } SER UCG	UAU } TYR UAC UAA } STOP UAG	UGU } CYS UGC UGA } STOP UGG } TRP	U C A G
C	CUU CUC } LEU CUA CUG	CCU CCC } PRO CCA CCG	CAU } HIS CAC CAA } GLN CAG	CGU CGC } ARG CGA CGG	U C A G
A	AUU AUC } ILE AUA AUG } MET or START	ACU ACC } THR ACA ACG	AAU } ASN AAC AAA } LYS AAG	AGU } SER AGC AGA } ARG AGG	U C A G
G	GUU GUC } VAL GUA GUG	GCU GCC } ALA GCA GCG	GAU } ASP GAC GAA } GLU GAG	GGU GGC } GLY GGA GGG	U C A G

68 Fill in the missing mRNA bases and the amino acid sequence that corresponds to the DNA base sequence below. [2]

DNA   CAC   GTG   GAC   TGA

mRNA   _____   _____   _____   _____

Amino acids _____   _____   _____   _____

Base your answers to questions 69 and 70 on the information below and on your knowledge of biology.

An investigation is carried out to determine the effect of exercise on the rate at which a person can squeeze a clothespin.

69 In this investigation, the independent variable is the

(1) control

(2) exercise

(3) rate of squeezing

(4) number of participants

69 ☐

70 Muscle fatigue occurs during this activity when

(1) carbon dioxide is used up in the muscle cells

(2) simple sugar is converted to starch in the muscle cells

(3) proteins accumulate in mitochondria in the muscle cells

(4) certain waste products collect in the muscle cells

70 ☐

71 Part of a laboratory procedure is shown in the diagram below.

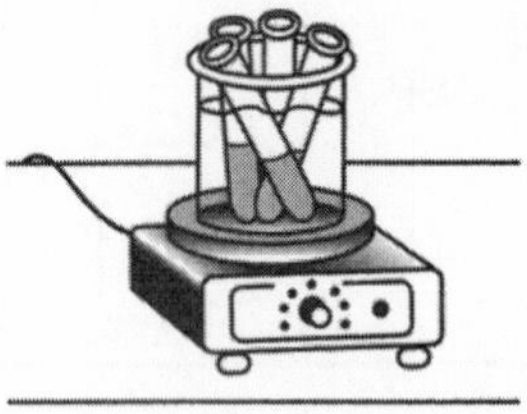

This setup would most likely be involved in a procedure to

(1) stain specimens while making a wet mount

(2) test for the presence of glucose using an indicator

(3) separate pigments in a mixture

(4) determine the pH of solutions

71 ☐

Base your answers to questions 72 and 73 on the information below and on your knowledge of biology.

A valuable medicine is obtained from a certain rare species of plant. Scientists are anxious to find another more abundant species of plant that is closely related to the rare one, and also produces the medicine.

Two newly discovered plant species, *A* and *B*, were studied and compared to the rare one. The results of the study are shown in the table below.

Species of Plant	Characteristics of Flowers	Shape of Leaves	Species Number of Chromosomes	Enzyme A Present	Enzyme B Present	Enzyme C Present
rare species	pink 5 petals	round	36	yes	yes	yes
species A	pink 5 petals	oval	34	no	no	yes
species B	white 5 petals	round	36	yes	yes	yes

72 Which newly discovered species is more closely related to the rare species? Support your answer. [1]

Species:_____

_____

_____

72 [ ]

73 Which procedure could also be carried out to help determine which newly discovered species is most closely related to the rare species?

(1) measurement of respiration rate in the plants

(2) chromatography of pigment extracts from the plants

(3) determination of the type of gas released by photosynthesis in the plants

(4) analysis of chemical bonds present in glucose in the plants

73 [ ]

74 The characteristics of four finches that inhabit the same island are represented in the chart below.

Characteristics Chart

Large Ground Finch	Warbler Finch
Beak: crushing Food: mainly plant	Beak: probing Food: 100% animal
Small Ground Finch	Large Tree Finch
Beak: crushing Food: mainly plant	Beak: grasping Food: mainly animal

Complete the table below using information in the characteristics chart and your knowledge of biology. [2]

Competes With the Large Tree Finch	Type of Finch	State *one* reason why it competes *or* does *not* compete with the large tree finch.
no		
yes		

74 ☐

75 Studies of the finches of the Galapagos Islands have shown that

(1) DNA will change to produce structures needed by birds to survive intense competition

(2) a bird's beak changes annually in response to the type of food that is most abundant each year

(3) natural selection occurs when there are scarce resources and intense competition

(4) the beak of a finch will change if the environment of the bird remains stable

75 ☐

Base your answers to questions 76 through 78 on the information below and on your knowledge of biology.

A student prepared four different red blood cell suspensions, as shown in the chart below.

Suspension	Contents
A	red blood cells in normal blood serum (0.7% salt solution)
B	red blood cells in 10% salt solution
C	red blood cells in distilled water
D	red blood cells in tap water

76 Which suspension would contain red blood cells that would appear wrinkled and reduced in volume?

(1) *A*

(2) *B*

(3) *C*

(4) *D*

76 ☐

77 The change in red blood cell volume is principally due to the movement of

(1) serum

(2) oxygen

(3) water

(4) salt

77 ☐

78 Which process is most likely involved in the change in red blood cell volume?

(1) active transport

(2) evaporation

(3) replication

(4) diffusion

78 ☐

LIVING ENVIRONMENT
JANUARY 2011

ANSWER SHEET

Student ....

Teacher ....

School .... Grade ....

Part	Maximum Score	Student's Score
A	30	
B–1	15	
B–2	10	
C	17	
D	13	
Total Raw Score (maximum Raw Score: 85)		
Final Score (from conversion chart)		

Raters' Initials

Rater 1 .... Rater 2 ....

Record your answers to Part A and Part B–1 on this answer sheet.

Part A

1	11	21 ....
2	12	22 ....
3	13	23 ....
4	14	24 ....
5	15	25 ....
6	16	26 ....
7	17	27 ....
8	18	28 ....
9	19	29 ....
10	20	30 ....

Part A Score

Part B–1

31	39 ....
32	40 ....
33	41 ....
34	42 ....
35	43 ....
36	44 ....
37	45 ....
38	

Part B–1 Score

LIVING ENVIRONMENT
JUNE 2011

Part A

Answer all questions in this part. [30]

Directions (1–30): For *each* statement or question, record on your separate answer sheet the *number* of the word or expression that, of those given, best completes the statement or answers the question.

1 Which phrase is an example of autotrophic nutrition?

(1) a cow eating grass in a field
(2) a mushroom digesting a dead log
(3) an apple tree making its own food
(4) a tapeworm feeding in the body of a dog

2 The ability of estrogen to affect certain cells depends directly on

(1) amino acids
(2) receptor molecules
(3) gametes
(4) nerve cells

3 By studying the chemicals in rare plants that grow only in rain forests, scientists hope to discover new life-saving medicines. Chances of finding such new medicines are reduced by

(1) predation by carnivores
(2) homeostasis in organisms
(3) recycling of materials in food webs
(4) loss of species due to human activities

4 When a species includes organisms with a wide variety of traits, it is most likely that this species will have

(1) a high proportion of individuals immune to genetic diseases
(2) a greater chance to survive if environmental conditions suddenly change
(3) less success competing for resources
(4) limitless supplies of important resources, such as food and water

5 Some diseases and their causes are listed below.

  A. Flu—influenza virus
  B. Lung cancer—smoking
  C. Cystic fibrosis—genes
  D. Dysentery—parasitic ameba

Which disease would individuals have the greatest difficulty preventing in themselves?

(1) A
(2) B
(3) C
(4) D

6 The diagram below represents the banding pattern for human chromosome 11, with some of the bands labeled.

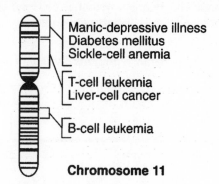

Manic-depressive illness
Diabetes mellitus
Sickle-cell anemia

T-cell leukemia
Liver-cell cancer

B-cell leukemia

Chromosome 11

The bands represent

(1) proteins
(2) genes
(3) starches
(4) enzymes

7 A liver cell can make enzymes that a heart cell can *not* make because liver cells

(1) digest large, complex molecules
(2) contain more DNA than heart cells
(3) use different genes than the heart cells use
(4) remove carbon dioxide from blood

8 As male children get older, some begin to closely resemble their fathers and have no resemblance to their mothers. Which statement best explains this observation?

(1) Several sperm fertilized the egg, so the fertilized egg contained more genes from their father.
(2) More genes are inherited from the sperm cell of their father than from the egg cell of their mother, so most traits will be like those of their father.
(3) More genes from their father are expressed in traits that can be seen, and more genes from their mother are expressed in traits that cannot be seen, such as blood type or enzyme function.
(4) Genes from their father are stronger than genes from their mother, so the genes from their mother are not expressed.

9 Which row in the chart below contains a cell structure paired with its primary function?

Row	Cell Structure	Function
(1)	ribosome	protein synthesis
(2)	vacuole	production of genetic information
(3)	nucleus	carbohydrate synthesis
(4)	mitochondrion	waste disposal

10 Which sequence represents the levels of biological organization from smallest to largest?

(1) organism → cell → tissue → organelle → organ system → organ
(2) organ system → organ → organism → cell → tissue → organelle
(3) organelle → organ system → cell → organism → tissue → organ
(4) organelle → cell → tissue → organ → organ system → organism

11 The diagram below represents division of a cell that produces two daughter cells.

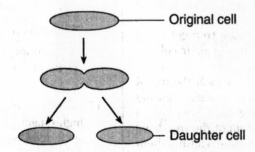

Which statement most likely describes the daughter cells produced?

(1) The daughter cells will pass on only half of the genetic information they received from the original cell.
(2) The daughter cells will each produce offspring that will have the same genetic information as the original cell.
(3) The daughter cells will each undergo the same mutations as the original cell after reproduction has occurred.
(4) The daughter cells will not pass on any of the genes that they received from the original cell.

12 Which concept is best represented in the diagram below?

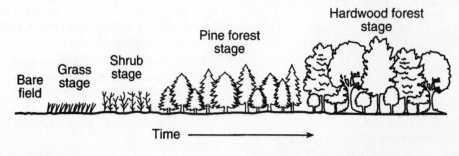

(1) random mutations
(2) ecological succession
(3) genetic engineering
(4) direct harvesting

13 The cells that make up leaves on a tree are genetically identical, yet the leaves often have different shapes and sizes, as shown in the diagram below.

Leaves of White Oak *(Quercus alba)*

Top leaves   Bottom leaves

Which statement best explains this difference in leaf appearance?

(1) The leaves at the top of the tree get more sunlight, causing the genes in their cells to be expressed differently.
(2) The genes in the cells of leaves at the top of the tree are destroyed by sunlight, causing the leaves to stop growing.
(3) The leaves near the bottom of the tree have more genes related to leaf size, causing them to grow larger.
(4) The genes in the cells of leaves near the bottom of the tree increase in number, causing them to grow even larger.

14 Selective breeding is a technique that is used to

(1) give all organisms a chance to reproduce
(2) produce organisms from extinct species
(3) produce offspring with certain desirable traits
(4) keep farm crops free of all mutations

15 On hot, dry days, guard cells often close microscopic openings in plant leaves, conserving water. This is an example of

(1) environmental factors causing gene mutation in plants
(2) finite resources acting as selecting agents for evolution
(3) a feedback mechanism for maintaining homeostasis
(4) differentiation in plants as a result of stimuli

16 The diagram below represents a portion of a DNA molecule.

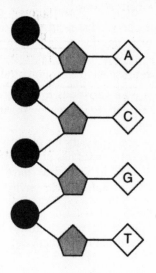

The letters represent different types of

(1) sugar molecules   (3) enzymes
(2) molecular bases   (4) proteins

17 Cotton plants produce seeds that contain high-quality protein. This protein could be used as a food source except that the seeds are poisonous to humans. Recently, scientists have inserted a section of DNA into the cotton plants that makes the cotton seeds nonpoisonous. The technique for this procedure is known as

(1) gene manipulation   (3) reproduction
(2) cloning   (4) direct harvesting

18 Which mutation in a fruit fly could be passed on to its offspring?

(1) a mutation in a cell of an eye that changes the color of the eye
(2) a mutation in a leg cell that causes the leg to be shorter
(3) a mutation in a sperm cell that changes the shape of the wing
(4) a mutation in a cell of the digestive tract that produces a different enzyme

19 Which process initially provides the link between an abiotic factor and the energy needs of an entire ecosystem?

(1) respiration   (3) decomposition
(2) photosynthesis   (4) predation

20 Buffalo grass is a species of plant found on the grazing prairies of Wyoming. It is a tough grass that has silicates (compounds containing oxygen and silicon) that reinforce its leaves. For hundreds of years, this grass has survived in an adverse environment. Which statement best explains the presence of this grass today?

(1) There are no variations in this grass species that help it to survive in an adverse environment.
(2) Silicates are necessary for photosynthesis.
(3) The current species has no mutations.
(4) The silicates in the grass have given the species an advantage in its environment.

21 The bud shown in the diagram below was produced by asexual reproduction.

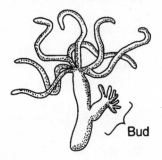

Bud

Which process is responsible for the formation of the bud?

(1) fertilization   (3) mitosis
(2) recombination   (4) meiosis

22 The temporary storage of energy in ATP molecules is part of which process?

(1) cell division   (3) protein synthesis
(2) cellular respiration (4) DNA replication

23 A function of white blood cells is to

(1) transport oxygen to body cells
(2) produce hormones that regulate cell communication
(3) carry glucose to body cells
(4) protect the body against pathogens

24 Competition for biotic resources can be illustrated by organisms fighting for a limited amount of

(1) air to breathe   (3) mates for breeding
(2) water to drink   (4) space for nesting

25 Many biological catalysts, hormones, and receptor molecules are similar in that, in order to function properly, they must

(1) interact with each other at a high pH
(2) interact with molecules that can alter their specific bonding patterns
(3) contain amino acid chains that fold into a specific shape
(4) contain identical DNA base sequences

26 If only one type of tree is planted in an abandoned field, the ecosystem will

(1) evolve quickly and become extinct
(2) be unable to reach dynamic equilibrium
(3) contain little genetic variability
(4) be unable to cycle materials

27 Which organisms directly help to reduce overpopulation in a deer herd?

(1) parasites and predators
(2) parasites and scavengers
(3) decomposers and predators
(4) decomposers and consumers

28 In the human body, oxygen is absorbed by the lungs and nutrients are absorbed by the small intestine. In a single-celled organism, this absorption directly involves the

(1) nucleus   (3) cell membrane
(2) chloroplasts   (4) chromosomes

29 An earthworm lives and reproduces in the soil. It aerates the soil and adds organic material to it. The earthworm is a source of food for other organisms. All of these statements together best describe

(1) a habitat
(2) autotrophic nutrition
(3) an ecological niche
(4) competition

30 Depletion of nonrenewable resources is often a result of

(1) environmental laws
(2) human population growth
(3) reforestation
(4) recycling

Part B–1

Answer all questions in this part. [13]

Directions (31–43): For *each* statement or question, record on the separate answer sheet the *number* of the word or expression that, of those given, best completes the statement or answers the question.

31 The bar graph below shows the height of a plant at the end of each week of a five-week growth period.

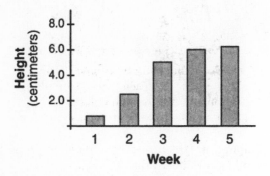

Which statement represents a valid conclusion based on the information in the graph?

(1) The plant was given water during the first three weeks, only.
(2) The plant will grow faster during the sixth week than it did during the fifth week.
(3) The plant grew fastest during the first three weeks, and then it grew slower.
(4) The plant grew slowest during the first three weeks, and then it grew faster.

32 A diagram frequently used in ecological studies is shown below.

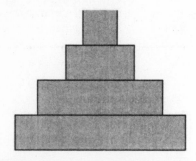

This diagram can be used to represent the

(1) dependency of animal survival on physical conditions in an ecosystem
(2) loss of energy from various groups of organisms in an ecosystem
(3) competition among species in an ecosystem
(4) mechanisms that maintain homeostasis in the plants in an ecosystem

33 A biologist formulates a hypothesis, performs experiments to test his hypothesis, makes careful observations, and keeps accurate records of his findings. In order to complete this process, the biologist should

(1) adjust the data to support the hypothesis
(2) eliminate data that do not support the hypothesis
(3) write a research paper explaining his theories before performing his experiments, in order to gain funding sources
(4) evaluate the findings and, if necessary, alter the hypothesis based on his findings, and test the new hypothesis

Base your answer to question 34 on the diagram below and on your knowledge of biology.

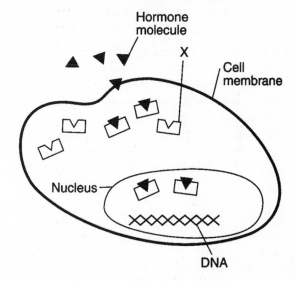

34 Structure X most likely functions in the

(1) transport of chemical messenger molecules into the cell nucleus
(2) extraction of energy from nutrients
(3) separation of cell contents from the outside environment
(4) digestion of large molecules

35 The arrows in the diagram below represent the movement of materials.

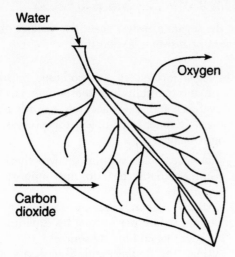

This movement of materials indicated by the arrows is most directly involved in the processes of

(1) respiration and replication
(2) photosynthesis and excretion
(3) digestion and recycling
(4) circulation and coordination

36 When using a compound light microscope, the most common reason for staining a specimen being observed is to

(1) keep the organism from moving around
(2) make the view more colorful
(3) determine the effects of chemicals on the organism
(4) reveal details that are otherwise not easily seen

Base your answers to questions 37 through 39 on the diagram below and on your knowledge of biology.

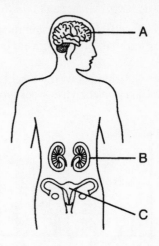

37 Failure of structure A to function properly would most directly disrupt

(1) autotrophic nutrition
(2) chromosome replication
(3) cellular communication
(4) biological evolution

38 Structure B represents

(1) cells, only
(2) cells and tissues, only
(3) an organ with cells and tissues
(4) a complete system with organs, tissues, and cells

39 Structure C is part of which body system?

(1) digestive   (3) circulatory
(2) reproductive   (4) nervous

40 Thrips are insects that feed on the pollen and flowers of certain plants. The size of a thrip population depends on the number of flowers available. Which graph best represents changes in a population of thrips if winter was longer than usual and the summer was too cool and dry for many flowers to bloom?

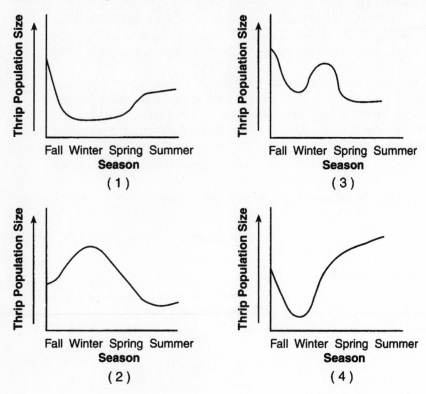

41 The percent of DNA that species A has in common with species B, C, D, and E are shown in the graph below.

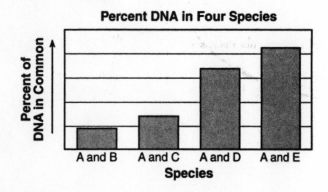

Which statement is a valid conclusion that can be drawn from this graph?

(1) Species A is closely related to species B, but is not related to species E.
(2) Fewer mutations have occurred in species B and C than in species A.
(3) Species A and E have the greatest similarity in protein structure.
(4) Environment influences the rate of evolution.

Base your answers to questions 42 and 43 on the passage below and on your knowledge of biology.

...Corals come in about 1,500 known species—from soft swaying fans to stony varieties with hard skeletons that form reef bases. They are made up of polyps, tiny animals that live in colonies and feed at night on microscopic plants and creatures. The coral's surface is the living part, with color infused by single-celled algae called zooxanthellae that live in polyp tissue. The algae act like solar panels, passing energy to the coral as they photosynthesize while feeding on the coral's waste.

Extremely sensitive, corals survive in a narrow range of temperature, sunlight and salinity. An uncommonly severe El Niño in 1998 raised ocean temperatures and changed currents, causing bleaching that devastated reefs worldwide. Scientists say parts of the Indian Ocean lost up to 90 percent of corals. The bleaching struck reefs around the Persian Gulf, East Africa, Southeast Asia and the Caribbean. Some recovered. Many died. ...

Source: Associated Press, December 2001

42 The relationship between the polyps and the zooxanthellae can best be described as
- (1) negative for both
- (2) neutral for both
- (3) positive for both
- (4) negative for one and positive for the other

43 The passage contains information concerning
- (1) limiting factors
- (2) reproductive methods
- (3) bacteria
- (4) competition

Part B–2

Answer all questions in this part. [12]

Directions (44–55): For those questions that are multiple choice, record on your separate answer sheet the *number* of the choice that, of those given, best completes each statement or answers each question. For all other questions in this part, follow the directions given and record your answers in the spaces provided.

44 State *one* way insect pests in an apple orchard can be controlled without using chemical pesticides. [1]

_____

_____

45 The tranquilizer thalidomide was once prescribed for pregnant women. When this drug was used between the third and sixth week after fertilization, serious deformities in the fetus occurred as the fetus developed. State why thalidomide would have a greater effect on development when used between weeks 3 and 6 than when used in late pregnancy. [1]

_____

_____

_____

46 Draw an arrow to indicate *one* part of the plant cell below that would *not* be found in an animal cell. The tip of the arrow must touch the part being identified. [1]

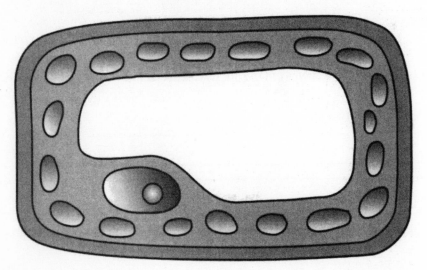

Base your answers to questions 47 and 48 on the information and chart below and on your knowledge of biology.

Body weight is considered to be a risk factor for diseases such as diabetes and high blood pressure. The Body Mass Index (BMI) chart can be used as a guide to determine if a person's body weight puts them at risk for such diseases. A portion of this chart is shown below.

Calculating Your Body Mass Index (BMI)

BMI	Healthy		Overweight					Obese			
	19	24	25	26	27	28	29	30	35	40	45
Height	Weight in Pounds										
5'4"	110	140	145	151	157	163	169	174	204	232	262
5'5"	114	144	150	156	162	168	174	180	210	240	270
5'6"	118	148	155	161	167	173	179	186	216	247	278
5'7"	121	153	159	166	172	178	185	191	223	255	287
5'8"	125	158	164	171	177	184	190	197	230	262	295
5'9"	128	162	169	176	182	189	196	203	236	270	304
5'10"	132	167	174	181	188	195	202	209	243	278	313
5'11"	136	172	179	186	193	200	208	215	250	286	322

Note: The answer to question 47 should be recorded on your separate answer sheet.

47 The BMI for a person who is 5 feet 9 inches tall and weighs 170 pounds is between

(1) 24 and 25
(2) 25 and 26

(3) 27 and 28
(4) 29 and 30

48 Is the person described in question 47 at risk for diseases such as diabetes or high blood pressure? Support your answer. [1]

_____

_____

Note: The answer to question 49 should be recorded on your separate answer sheet.

49 The diagram below represents a food web.

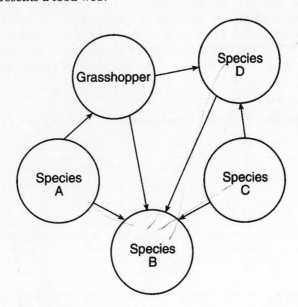

Which species would most likely be a decomposer?

(1) A

(2) B

(3) C

(4) D

Base your answers to questions 50 and 51 on the passage below and on your knowledge of biology.

Plants of the snow lotus species, *Saussurea laniceps*, are used in Tibet and China to produce traditional medicines. These plants bloom just once, at the end of a seven-year life span. Collectors remove the taller blooming plants, which they consider to have the best medicinal value. Some scientists are concerned that the continual selection and removal of the tall plants from natural ecosystems may result in a change in the average height of the snow lotus in future populations.

Note: The answer to question 50 should be recorded on your separate answer sheet.

50 The removal of the taller plants is an example of

(1) genetic engineering

(2) direct harvesting

(3) selective breeding

(4) asexual reproduction

51 State *one* way that the removal of the taller snow lotus plants from ecosystems interferes with the process of natural selection. [1]

_____

_____

_____

Base your answers to questions 52 through 55 on the information below and on your knowledge of biology.

A biology class conducted an experiment to determine the rate of respiration of yeast in bread dough at various temperatures.

Bread dough will rise due to the production of carbon dioxide by the yeast present in the dough.

An equal amount of dough was placed in the bottom of each of five graduated cylinders. Each cylinder was then placed in a different water bath to maintain a particular temperature. A diagram of the setup is shown below.

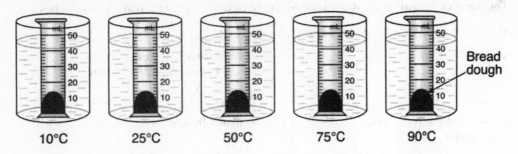

The amount of expansion of the dough in each cylinder was measured after 15 minutes. The results are shown in the data table below.

The Effect of Temperature on Yeast Respiration

Temperature of Water Bath (°C)	Change in Volume of Bread Dough (mL)
10	4
25	11
50	20
75	25
90	2

Directions (52–53): Using the information in the data table, construct a line graph on the grid, following the directions below.

52 Mark an appropriate scale, without any breaks, on the axis labeled "Temperature of Water Bath (°C)."   [1]

53 Plot the data from the data table. Surround each point with a small circle and connect the points.   [1]

Example: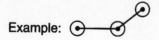

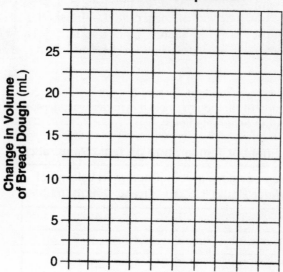

The Effect of Temperature on Yeast Respiration

Change in Volume of Bread Dough (mL)

Temperature of Water Bath (°C)

54 At which temperature did yeast cells produce the *least* amount of gas in 15 minutes?   [1]

_____°C

55 Identify the independent variable in this investigation.   [1]

_____

_____

Part C

Answer all questions in this part. [17]

Directions (56–72): Record your answers in the spaces provided in this examination booklet.

Base your answer to question 56–59 on the information below and on your knowledge of biology.

Many people have a sensitivity to peanuts. The symptoms can include watery, itchy eyes and difficulty breathing. This allergic reaction can be mild, severe, or fatal.

56–59 Discuss why an individual can have a sensitivity to peanuts. In your answer, be sure to:

- identify the human system that is responsible for this sensitivity to peanuts [1]
- identify the specific type of molecule that triggers an allergic reaction [1]
- state *one* reason why a person could be allergic to peanuts, but *not* be allergic to walnuts [1]
- describe how this reaction is similar to the rejection of a transplanted organ [1]

_____

_____

_____

_____

_____

_____

_____

_____

60 State *one* way the decision of high school students to drive to school rather than ride a bus to school can have a *negative* environmental impact on future generations. [1]

_____

_____

Base your answers to questions 61 through 64 on the passage below and on your knowledge of biology.

Dandelions are weeds that are very common in many grassy areas of New York State. Dandelion flowers first open up in a bright-yellow stage, and later turn a fluffy white when they are ready to release their seeds. The seeds are carried by the wind, and can sometimes travel great distances before landing and growing into new plants. The stems of dandelions are usually very long, typically about 20–30 centimeters (cm), and stand high above the surrounding grass.

A science teacher in Niagara County discovered an area in her lawn where nearly every dandelion had a stem less than 1 cm long. These short dandelions were replacing large amounts of grass in the lawn surrounding her house. They were growing much more thickly than the taller dandelions in other nearby areas. The short dandelions appeared to be growing very successfully in one area of her lawn, but did not appear to have spread to other areas of her lawn. The science teacher noticed that every time she mowed her lawn, the short dandelions were left untouched by the mower blades, and that their numbers were steadily increasing.

61 State *one* possible cause of the genetic variation in dandelion height.   [1]

_____

62 State *one* possible explanation for the fact that the short dandelions had not yet spread to other areas of her lawn.   [1]

_____

_____

63 State *one* possible reason why the amount of grass was decreasing, while the number of short dandelions was increasing in the lawn of the science teacher.   [1]

_____

_____

64 State *one* possible advantage the short dandelions may have over the tall dandelions in this yard.   [1]

_____

_____

Base your answers to questions 65 and 66 on the information below and on your knowledge of biology.

The New York State Department of Health issues health advisories on eating specific fish. Some of these fish contain toxic chemicals that were passed through the food chain and accumulate in the tissues of these fish. The advisories include general advice on fish caught in specific bodies of water. The advisory tells people how to minimize their exposure to toxins in these fish in order to reduce health risks associated with them.

65 Different species of fish are found in different positions in a food chain, depending on what they eat. Explain how the position of a fish in a food chain would affect the amount of toxins present in the tissues of the fish. [1]

_____

_____

66 State *one* action that could be taken to reduce the amount of toxins present in the environment inhabited by these fish. [1]

_____

_____

_____

Base your answers to questions 67 and 68 on the information below and on your knowledge of biology.

The diagram below represents some stages in the life cycle of humans. The numbers in the diagram represent various processes in the cycle.

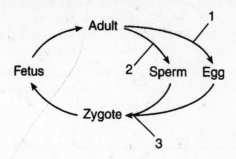

67 State how processes 1 and 2 affect the amount of genetic information provided by a parent to its offspring. [1]

Each of the gametes have a haploid number of the regular number of chromosomes that are combined to form the offspring.

68 State how process 3 affects the amount of genetic information an offspring receives. [1]

The zygote is a diploid cell that begins to divide to create different body systems for the baby

69 Suggest *one* way that doctors or patients can help to reduce the chances of bacteria becoming resistant to an antibiotic. [1]

_____

_____

Base your answers to questions 70 through 72 on the information below and on your knowledge of biology.

In the 1980s, global deforestation was estimated at 17 to 20 million hectares per year, an area the size of Great Britain. Today, the area affected by deforestation has decreased significantly in some regions of the world through the use of sustainable forest management. However, there are still regions of the world affected by wide-scale deforestation, because of the short-term economic benefits. The harmful effects of deforestation on regional and worldwide climate and ecology continue as forest areas are destroyed.

70 State *one* short-term economic benefit of deforestation.   [1]

_____

_____

71 Explain how deforestation decreases biodiversity.   [1]

_____

_____

72 Explain how wide-scale deforestation may contribute to global warming.   [1]

_____

_____

Part D

Answer all questions in this part.   [13]

Directions (73–85): For those questions that are multiple choice, record on your separate answer sheet the *number* of the choice that best completes the statement or answers the question. For all other questions in this part, follow directions given in the question and record your answer in the spaces provided.

Note: The answer to question 73 should be recorded on your separate answer sheet.

73 A model cell setup is represented in the "Initial State" diagram below.

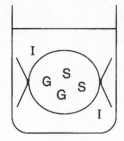

Initial state

Which diagram indicates the areas where each of these substances would be located after 20 minutes?   [1]

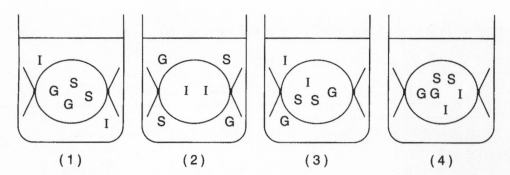

Note: The answer to question 74 should be recorded on your separate answer sheet.

74 Which factor most likely contributed to the evolution of Galapagos Islands finches with different beak shapes?

(1) similar climates on the different islands
(2) competition between the finches for food
(3) cloning experiments carried out by native people on the islands
(4) increased rate of asexual reproduction

Base your answer to question 75 on the finch diversity diagram below and on your knowledge of biology.

Variations in Beaks of Galapagos Islands Finches

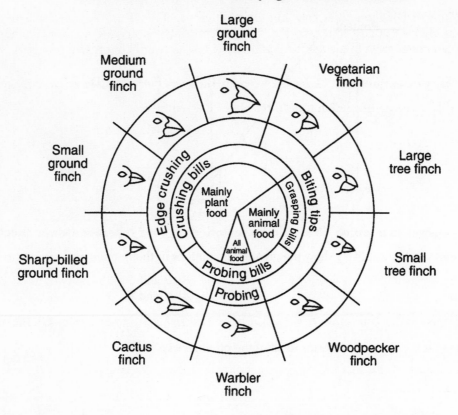

From: *Galapagos: A Natural History Guide*

Note: The answer to question 75 should be recorded on your separate answer sheet.

75 Warbler finches are classified as

(1) producers

(2) herbivores

(3) carnivores

(4) decomposers

Base your answers to questions 76 through 78 on the diagram below and on your knowledge of biology. Letters *A* through *E* represent different species of organisms. The arrows represent long periods of geologic time.

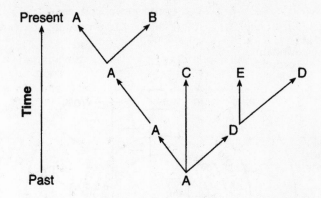

Note: The answer to question 76 should be recorded on your separate answer sheet.

76 Which species would most likely show the greatest similarities in their amino acid sequences?

(1) *A* and *E*   (3) *B* and *D*
(2) *A* and *B*   (4) *C* and *E*

77 Which species is the common ancestor to all of the other species? [1]

_____

78 Identify *one* species that was *not* able to adapt to its environment. Support your answer. [1]

_____

_____

_____

_____

Base your answers to questions 79 through 81 on the information and diagram below and on your knowledge of biology. The diagram represents some of the steps in a procedure used in a specific laboratory activity.

Samples of DNA from an eye-color gene of four individuals, W, X, Y, and Z, were cut into pieces using a type of chemical. The results of this procedure are shown below.

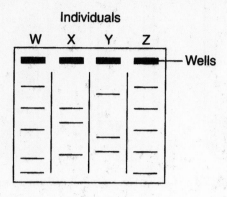

79 Identify the specific type of chemical used to cut the DNA in this procedure. [1]

_____

80 Which *two* individuals have DNA base patterns for this gene that are the most similar? Support your answer. [1]

Individuals: _____ and _____

_____

_____

Note: The answer to question 81 should be recorded on your separate answer sheet.

81 The diagram represents the results of the procedure known as

(1) cloning
(2) chromatography

(3) gel electrophoresis
(4) protein sequencing

Base your answer to question 82 on the results of an experiment using plant pigments represented below and on your knowledge of biology.

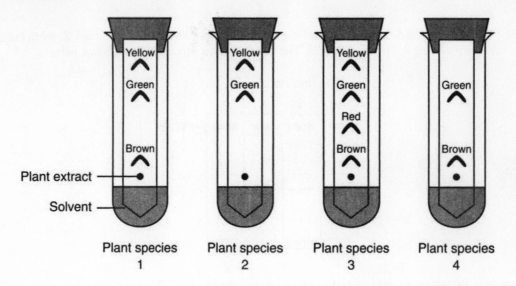

Note: The answer to question 82 should be recorded on your separate answer sheet.

82 Which phrase could be used to describe this technique?

(1) the use of chromatography to separate molecules in a mixture
(2) the use of cut leaves to observe certain colors
(3) using indicators to determine pH
(4) using dichotomous keys to identify plants

_____

Base your answers to questions 83 and 84 on the information below and on your knowledge of biology.

A student checked her pulse rate three times during one day. When she first woke up, her pulse rate was 54 beats per minute (bpm). As she walked to her first-period class, it was 71 bpm. Later, she recorded 98 bpm after playing in a basketball game.

83 State *one* reason for the change in her pulse rate throughout the day. [1]

_____

_____

84 State *one* way a change in pulse rate helps to meet the needs of the body. [1]

_____

_____

85 The diagram below shows a student heating some test tubes with chemicals in them during a laboratory activity.

Explain why putting stoppers in the test tubes could be dangerous. [1]

_____

_____

_____

LIVING ENVIRONMENT
JUNE 2011

Part	Maximum Score	Student's Score
A	30	
B–1	13	
B–2	12	
C	17	
D	13	

Total Raw Score
(maximum Raw Score: 85) ☐

Final Score
(from conversion chart) ☐

Raters' Initials

Rater 1 . . . . Rater 2 . . . .

ANSWER SHEET

Student . . . .

Teacher . . . .

School . . . . Grade . . . .

Record your answers to Part A, Part B–1, Part B–2, and Part D on this answer sheet.

Part A

1 . . . .   11 . . . .   21 . . . .
2 . . . .   12 . . . .   22 . . . .
3 . . . .   13 . . . .   23 . . . .
4 . . . .   14 . . . .   24 . . . .
5 . . . .   15 . . . .   25 . . . .
6 . . . .   16 . . . .   26 . . . .
7 . . . .   17 . . . .   27 . . . .
8 . . . .   18 . . . .   28 . . . .
9 . . . .   19 . . . .   29 . . . .
10 . . . .   20 . . . .   30 . . . .

Part A Score
☐

Part B–1

31 . . . .   38 . . . .
32 . . . .   39 . . . .
33 . . . .   40 . . . .
34 . . . .   41 . . . .
35 . . . .   42 . . . .
36 . . . .   43 . . . .
37 . . . .

Part B–1 Score
☐

Part B–2

47 . . . .   50 . . . .
49 . . . .

Part B–2 Score
☐

Part D

73 . . . .   76 . . . .
74 . . . .   81 . . . .
75 . . . .   82 . . . .

Part D Score
☐